Also in the Variorum Collected Studies Series:

ROY M. MacLEOD
Public Science and Public Policy in Victorian England

MAURICE CROSLAND
Studies in the Culture of Science in
France and Britain since the Enlightenment

A.W. SKEMPTON
Civil Engineers and Engineering in Britain before 1840

A. RUPERT HALL
Science and Society: Historical Essays on the
Relations of Science, Technology and Medicine

A. RUPERT HALL
Newton, his Friends and his Foes

P.M. HARMAN
After Newton: Essays on Natural Philosophy

R.W. HOME
Electricity and Experimental Physics in 18th-Century Europe

A.J. TURNER
Of Time and Measurement: Studies in the History of
Horology and Fine Technology

GERARD L'E. TURNER
Scientific Instruments and Experimental Philosophy, 1550–1850

JOHN HEDLEY BROOKE
Thinking about Matter:
Studies in the History of Chemical Philosophy

ROBERT FOX
Science, Industry and the Social Order in
Post-Revolutionary France

ROBERT FOX
The Culture of Science in France, 1700–1900

TREVOR H. LEVERE
Chemists and Chemistry in Nature and Society, 1770–1878

ALLEN G. DEBUS
Chemistry, Alchemy and the New Philosophy, 1550–1700

COLLECTED STUDIES SERIES

Science for All

Professor William H. Brock

William H. Brock

Science for All

Studies in the History of Victorian Science and Education

VARIORUM
1996

Published by VARIORUM
Ashgate Publishing Limited
Gower House, Croft Road,
Aldershot, Hampshire GU11 3HR
Great Britain

Ashgate Publishing Company
Old Post Road,
Brookfield, Vermont 05036–9704
USA

ISBN 0–86078–542–4

British Library CIP Data
Brock, William H.
Science for All: Studies in the History of Victorian Science and Education. — (Variorum Collected Studies Series: 518)
I. Title II. Series
509.034

US Library of Congress CIP Data
Brock, William H.
Science for All: Studies in the History of Victorian Science and Education / William H. Brock
p. cm. — (Variorum Collected Studies Series: CS518)
Includes bibliographical references and index.
ISBN 0–86078–542–4 (hc : alk. paper)
1. Science—Great Britain—History—19th century. 2. Science—Social aspects—Great Britain—History—19th century.
3. Science—Study and teaching—Great Britain—History—19th century. 4. Great Britain—History—Victoria, 1837–1901. I. Title.
II. Series: Collected Studies: CS518.
Q127.O5B76 1996 95–41842
509.41'09'034–dc20 CIP

The paper used in this publication meets the minimum requirements of the American National Standard for Information Sciences - Permanence of Paper for Printed Library Materials, ANSI Z39.48-1984. ™

Printed by Galliard (Printers) Ltd
Great Yarmouth, Norfolk, Great Britain

COLLECTED STUDIES SERIES CS518

CONTENTS

EDUCATION

This volume contains x + 332 pages

PREFACE

The compilation of a volume of one's published essays is an opportunity for academic reflection. While there is, I trust, a logic in the way the nineteen essays are grouped into the three sections of the patronage of science, science and culture, and science education, the reader may suspect there was no such logic in their composition. Serendipity and external commissions undoubtedly play an important role in the trajectory of scholarly research, but the fact that teaching and research go hand in hand is no mere rhetorical gambit, at least not in my experience as a University teacher since 1960. All of the essays grouped in Section 3, for example, arose from teaching on a Diploma (later MA) course on the History of Education in the School of Education at the University of Leicester between 1968 and 1980. Teaching with inspiring colleagues such as Professors Brian Simon, A. Jack Meadows, Malcolm J. S. Seabourne, David A. Reeder and Donald K. Jones, led specifically to my interest in the educational work of the chemist, Henry Armstrong. In 1973 I published an edited selection of his writings on heurism as *H. E. Armstrong and the Teaching of Science 1880-1930* (Cambridge, 1973). Essay XI in the present collection investigated how Armstrong's practical approach to teaching science was worked out in the chemistry curriculum of Finsbury Technical College, while XIX, the hitherto unpublished title essay, elaborates and extends some of the themes of the introduction to the 1973 book.

Among Armstrong's colleagues at Finsbury College, who shared his enthusiasm for teaching science practically, were the electrical engineers, William E. Ayrton and John Perry. Both men had gained their teaching experience in Japan at the Royal Engineering College in Tokyo. This extraordinary overseas experience formed the subject of my Presidential Address to the British Society for the History of Science in 1980, reprinted as item XVI. In Tokyo, while trying to teach electrical engineering to Japanese students, Ayrton and Perry introduced the use of squared, or graph, paper for the analysis of functional relationships in practical classes. A lecture on this subject given to the British Society for the History of Mathematics in 1980 was subsequently developed with my then doctoral student, Michael A. Price, as essay XIII.

The remaining essay within the history of science education section, on the social history of mathematics (XII), arose from an intention in the 1960s to write a biography of the chemist and science educator, Sir Edward Frankland (1825-1899). While exploring the rich archives of the Royal Institution for Frankland material during a sabbatical leave in 1968, I began to read the fascinating journals of the physicist John Tyndall, and of the mathematician Thomas Archer Hirst. The latter's journals were edited by Roy MacLeod and myself over a period of several years and eventually published by Mansell in a microfiche edition as *Natural Knowledge in Social Context. The Journals of Thomas Archer Hirst FRS* (1980). Hirst's involvement in the creation of the Association for the Improvement of Geometrical Teaching, led directly to the writing of essay XII, as well as to a joint essay with Roy M. MacLeod on the scientists' Declaration of 1864, which MacLeod will include in the second of his Variorum collections. During the 1970s, working under the auspices of the then Social Sciences Research Council, MacLeod gathered a team of historians of science together to investigate the significance of the Natural Science Tripos examination system which the University of Cambridge introduced in 1851. The question of the education of candidates before they went to Cambrdige inevitably came up, and essay VIII was an attempt to show how schools were influenced by Cambridge University's new-found interest in examinations.

My doctoral thesis (1966) had been concerned with the chemical, as opposed to, medical career of William Prout (1785-1850), who was one of the authors of the eight Bridgewater Treatises composed to demonstrate the existence of God from the evidence of design in Nature. Essay V on the Bridgewater Treatises was a spin-off from this graduate work. In retrospect, I recognize it as an adumbration of my interest in Victorian Studies, where the theme of Science and Religion has been an important element in both my undergraduate and postgraduate teaching. In 1966 I became a member of the Victorian Studies Centre which Leicester had launched as an interdisciplinary approach to postgraduate teaching of nineteenth-century history, literature and culture. Between 1974 and 1990 I was to direct the activities of the Centre, which included the organization of conferences on Victorian themes. Item VII was written for such a conference on the theological, literary and scientific work of the clergyman-novelist, Charles Kingsley, which I organised in 1974. Two other cultural essays from my 'Victorian' period, VIII and IX, arose from conferences on Alexander von Humboldt arranged jointly by the German Historical Institute in London and the Royal Society in 1983, and one on the remarkable Anglo-German chemist, August Wilhelm von Hofmann arranged jointly between the Deutsche Chemische Gesellschaft and the Royal Society of Chemistry in the recently re-united Berlin in 1992.

Although my biography of Frankland was never started (it has subsequently been brilliantly executed by Professor Colin A. Russell), it did lead to teaching engagements with the Open University during its early formative years. Unpublished item XV is a slightly adapted version of a radio script made for the AST281 'Science and Technology' course in 1973 under the auspices of Colin Russell and the producer, Mary Hoskins, with whom I also enjoyed making a television film on Frankland and his role in nineteenth-century science teaching. Frankland, together with Tyndall and Hirst, were all teachers at Queenwood College in Hampshire in the 1840s and 1850s. In the unpublished essay XVII I have focused less on the College's fame as an early institution for science teaching than on the school itself and its remarkable headmaster, George Edmondson. All of my long-term interest in the history of science education is effectively summarised in essay X, which was commissioned as a subject entry for the *Companion to the History of Modern Science* (1990), a volume edited by colleagues at the University of Leeds, which is proving invaluable in teaching the history of science.

Besides the history of science education, my other primary research interests have always been the history of chemistry, and the development of science periodicals. The former interest, now encapsulated in my *Fontana History of Chemistry* (1992; American edition, *Norton History of Chemistry*, 1993), had begun with research on nineteenth-century chemists' scepticism towards the atomic theory of matter. This led to my first book, *The Atomic Debates* (Leicester, 1967). During the course of this research, which focused on the work of Sir Benjamin Collins Brodie, I discovered the short-lived journal, *The Laboratory*. Among other fascinating contents, this contained some reminiscences by Faraday's erstwhile laboratory assistant, William Bollaert. An invitation to address the Library Circle of the Royal Institution, when I was spending study leave there in 1968, led to essay VI. My work in the history of chemistry is reflected in the present collection of essays by items III and IV. Together, these two essays offer some insights into the social organisation of British chemistry in the mid-nineteenth century, as well as bearing directly on the patronage of science. The latter issue was the subject of an important conference at the Royal Society which was organized by Gerard L'E. Turner in 1972 for the British Society for the History of Science in conjunction with the History journal, *Past and Present.* Although I was not an invited speaker, the late Trevor Aston kindly suggested that I should work up my 'from the floor' comments on other papers. The result was essay I, which I have been pleased to find frequently cited in subsequent studies of the patronage of science. Its companion piece, essay II, was composed in celebration of the 150th anniversary of the foundation of the British Association for the Advancement of Science in 1981, and refers specifically to the Association's role in patronising nineteenth-century scientific research.

An article on Victorian commercial science journals published in A. J. Meadows, ed., *Development of Science Publishing in Europe* (Elsevier, 1980) subsequently led to a commision by Taylor & Francis, Britain's oldest surviving science publishing house, to write their history. In the event, *The Lamp of Learning* (1984), which was written jointly with A. J. Meadows, was a general history of science publishing since the 1780s. It led me subsequently to publish studies of *Chemical News* (in the *Bulletin of the History of Chemistry*, 12(1992), pp. 30-35), an analytical bibliography of journals of microscopy for the 150th anniversary of the Royal Microscopical Society (*Journal of Microscopy*, 155(1989), pp. 248-266), an analytical historical survey of chemical periodicals in my *Fontana History of Chemistry*, and a bibliographical guide to the whole subject in J. Donn Vann and Rosemary T. Van Arsdel, *Victorian Periodicals and Victorian Society* (University of Toronto Press: Toronto, 1994). In the present volume, this long-term research interest in periodicals is demonstrated by an unpublished essay, XIV, on the *English Mechanic* that was developed at various meetings of the History of Education Society, the Periodicals Group of the University of London, and the British Society for the History of Science between 1980 and 1994. Serendipity certainly played its role here, for had the University of Leicester not acquired a run of *English Mechanic* in the late 1970s, this piece would never have been written.

When the enterprising James Wilson of Rugby School began teaching science in the 1860s he reflected:

> At present science is only taught by clever and enthusiastic men who could teach anything. But when it is taught by everybody there will be another story.

The essays gathered together in this volume and written over a period of thirty years offer support for Wilson's observation. Making science available to all has proved a tricky problem and is still not resolved to educationists' satisfaction. It is the belief of most historians of science, myself included, that our discipline has much to offer in the diffusion of scientific understanding in society. Perhaps these essays in the history of Victorian science and education will also cast some light on science, culture and education as we move collectively into a twenty-first century in which 'Science for All' will remain on the agenda of priorities.

Leicester, 1995 WILLIAM H. BROCK

ACKNOWLEDGEMENTS

I wish to thank the following editors and publishers for their permission to reproduce the articles included here: Professor G. L'E. Turner and Kluwer Academic Publishers, Dordrecht, The Netherlands (I); Professor R. M. MacLeod, Dr P. M. Collins and Science Reviews Ltd, Northwood, Middlesex (II); Taylor & Francis, London and Basingstoke (III, IV, IX, XII); The Royal Society (V); Professor Peter Day, FRS and the Royal Institution of Great Britain (VI); Professor J. C. Amalric and *Cahiers Victoriens & Edouardiens*, Université Paul Valéry, Montpellier (VII); Professor C. Meinel, Dr H. Scholz and Verlag Chemie Verlagsgesellschaft GmbH, Weinheim (VIII); Drs R. C. Olby, G. N. Cantor, J. R. R. Christie, M. J. S. Hodge and Routledge, London (X); the History of Education Society (XII); Kluwer Academic Publishers, Dordrecht, The Netherlands (XIII); the British Society for the History of Science (XVI); and Professor R. M. MacLeod and Studies in Education Ltd, Nafferton, East Yorkshire (XVIII). Essay XIII was originally written and jointly published with Dr M. A. Price. I am grateful to him for allowing me to reprint the essay here. Acknowledgement is also gratefully made to the Open University for permission to adapt the script of a radio broadcast in the course AST 281, 'Science and the Rise of Technology since 1800'delivered in 1973(XV); and Cambridge University Press for permission to adapt seventeen paragraphs from my Introduction to *H. E. Armstrong and the Teaching of Science* (1973) in item XIX.

PUBLISHER'S NOTE: *The articles in this volume, as in all others in the Collected Studies Series, have not been given a new, continuous pagination. In order to avoid confusion, and to facilitate their use where these same studies have been referred to elsewhere, the original pagination has been maintained wherever possible.*

Each article has been given a Roman number in order of appearance, as listed in the Contents. This number is repeated on each page and quoted in the index entries.

I

The Spectrum of Science Patronage

> Science in England does everything but pay. You may earn praise but not pudding! T. H. Huxley

The development of government participation 'in the support of research' has been described as 'one of the most significant characteristics of nineteenth-century science'.[1] Nevertheless, it may well be asked whether *ad hoc* patronage from outside the State sector was not only more widespread than historians have noticed, but also of equal, if not greater, importance than government patronage in the professionalization and institutionalization of science.

Of course, such a sweeping assertion demands qualification. Any statement concerning the amount or extent of private individual patronage, as opposed to patronage by the State, would be extremely difficult to quantify. As Professor Joad might have said: 'It depends what you mean by *science*'–the patronage of science teaching, of research, or of science in industry, or all of these things? Granted, as J. B. Morrell has demonstrated, that as scientific research became more and more expensive after, say, the 1860s, State patronage became the inevitable long-term economic solution. Nevertheless, it is worth drawing attention to the range and significance of individual patronage of science teaching, research and industrial science during the nineteenth century. In so doing, I shall view the matter not through the eyes of patrons, but through those of the patronized—the seekers of patronage—the scientific workers themselves. It is clear, for example, that the ability of American universities to innovate fresh developments in science teaching and research in both the nineteenth

Reprinted by permission of Kluwer Academic Publishers.

and twentieth centuries has been directly related to their openness to outside sources of support.[2]

This essay arose from a comment on the papers of D. S. L. Cardwell, R. Fox, and J. B. Morrell,[3] which together demonstrate the important roles that self-help, voluntaryism and individualism have played in the patronage of science in the nineteenth century. In the case of Great Britain their role was positive, supplementing and complementing the State's expedient and utilitarian patronage, which is considered at length by R. M. MacLeod in Chapter IV; in the case of France, their virtual absence, together with the standardization and bureaucratic character of the State's patronage, helps to explain the 'decline' of France as a scientific force in Europe later in the century. Other recent studies, notably by Miller,[4] have revealed the extent and significance of the *ad hoc* philanthropic provision of endowments and patronage in the United States of America. Just as Charles Babbage has seriously misled historians of ideas concerned with the 'decline' of science in Great Britain in the 1830s, has he not equally misled the historians over the role of State support for science in France (as Gerrit Moll hinted), and, by implication, Germany? Or, to put it more generally, have not historians been misled by those Victorian advocates of 'declinism' and of State support for science who looked to Germany for a model, as if it were the ideal and only valid system of patronage? Should not administrative, educational and science historians regard the German system of patronizing teaching and research as a nineteenth-century anomaly, rather than holding it up as a model simply because it proved influential in the development of twentieth-century systems of research support and allocation, or because it offered Victorian scientists a further method of academic professionalization?

Victorian men of science in Great Britain undoubtedly found it expedient to use the threat of industrial recession and loss of scientific supremacy, because they were concerned to professionalize science; but it is becoming increasingly doubtful whether Germany's provisions for, and investment in, university education and research was the most crucial factor in her success between 1860 and 1914.[5]

It is not my intention to provide pat answers to these admittedly rhetorical questions, but to question the assumption that State support for science was the ideal method of patronage, and that in its absence, the fullest development of a science-based society was retarded. It should be noted, for example, that patronage of the sciences in nineteenth-century Germany did not come exclusively from the State. Although Liebig's laboratory at Giessen was later held up as an example of the beneficient effects that would follow from State patronage of science,[6] it was, in the first instance, purely a privately-financed speculation on his part in order to recruit apothecary students who would qualify for a professional licence only if they had attended practical classes. Even in Germany, then, private individual effort was expected and considered normal—especially in the most expensive of sciences, astronomy and chemistry. Following the opening of the spacious and costly (£10,000) chemical laboratories at Leipzig by the Saxon government in 1872, the Professor of Chemistry, Kolbe, could appeal for further State aid, and point out that a professor's individual financial support and subsidy was still necessary to make the system work; a system whose benefits, he hastened to stress, were the material prosperity of the nation.[7] We see the promotion of individualism most clearly in the German *Privatdozent* system, which gave a young scholar the opportunity to seek a patron—the State—by supporting himself through teaching and research. But a private income and a relatively inexpensive research programme were really required in this search for patronage.[8]

Undoubtedly, one of the reasons for the relatively smooth relationship between research, the professionalization of science, and the development of industry in Germany was financial. But the historians' view of Germany as a country that promoted science exclusively by lavish State patronage must be qualified. It is not so much that Germany was anomalous in the extent of her State patronage, but that State support for science was only one of many varieties of patronage at work in the nineteenth century. There was a 'spectrum' of patronage, the colours of which were more vivid and intense in some countries than others, or in certain periods than in others. German conditions promoted State patronage of an intensive kind, but other colours of the spectrum, notably industrial and commercial patronage, were also present.

In the analysis of this spectrum, the questions we are faced with are: How did a nineteenth-century scientist earn his living, or how did he finance his pursuit of science? What were his sources of income? What career incentives were there to become a full-time man of science? What patrons could a scientific tyro find? Patronage in its original ecclesiastical and civil contexts meant the placement of clergy in livings and 'the recruitment of public servants by private recommendation',[9] but as Richard Johnson has carefully explained in connexion with his study of the British Education Department, the nature of civil (though not ecclesiastical) patronage altered considerably in the Victorian period. He explores the nature of this change by distinguishing two aspects of patronage:

> First, patronage requires some personal link, however indirect, between patron and client, and systems of patronage may vary according to the nature of this Nexus. Secondly, patronage may, in practice, be exercised more or less functionally for the job in hand. The criteria used in placing a man may be intrinsic to the job (e.g. appropriate expertise) or may rest on extrinsic considerations (e.g. the needs of the client or the private and political interests of the patron).[10]

Although Johnson's distinctions are particularly helpful in exploring the changing nature of Civil Service recruitment in Victorian Britain, they may also be usefully applied outside the State sector, where personal links were just as significant, and where there were increasing needs and opportunities for scientific and medical 'expertise'.[11] We may delineate at least six areas of nineteenth-century patronage that, in practice, applied in varying degrees to all scientifically-developed and developing countries: State patronage; industrial, commercial and legal patronage; corporate patronage; proprietorial patronage; publishers' patronage; and the most fundamental and general phenomenon, inter-scientific patronage. Each of these will be briefly examined—briefly, because what has become most apparent in making these remarks is how little work has been done in these areas by historians of science.

The Spectrum of Science Patronage

State Patronage

In 1869, two-thirds of the income of the ten Prussian universities came from the State, the remainder being more or less supplied by students' fees.[12] However, in Britain and France, as Morrell and Fox have argued,[13] government patronage was usual only when there were obvious commercial, military or national benefits. Babbage, writing as splenetically in 1851 as he had in 1830, observed that State patronage could be both financial and honorary.

> What are the salaries of the highest offices to which the most successful [men of science] may aspire? What are the honorary distinctions which the most eminent can attain?
>
> Offices of a strictly scientific nature are few, and their salaries are generally of small amount: amongst them are—
>
> A few of the professorships at our universities (i.e. Regius chairs)
>
> The Astronomer Royal
>
> The Astronomers of some of our Colonial observatories
>
> The Master of Mechanics to the Queen
>
> The Conductor of the Nautical Almanac
>
> The Director of the Museum of Economical [*sic*] Geology and of the Geological Survey
>
> Various officers of the same institutions
>
> Some of the officers in the Natural History department of the British Museum. The most valuable of these, that of Astronomer Royal, receives about £1,300 a year, including a pension of £300.
>
> Thus there is amongst this class one solitary prize of at the utmost £1,300 a year, and that is confined to one department of science. Offices for which men of science are at least as fit as any other persons, are numerous, though they are very rarely attained by those who pursue it.[14]

However, it should be clear from the 'less obscure' government institutions cited by Morrell that the picture painted by Babbage, while undoubtedly correct concerning 'the highest offices' or the patronage of scientists for non-scientific roles, was far too gloomy. Indeed, it is worth going further than the parliamentary estimates showing the 'regular expenditure by the state on its chief scientific institutions' cited by Morrell (Table 1), and enumerating the scientific employment which several government departments and institutions were able to give, or were increasingly forced to give, by the altering nature of government administration. Needless to say, these aspects of government have received very little attention from historians of science,[15] while the employment and use of scientists has only been of incidental interest to administrative historians.[16] Obscure though some of these avenues of patronage may have been, they nevertheless offered livelihoods, or supplementary support, to many ordinary men of science; not merely to the innovators that Babbage was exclusively concerned with, but Bacon's 'rank and file'.

The Mint. In 1861, the Assay Office of the Royal Mint[17] was reorganized under the leadership of a new scientific Master of the Mint, Sir John Herschel. His predecessor, R. L. Shiel, had already recommended that more than one assayer should be employed in testing coins, on a fixed scale of fees, and from different laboratories, in order to produce independent analyses and avoid fraudulent collusions. Herschel approved this plan to employ 'non-resident assayers' to rate ingots as they came from the melting pots, and in verifying the composition of coins before they were issued. As a result, the chemists W. A. Miller and T. Graham were appointed non-resident assayers, A. W. Hofmann assuming Graham's place when the latter succeeded Herschel as Master in 1856. When Hofmann returned to Germany in 1865, he was replaced by John Stenhouse. Thus, between 1850 and 1870, at least, the Mint patronized and, *ipso facto* subsidized chemical research.

Non-resident assayers received 2*s*. 6*d*. (12½p) (increased by Graham to 4*s*. 6*d*. (22½p)) per assay, but in busy times, such as in 1860 when bronze replaced copper for coinage, and much of the Mint's work had to be sub-contracted, the fee was raised to 10*s*. (50p) per assay. The fees earned by the non-resident assayers were, therefore, very considerable indeed.[18] During the decade 1855 to 1865, Hofmann earned a total of £6,341 from such assay work.[19] This produced a fairly steady average annual income of over £600, from which he was able to pay two assistants at the Royal College of Chemistry (at £50 per annum each) and probably make a clear 'profit' of £500 per annum. Moreover, the assistants, who performed the actual assays under Hofmann's direction, were free at other times for 'purer' chemical researches. It may be fairly claimed, therefore, that the Mint's need for 'experts' after 1851, assisted and promoted the ordinary scientific investigations of these experts. This was not a unique British phenomenon. In France, in 1848, the brilliant organic chemist Laurent replaced E. M. Peligot as assayer in the Paris Mint, where he was able to convert a damp cellar into a private laboratory. Indeed, the need for State assayers is probably one of the oldest forms of State patronage of chemistry and metallurgy.

Inland Revenue and Excise Department. Like the Mint, the Commissioners of the Inland Revenue (and their equivalents in other countries) have long had a symbiotic relationship with science, because of their commitment to standards and standardization for purposes of taxation and financial control. For example, in 1805, Thomas Thomson and two other chemists were asked to investigate the relative quantities of malt which could be made from English and Scottish barley. During this inquiry, Thomson invented a saccharometer which the Scotch Excise Bond adopted, and for which he was rewarded. In 1820, the English Excise Board paid Thomson £200 for four months' work on the strength of ales and porter—which is to be contrasted with his current government salary, as a Regius Professor at Glasgow, of £50 per annum.[20]

Intermittently, throughout the nineteenth century, scientists were employed as examiners by the Inland Revenue on an *ad hoc* basis in order to solve fiscal problems. Between 1854 and 1856, Hofmann, together with Graham and the pharmacist Theophilus Redwood, were commissioned to study the denaturing of spirit, and its commercial use prior to, and following, legislation for the sale of duty-free industrial alcohol in 1855 (18 & 19 Victoria, cap. 38). Their report suggested that crude wood naphtha had few advantages for commercial users, and many disadvantages. They therefore recommended that a denatured spirit composed from 'a mixture of spirits of wine with ten per cent of purified wood-naphtha or methylic spirit . . . may be designated *methylated spirit* for convenience and freed from fiscal control'.[21] It is interesting to notice how, having been patronized by the State, scientists, *qua* 'experts' were extremely well placed to advise on fiscal policy. In the present example, Hofmann, Graham and Redwood saw the creation of duty-free alcohol as an act of fiscal enlightenment:

> The command of alcohol at a low price is sure to suggest a multitude of improved processes and of novel applications, which can be scarcely anticipated at the present moment. It will be felt far beyond the limited range of trades now more immediately concerned in the consumption of spirits; like the repeal of the duty on salt, it will at once most vitally affect the chemical arts, and cannot fail ultimately to exert a beneficial influence upon many branches of industry. The same measure also practically removes one of the last anomalies in the duties imposed for revenue, alcohol having, as a raw material of manufacture, a claim to exemption from duty according to sound principles of taxation.[22]

Earlier, in 1852, the same team had advised the Inland Revenue on the original gravities of worts used, before fermentation, in the brewing of beers, in order to help quantify a taxation qualification permitted under 10 Victoria, cap. 5.[23] The need for scientific knowledge in dictating standards of fiscal control was responsible for the Inland Revenue paying for its recruits to attend courses at the Royal College of Chemistry, the Royal School of Mines, and the Chemistry Department at University College.[24] Its recruitment and educational policy would obviously repay further attention by administrative and science historians, especially since, although it certainly sent recruits to University College in the 1850s and 1860s, Williamson could complain in the 1870s that when it came to government science appointments, recruits from the government-supported schools were preferred.[25]

The Admiralty and the War Department. Morrell has drawn attention to the way the Admiralty supported astronomical and hydrographical research. In fact, the Admiralty and the War Department were the biggest government departmental spenders. By their very nature they were concerned with an increasing number and variety of scientific and technical problems; hence, as Gillian Sutherland has commented, they were the departments which had to face the administrative problems of working with 'experts' long before other departments.[26] There is evidence that the Treasury—the overlord of the Admiralty and the War Department—found it extremely difficult to control, or *not* to patronize experts and permit professional/technical expenditure. This, as Wright has found, applied with particular force because these departments (and the Medical Department) 'could argue with some justification that the Treasury could not make informed judgements on subjects about which it knew very little'.[27]

Clearly, science had much to gain from such 'enforced' patronage; unfortunately neither the Admiralty nor the War Department has received any direct attention from science historians.[28] For example, enormous sums of government money were spent on the health of the Army and Navy, drugs and chemicals being supplied directly by the Society of Apothecaries' laboratory in London. Consequently, the Society's operations were of considerable size and scope, offering employment to a large number of pharmacists and workmen. Between 1805 and 1810, for example, the Royal Navy patronized the Society to the extent of £24,917 per annum for medicines.[29]

One of Hofmann's most distinguished English pupils was Sir Frederick Abel (1827–1902), who made his career as a chemist of the War Department from 1854 to 1888 before becoming the first Director of the Imperial Institute. At the Woolwich Arsenal, he and a small group of chemists developed the study of explosives, including cordite. The development and financial status of this research group, together with its direct, but sometimes less than honest, competition with Nobel's activities, cries out for further research.[30]

Chemistry was not the only subject sponsored by the War Department. In his report to the Devonshire Commission in 1872, John Anderson, Superintendent of Machinery to the Department (later the Director-General of Ordnance Factories) reported that it had spent three million pounds on engineering works during his thirty years' service.[31] While most of this enormous sum of money had gone on the cost of raw materials for munitions, much undoubtedly went on the salaries of those employed on research and development.

Medical. The nineteenth century was the century of sanitation, a movement sponsored and produced initially by engineering and latterly by science. Insofar as the whole sanitary movement involved the collection of evidence, the establishment of local Boards of Health from 1848 onwards, the appointment of scientific authorities on health, like John Simon to the Privy Council, and the creation of the Local Government Board in 1871, were all to provide many new career opportunities for physicians, general practitioners, chemists, civil engineers and pharmacists. From this 'emergency' patronage of experts came Medical Officers of Health, Public Analysts, and Borough Surveyors and Engineers. As Maurice Wright has pointed out, not the least interesting feature of Lambert's brilliant biography of Simon is the light it throws on 'the astonishing growth of the Medical Department from a small subordinate part of the Privy Council with a staff of one to an independent department with an annual expenditure of £16,000'.[32]

Both the sanitarians' idea that dirt produced disease, and the germ theory of the 1870s and 1880s, stressed the role of water in the improvement of public health and the alleviation of epidemic conditions. Throughout the Victorian period, both in Britain and on the Continent, the chemical analysis of water (which had been already promoted by the growth of spa towns in the eighteenth century) was important. Physicians, and, increasingly, professional chemists, were patronized by both governments and individuals to examine water supplies. Payment, which seems to have ranged from 5*s.* (25p) to 7*s.* 6*d.* (37½p) per analysis, soon came to provide a useful supplementary income for these 'experts' who, in their turn, frequently became professional water analysts.

Sometimes the State's concern and patronage could lead to concern and patronage by industry. The discharge of sewage into the Thames frequently rendered it foul and malodorous in hot weather. In 1859, the Metropolitan Board of Works asked Hofmann and Frankland to conduct research on the deodorization of the river. After much work, in which W. A. Miller was also involved, they recommended treatment of the stinking river with Dales's muriate of iron, a concentrated solution of ferric chloride.[33] It was claimed that the malodorous parts of the river would then form a harmless mud with the solution. However, this deduction was challenged by other chemists (Odling and Letheby) who argued that the mud would have to be dredged at additional cost, that the deodorization could be done more cheaply without the production of mud by the use of calcium chloride, and that, in any case, Dales's solution contained arsenic! During the scientific skirmish that followed, Dales employed two other chemists to defend their preparation against the arsenic allegation, while the Board of Works quietly dropped the matter.

Earlier, in 1850, the General Board of Health commissioned a report on the supply of water to the Metropolis. Analytical data was collected from Lyon Playfair (then chemist to the Museum of Practical Geology), Richard Phillips and Hofmann.[34] As a result, the Home Secretary, Sir George Grey, requested Hofmann, Graham and Miller to provide a scientific analysis of the data collected by the Board of Health and their recommendations.[35] Having been drawn into this field, Hofmann made many other chemical studies of London water supply, drainage and sewage disposal.[36] He was therefore an ideal choice of the Lords Commissioners of the Treasury to take over the monthly analyses of Thames water from R. D. Thomson in 1864. Hofmann was in turn replaced by Edward Frankland, who built a second career as the unofficial chief government scientist on the potability and supply of water and the pollution of British rivers.[37]

Although English examples have been chosen, patronage of science by government departments interested in public health, the supply of water and the disposal of sewage, could be paralleled in France, Germany and America. It is also clear that it would be extremely useful if historians possessed a breviate of Parliamentary Papers concerned with scientific and technical matters, and a list of the scientific experts who prepared reports for government departments, or who sat on, or gave evidence to Royal Commissions.[38] Was this expertise, and patronage of expertise, thinly or widely spread? Was there really an inner core of science-experts who were regularly patronized, as cursory examination suggests?

Education and Examinations. Augustus De Morgan complained in

1856 of the 'general mania for examining everybody by means of written answers to printed questions'.[39] Cardwell has called this method of raising standards by competition which swept Britain in the 1850s a 'remarkable social phenomenon', since its tentacles affected 'all classes of the community above the very humblest'. But the social effects were not confined to the examinees; what about the need for examiners? Cardwell enumerates ten examinations, of which three were established by the State, namely Board of Trade examinations for Masters and Mates (1850), competitive examinations for entry to the Indian Civil Service (1854), and Department of Science and Art Examinations (1859).[40] Even this ignores the military examinations, and Woolwich and Sandhurst founded in 1857[41] that became significant for patronizing London scientists like Tyndall, Frankland and Hirst. Indeed, all the examinations mentioned involved some science papers, thus offering scientists employment in setting and marking papers, or, in the case of the Department of Science and Art, offering a new career, that of a science teacher paid by the results his classes obtained in examinations.[42]

In his complaint against the examination system, Henry Armstrong observed:

> The discovery has . . . been made that examinations not only afford remunerative occupation, but also that if carried out on a sufficiently large scale, they can be made remunerative—indeed we may class examining among the new industries discovered in modern times.[43]

But admission that an examinations industry was created in Victorian Britain, must not disguise the fact that the remunerations received by men of science who acted as examiners were a not insignificant form of State, and private, patronage. What sort of fees were earned by examiners in Victorian England?

Although government assistance to British universities did not begin on a large scale until 1889, we may note that by the 1870s the University of London (a purely administrative and examining organization) spent £2,000 per annum on examinations, of which £1,700 was underwritten by the government. Nearly £1,000 was spent on science examinations as follows:[44]

2 examiners in chemistry, £175 each	£350
2 assistants in chemistry, £25 each	£50
2 examiners in experimental philosophy, £100 each	£200
2 examiners in botany, £75 each	£150
2 examiners in geology/palaeontology, £75 each	£150
Expenses of practical examinations	£60
	£960

Letters and journals of the period reveal the keen competition among scientists for examinerships. For example, the geometer T. A. Hirst was disappointed to find that J. J. Sylvester, with whom he dared not compete or offend, had applied for the vacant London University examinership in mathematics in 1865.[45]

The really big government patron, however, was undoubtedly the Department of Science and Art, which between 1853 and 1870 spent £51,884 on the preparation and marking of papers.[46] Although it must be borne in mind that half of these examinations were in drawing and applied art, a goodly proportion of this sum gave supplementary earnings, or even 'double salaries', to a group of London scientists who were teaching at the D.S.A.'s Metropolitan School of Science, namely Hofmann, Frankland, Tyndall, Huxley, Ramsay, and others within their immediate circle like Hirst and Debus. Indeed, it is noticeable that members of the X-Club were very prominent in the receipt of this government patronage.[47]

Honours and Pensions. When Babbage reviewed the state of British Science in 1851, he still complained, as he had done in 1830, about the absence of honorary distinctions for scientists and the State's failure to patronize science by an honours system.

> During the eleven years of the present reign [of Victoria], one solitary instance is to be found of a baronetcy given for science [J. W. Herschel], and that too occurred only at a festival (the coronation) at which baronetages and peerages were showered upon those whose sole claim was founded on the mere support of party.
>
> During the same interval, about half a dozen of those who cultivate science, have been knighted . . . When this is compared

> with the most successful prizes in the Army, the Navy, the Church, or the Bar, it shows at once the inferior position occupied by science.[48]

Babbage's reflection on the extent of political patronage compared with scientific patronage is fair. On the other hand, as MacLeod has shown, Peel had attempted to make the Crown civil list of pensions apply to scientific services to the State;[49] indeed, Peel would have liked to have developed the system into one of 'Fellowships' for young men, at the peak of their creativity, who lacked financial means for research.[50] Such ideals were soon dashed by political necessities. Consequently, the civil list remained an extremely minor form of State patronage of science throughout the nineteenth century. Only about thirteen per cent of the sum available as pensions money went to scientists. Moreover, the distribution of pensions remained élitist, only about six per cent of the total awarded between 1772 and 1914 being granted to scientists, and these usually when they were well beyond the active stage of their careers.[51]

Industrial, Commercial and Legal Patronage

The patronage shown by the State towards men of science through official inquiries could also come from industry, commerce and the law. Industrial consultancy, or employment and retainment as a scientific expert in patent actions or murder trials, became a lucrative sideline, or even the principal activity, for many scientists. Frederick Penny, teaching chemistry in the unremunerative chair at Anderson's University, Glasgow, where classes were small and the fees low, found patronage by exploiting his brilliant analytical talents in legal and commercial consultancy, instead of fulfilling the scientific promise he had shown when he tackled the problem of Prout's hypothesis in London.[52]

In a letter to Liebig written in 1853, Hofmann stressed that, financially, he was more than happy with life in England, where it was possible to earn between £8,000 and £9,000 per annum by acting as a scientific witness in a law court. Hofmann himself took part in the Medlock patent trial of 1860, which concerned red and purple dyes, Gerard's patent for aniline blue in 1861, 'Paraffin' Young's patent trial concerning paraffin and paraffin oils in 1864, and Fox Talbot's action in 1856 against Laroche concerning processes of photographic development.[53] Friends and colleagues were frequently ranged against each other by the industrialists, patentees and lawyers concerned. Faraday found himself opposing his older colleagues, Humphry Davy and W. T. Brande, in an insurance case (Severn & Kings, sugar bakers versus Imperial Insurance Co., 1820), while on more than one occasion it would seem to the historian that all the chemists in London were being shared by aggrieved parties in a patent action.[54]

Industrial consultancy became increasingly profitable in the nineteenth century. Perhaps the best-known case is that of Fleeming Jenkin (1833–85), who, after much experience in engineering design, formed his own consultant engineering partnership in London in 1861, and became with William Thomson (Lord Kelvin) chief consultant for Great Britain's cable-laying expeditions. Industrial and commercial patronage, of course, provided most engineers with their careers. However, in the case of a more academically-inclined engineer, such as Jenkin, it compensated him for the lack of patronage towards engineering shown by the University of Edinburgh.[55]

Such industrial, commercial and legal patronage was not, of course, in any way unique in Great Britain. One thinks of Liebig's investigations of 'spontaneous combustion', for example, while Germany's industrial prowess owed as much, if not more, to her commercial patronage of scientific expertise, than to the State's patronage of scientific higher education. Nevertheless, the work done by nineteenth-century scientists in such advisory, consultant or legal capacities has been usually neglected by historians, or relegated to footnotes.[56] But within the social context of the time, this kind of patronage is of considerable significance, both to our theme, and for the light it throws on the relationship between science, commerce, technology and industry in the nineteenth century.

Corporate Patronage by Learned Societies

By corporate patronage is meant the patronage that comes to individual scientists through their membership of, and participation in the affairs of, learned societies, or to science generally through the dispensation of funds, endowments, medals and honours by such societies. One of the principal reasons for the proliferation of learned societies in the nineteenth century, was publication.[57] To subsidize a costly scientific paper which advanced or enhanced the reputation of a scientist was, and still is, an important form of patronage. Following the reform of the Royal Society in the 1840s, which rendered Fellowship a hallmark of honour and distinction, a refereeing policy was instituted, and papers submitted by non-Fellows had not merely to be sponsored by a Fellow, but usually reported on by a Fellow expert in the field. It would be interesting to make a study of the rich collection of the Royal Society's referees' reports.[58] Was there a nucleus of Fellows or 'patrons' who refereed? Did the Royal Society and other societies (like the Chemical Society) foster new areas of study by patronizing the work of newcomers? Or, how far was the success of journals like *Chemical News* and *Nature* due to a failure by the Royal Society and the Chemical Society to patronize the younger men of science who were working in new or unfashionable fields? Only fundamental archival research can answer these questions in historical sociology; they are raised here only to demonstrate that 'recognition' by membership of a peer group and publication within the group's journal is a form of patronage.

Before the development of government research grants, corporate patronage could also be significant financially. The British Association really did *advance* science by the annual grants it distributed from its subscriptions, especially during its first two decades of existence. Between 1833 and 1850 it dispensed £1,500 and supported some 300 research projects. As Morrell has illustrated,[59] provincial scientists like P. G. Tait, who were not Fellows of the Royal Society, depended considerably on such support, which they obtained more easily than from the Royal Society. The latter's government grant of £1,000 per annum, which began to operate in 1850, has been studied as a system of patronage by R. M. MacLeod, who has underlined its importance not only as a 'working model for the first system of professional grants to be launched by the British Government' after the First World War, but because its dispensation of about £179,000 to 938 men and for 2,316 different projects between 1850 and 1914 was 'a landmark in the extension of State support' for fundamental research.[60] However, although technically a State grant, it was administered by the Royal Society, whose patronage it became. Inevitably, the Royal Society tended to patronize its own Fellows rather than other men of science; or even to patronize metropolitan Fellows and neglect those living in the provinces. There is a law of patronage—it shall not be equally distributed.

The Royal Society's earliest instrument of patronage was not financial, but the award of two gold Royal Medals worth twenty guineas each. These medals, which began to be distributed in 1826, were intended to be awarded 'in such a manner as shall by excitement and competition among Men of Science seem best calculated to promote the objects for which the Royal Society was instituted'.[61] MacLeod has shown that, like the civil list, the system of Royal Medals did not always patronize the most deserving. Often the Council of the Royal Society failed to reward the most recent scientific discoveries; it often excluded non-Fellows of the Royal Society by 'defining' the award solely in terms of papers which had appeared in the *Philosophical Transactions*, and, in general, honoured a smaller group of scientists than was truly active at any moment.

Few other learned societies were able to patronize their membership financially, or by honours, until the 1860s and 1870s, and for the means they were dependent on the inter-scientific patronage which is considered below.

Proprietorial Patronage

Institutions, other than learned societies, which relied on aristocratic, commercial, and limited industrial support, sometimes fostered scientific research and so patronized science. In London, the Royal Institution, the London Institution, the Society of Arts, the Royal College of Chemistry, and the Civil Engineering College at Putney, were all financed by private subscriptions, and possessed at one time or another a proprietorial share system. Although less prosperous and efficient than it had been in the eighteenth century, the Society of Arts continued to patronize the applications of science and art to industrial purposes. This it did through industrial exhibitions, the support of scientific education, endowed lectures, and prizes and medals for a whole range of inventions and improvements, from optical glass to agriculture.[62] Recent studies of the early years of the Royal Institution[63] and of the Royal College of Chemistry[64] have demonstrated the importance of proprietorial 'say' in the structure and activities of such institutions. The private patrons who created these institutions had a definite 'image' (or rather, a diverse set of 'images') of the role of science in the advancement of agriculture, of commerce and, usually, society itself, and their patronage was as much idealistic as utilitarian. Nevertheless, the limited facilities these patrons offered, were to provide niches for some of the most creative British scientists: Davy, Faraday and Tyndall at the Royal Institution, Grove and Armstrong at the London Institution, Hofmann at the Royal College of Chemistry, and Frankland (briefly) at the Putney College.

Proprietorial patronage could be impressively enthusiastic. When, in 1845, the Royal College of Chemistry needed £2,000 for its building fund, Hofmann reported to Liebig:

> Every method is being tried to collect the missing money by the Council members, especially by Sir James Clarke [the Queen's Physician], and Mr Bingham Baring [Lord Ashburton]. At the last Council meeting Dr Gardiner [Secretary of R.C.C.] was authorized to send out 10,000 letters to as many people, and not less than 170 agents were nominated for canvassing the country. Dr Gardiner is on his feet night and day to collect the money in the shortest time.[65]

Such canvassing methods did not always work (as they did in this case), and the promise of science seen in the proprietors' images of science seems to have been found wanting by about 1850. The Royal College of Chemistry failed financially because the proprietors lost faith in the success of their patronage of science. The London Institution, in contrast to the successful Royal Institution, never succeeded in formulating a coherent and continuous system of patronage, and ceased to patronize science altogether after Armstrong resigned in 1884; while the attempts to create a new middle-class 'University' for Wales at Neath (Gnoll College), which was to have provided half a dozen scientific chairs in the 1850s, met with dismal failure.[66]

I

The Spectrum of Science Patronage

Publishers' Patronage

Publication of books could be fairly lucrative for men of science, even before the present system of royalty payments to authors began during the late Victorian period. Contracts were usually for a lump sum of money in payment for the first impression of a book, with further payments for succeeding impressions or new editions, for overseas sales and translations. Unfortunately, Victorian biographers are usually reticent concerning the financial success of their subjects' books. It is clear, however, that certain publishers, notably Longmans, Taylor & Francis, John Murray and Macmillan, were able to offer considerable patronage to scientists.[67]

John Murray offered Charles Lyell 400 guineas for his unfinished *Principles of Geology*, for sale in two volumes in an impression of 1,500 copies at 30*s*. (£1.50) a copy, this representing half his profit. In the event, Lyell received 300 guineas for volume I, 160 guineas for volume II, and £370 for volume III, together with £210 for the second edition of the first two volumes.[68] Thus, by 1833, Lyell had earned over £900 from his writing. Darwin's *Origin of Species* was also published by Murray on the same sort of generous terms, that is, a half share of the profits.

The scientific printing clubs which began seriously in the 1840s also offered moderate patronage. Based upon the idea of subscription publishing, their interest in foreign scientific works meant they were ready to employ translators. In 1848, Playfair promised to furnish Frankland with some translation work for the Cavendish Society when Frankland returned from Germany.[69] Although Frankland did not take up this offer, the printed reports of the Cavendish Society show that Graham received £104 for his *Chemical Reports and Memoirs* (1848), and Henry Watts £195 for the first two volumes of his heroic translation of Gmelin's *Handbook of Chemistry* in 1849.

From the 1850s, textbook writing, article writing and translation became significant sources of income for men of science. The *Philosophical Magazine* and the *Lancet* paid well for translations, and scientists like Hofmann, Tyndall, Francis and Hirst, or Gerhardt in France, first launched themselves through this patronage by commercial publishers. Some scientists, like William Crookes, who never held an academic position, earned their livelihoods entirely as consultants and journalists. Crookes's contribution to science journalism was considerable; he was editor of the *Liverpool Photographic Journal*, the *Journal* of the London Photographic Society, *Chemical News*, and the *Quarterly Journal of Science*.

During the 1860s, schools also had the problem of finding suitable textbooks; for this was an area of publishing where previously there had been little demand. Publishers like Alexander Macmillan, who was closely involved with Huxley and his fellow-scientists, recognized this as an area of need, and commissioned many leading men of science, rather than teachers, to write elementary science texts. Consequently, by the 1870s, up-to-date and readable school, college, and university science textbooks had become available, and textbook writing had become a significant form of patronage.

Interscientific Patronage and Individual Philanthropy

We may distinguish financial and non-financial systems of mutual support and patronage between men of science. Corporate systems of mutual financial aid, and honours, such as that administered by the Royal Society and the British Association, have already been mentioned. But the fuller development of this system was ultimately dependent on individual scientific generosity. Individuals whose financial circumstances placed them above their peers, or who made their fortunes as young men, frequently played the role of science patrons. In chemistry and astronomy, for example, Warren De La Rue (1815–89), who made his fortune in printing, was noted for his encouragement of younger scientists, and for the way 'he gave money as well as his own time freely for the advancement of pure science'.[70]

The physicist, J. P. Gassiot (1797–1877), who, like De La Rue, helped to found the Chemical Society and support the London Institution, made his wealth from the wine trade. When the British Association took over the Kew Observatory in 1842, he presented it with instruments; in 1859 he created the Royal Society's Scientific Relief Fund, and in 1871 he made over securities to the Royal Society for making magnetic and meteorological observations.[71] Cardwell has mentioned the examples of endowments to Owens College by C. F.

Beyer and J. Whitworth (whose National Scholarships endowed in 1868 were a stimulus to the development of British technical education).[72] Probably the best-known example of private patronage in Britain is, however, Ludwig Mond's gift of £100,000 for the endowment of the Davy-Faraday laboratories at the Royal Institution in 1896.[73] He also gave £16,000 to the Royal Society for the continuation of the *Catalogue of Scientific Papers*.

By examining the histories of individual learned societies it would be possible to find many examples of individual generosity, either by way of financial endowments for research, as with G. D. Longstaff's gift of £1,000 to the Chemical Society in 1876, or patronage by medal, as with the same Society's 'Longstaff Medal' in 1880.[74]

Non-scientists also act as patrons, and nowhere is this philanthropic phenomenon seen more clearly than in America. Here Andrew Carnegie endowed the Mount Wilson Solar Observatory the traction-magnate C. T. Yerkes endowed the Yerkes Observatory in Wisconsin; while an astronomer, George Ellery Hale (1868–1938) was to become famous not merely as an astrophysicist, but as an administrator with a flair for finding rich patrons for new observatories.[75]

Equally important was the freemasonry of inter-scientific patronage, which is seen most obviously by the historian in the patronage of younger men by Laplace and Berthollet at Arcueil,[76] the mutual assistance and group pressure exerted by invisible colleges and cliques such as the X-Club,[77] the American Lazzaroni,[78] and the nebulous Cambridge network.[79] A great deal of material that would trace and delineate this kind of patronage remains unexploited; for example, the sponsorship forms for election of Fellows of the Royal Society, and the (usually) printed testimonials of scientists. Concerning the latter, it is not unusual to find one scientist being asked to supply testimonials for two or even more candidates for a job. In 1850, both Frankland and Stenhouse approached Hofmann for testimonials to support their candidature for the new chemistry chair at Owens College, Manchester. Hofmann expressed his dilemma to Liebig:

> Stenhouse's case seems of greater importance inasmuch as his success in Manchester would be of help for a future career in

I

> Glasgow (Regius Professorship). He has done a lot of work and although somewhat deficient in theory . . . one cannot deny that he has a good chemical nose. He gave an agreeably pleasant lecture to the Royal Institution some days ago. I know Frankland only from his work which certainly commends him highly.[80]

Clearly, scientific status measured by publication output weighed in Hofmann's choice of patronage, but so did extra-scientific factors like his personal acquaintance with Stenhouse,and not with Frankland. In practice, 'out of good conscience', Hofmann decided to give them both a good testimonial. Ironically, his correspondent, Liebig, was in an even more difficult position since he was asked for testimonials from Stenhouse, Frankland and Angus Smith, all of whom had worked at Giessen. For Liebig, Frankland was the best, though not wishing to offend the others, he wrote diplomatic testimonials for them.[81]

Focusing the Spectrum

The spectrum of patronage sketched in the six preceeding sections can be analysed from a different perspective, that of an individual scientist's career. How did a young man who was determined upon a scientific career find patrons? Having successfully launched himself on such a career, how did he maintain a relationship with his patron or patrons, and attract new forms of patronage? Was the mark of a successful scientist determined by the number of wavelengths of the patronage spectrum he absorbed? Or is this to confuse worldly financial success with intellectual or practical distinction?

Hofmann's name has been mentioned frequently. He was both financially successful and an innovating chemist. How does he reflect the spectrum we have analysed? His career at the Royal College of Chemistry was launched in 1845 by the patronage of Liebig, whose assistant he had been at Giessen, where he obtained his Ph.D. in 1841. Despite the College's financial insecurity, caused largely by the disillusionment of its proprietors with their patronage, the College proved an important training school for British chemists. Hofmann stayed in England for twenty years, accepting new patrons in 1853, when the College was absorbed by the government's School of Mines. He returned to Germany in 1865 as Professor of Chemistry at the

University of Berlin, partly disillusioned, one suspects, by the failure of his State patrons to create a technical university around his laboratory, as his Royal patron, Prince Albert, had hoped.[82]

Hofmann, who made a fortune from his subject in both England and Germany, was one of chemistry's most cosmopolitan and placcable characters. When he arrived in London he immediately admired the seriousness of purpose of the proprietors' patronage of the Royal College of Chemistry. Accordingly, he attended Brande's and Graham's lectures at the Royal Institution and University College in order to improve his own. Life in London proved ambivalent. It was 'a peculiarly gay, endlessly varied confusion, stimulating and refreshing for the fortunate and hopeful, but discouraging for the depressed'.[83] If the finances of the College and his patrons were desperate, Hofmann found there were large sums of money to be made from private and government patrons who needed chemical expertise to solve industrial, commercial, or legal problems; as he advised Liebig cynically, do not 'lift up your hand without first making sure that it is counterbalanced with gold'.[84] This cynicism and opportunism were perhaps visible publicly, for T. A. Hirst recorded in his private journal that 'by robbery and intrigue [Hofmann] manages to seek an income of 12 or 14 hundred pounds a year out of England'.[85] Was Hirst merely jealous, or was he being over-altruistic? Certainly the private papers of Victorian scientists suggest that the need to find patrons was more important than altruism. Such papers also reveal the sound commercial sense of most scientists when driving a bargain with patrons. Walton, his English publisher, agreed to pay Liebig £150 for the third enlarged edition of his *Familiar Letters on Chemistry*. (This was for an edition of 4,000 copies.) However, through the agency of Hofmann, by publishing the four new 'Letters' of this edition in Wakley's *Lancet* beforehand, Liebig earned another £150. When Hofmann and Graham investigated rumours that the Burton-brewed beers of Allsopp and Bass were contaminated by strychnine, Hofmann arranged for Liebig to be paid £100 for writing an open letter in praise of the English brew. His main chemical test 'consisted in my drinking a bottle with great enjoyment'.[86]

Hofmann, then, found patronage from the State in England (and Germany), he was patronized by the legal profession and by commerce, he was patronized by publishers and editors for his ability in translation, he was patronized by the Chemical Society and the Royal Society, by private institutions, by proprietors like Ashburton and Clarke, and by scientists like Liebig. Hofmann, at least, seems to reflect all the colours of the patronage spectrum.

It may well be thought, however, that Hofmann's experience of patrons was atypical, either because chemistry provided greater and wider opportunities for patronage, or because Hofmann was an influential and distinguished figure. This is probably true. All the same, lesser mortals, although not always successful in these bids for patronage, seem to have been clear about the variety of patronage available. Lesser mortals also reveal the great importance of interscientific patronage.

Consider Thomas Archer Hirst (1830–92), Hofmann's jealous critic. Hirst, a representative mathematician, is not at all well-known to historians, but we can learn a good deal about him and his friends because he kept a Journal for most of his life.[87] A Yorkshireman, Hirst received a slight acquaintance with mathematics and science at the West Riding Proprietory School (later absorbed by Wakefield Grammar School) before being apprenticed to a railway surveyor and civil engineer at Halifax in 1845. There is little doubt that Hirst would have become an engineer had not he become the friend of John Tyndall, then a fellow surveyor during the 'railway mania' of the 1840s. Tyndall became Hirst's mentor, extending his intellectual horizons, and awakening new ambitions in him. 'There is no person of whose acquaintance I feel more proud', Hirst wrote on 15 August 1848, 'No friend for whom I have more regard, and for whose abilities I have more esteem. May that acquaintance, from which I have already reaped so much, and from which I hope to reap more, continue . . .'

It was through Tyndall's influence that Hirst decided to study mathematics with Stegmann at the University of Marburg between 1850 and 1852, and by the same 'angel of mercy, guiding star and helpmate' he became a mathematics teacher at the proprietorial Queenwood College in Hampshire, from 1853 until 1856, at the considerable salary of £200 per annum. 'The profession of schoolmaster is no drudgery', he decided, 'but when properly undertaken [it is] a noble task, and a healthy discipline' (14 August 1853). Unfortunately, it left little time for original mathematical work; moreover, the tragic death of his Irish wife in 1857 left him lonely, and cut off from his London friends, especially from Tyndall, who had received the patronage of the Royal Institution in 1853. Nevertheless, until 1860, Hirst devoted himself entirely to mathematics, studying with Chasles in Paris and Cremona in Rome, supporting himself on a small income derived from his parents' and wife's estates. Returning to London in 1860, he became a mathematics teacher at University College School through the influence of a fellow mathematician, J. J. Sylvester. An abundance of patronage now followed. In April 1861, through the efforts of another fellow student at Marburg and teacher at Queenwood, Edward Frankland, he was elected to the Royal Society. At about the same time, Ballière, the publishers, commissioned him to translate Ganot's *La physique*, on the strength of recommendations from William Francis, for whose *Philosophical Magazine* he had already translated several German and French mathematical and physical articles. However, finding that the translation of Ganot interfered with mathematical research, he passed the commission over to another Marburg friend, Edmund Atkinson. In September 1862, Longman asked Hirst to revise all the mathematical entries in Brande's *Dictionary of Arts and Sciences*, which he did grudgingly for the money. In 1864, after having written an article for *The Reader*, he commented, 'I must avoid such temptations in the future, it is inconvenient to be poor but to me the moral effect is bad of writing for money unless I limit myself to translation or a purely mechanical reproduction of other views' (9 October 1864). In fact, Hirst's translations of Wiedemann, Kohlrausch, Verdet, Poinsot and, especially, of Clausius,[88] were important for the development of science.

In July 1864, with no other monetary prospect in view, Hirst resigned his teaching post at University School, in order to devote himself more fully to geometry. But in December he collected testimonials from Cayley, Salmon, Sylvester, Smith, Stegmann, Birkhardt, and Cremona in support of his application for the vacant chair of mathematics at Cork. The drawbacks were acute, however. Although 'the position and emolument would be acceptable enough, . . . the price at which I shall have to purchase both is great. It will involve my leaving London and changing entirely my friends and my life. To leave London is the greatest obstacle. For me it is the centre of all scientific interest, and there dwells, and will always dwell, my constant and true-hearted friend, John Tyndall' (1 January 1865). In the event, it was Tyndall who advised against taking the Irish post; but Tyndall did not demur a few months later when Roscoe suggested that Hirst should apply for the vacant chair of mathematics at Owens College, which was worth £400 per annum. Once again, however, the strings of interscientific patronage pulled in a different direction. This time it was A. W. Williamson, the Professor of Chemistry at University College, whose father-in-law, A. W. Key, was headmaster of the College's school and Hirst's former employer, who prevented Hirst's loss to Manchester. It was Williamson's hope that the chair of Experimental and Mathematical Physics could be made into two chairs, the former to be filled by Tyndall (it went to G. C. Foster) and the latter by Hirst. From 1865 to 1866, Hirst was therefore Professor of Mathematical Physics, then in 1867, in succession to De Morgan, Professor of Pure and Applied Mathematics, and from 1868 to 1871 he also acted as Secretary to the peripatetic British Association for the Advancement of Science.

The X-Club had formed in 1864, and its members, who included Hirst, gave one another mutual support and patronage. Never robust in health, Hirst resigned his position at University College in 1870, and through Lubbock's assistance he took the apparently inferior administrative position of Assistant Registrar of the University of London. One is tempted to call this a sinecure; the post was dull and routine, well below his intellectual abilities; on the other hand, the hours were not onerous. In 1873, through the influence of his fellow 'Xers', Huxley and Spottiswoode, he accepted the much more onerous post of Director of the Royal Naval College at Greenwich, newly created by the Admiralty. The salary was £1,200 per annum, together with a free house. The Admiralty's intention was to establish a naval university which would instruct engineers and naval dockyard personnel, together with executive officers, in mathematical and physical science. As Huxley noted, the Admiralty needed 'a man of scientific eminence to give the institution dignity with the public; a man who can understand and attract young men and get them to enter it'.[89]

Goschen was at this time First Lord of the Admiralty, and, according to Tyndall, the Minister was extremely sympathetic towards Hirst's mathematical studies, and willingly allowed him 'to arrange matters as to reconcile the prosecution of those studies with his duties as Director'.[90] Hirst retired in 1883 on a fairly generous pension of £417 per annum.

Hirst, then, although he ended his career with State patronage, and met with publisher's, corporate, and institutional patronage during the course of his career, is a good example of the force of interscientific patronage. It would, no doubt, be possible to look at other scientists' careers and find that they exemplify to a greater degree State patronage (e.g. G. B. Airy), or industrial patronage (e.g. F. Penny), or corporate patronage (e.g. E. Sabine), or institutional patronage (e.g. J. Tyndall) or publisher's patronage (e.g. J. N. Lockyer). But all careers will reveal a degree of interscientific patronage, since this is a fundamental characteristic of professionalization. The spectral categories of patronage which have been discussed are by no means exhaustive, or, possibly, the only means of analysis. The role of Army Officers, like Donnelly, Strange, and Abney, in science education and administration, may suggest that the Army (and *ipso facto* the Navy) should be examined as patrons separate from their role as government patrons.[91] This metamorphosis of patronage, or our inability to classify forms of patronage in any rigid manner, is also encompassed by the spectral metaphor. The categories of patronage are blurred at the edges; there is a horizontal distribution and diffusion of patronage resembling what Clark has called, for the institutionalization of higher education, a 'combined process model'[92] in which innovation and the creation and distribution of systems of patronage develop concurrently both inside and outside universities, scientific societies, proprietorial institutions, publishers' offices, government departments, and industry. The government grant to science which began in 1850 was a form of State patronage; but it was administered by the Royal Society, which, in turn, became a corporate patron; while its choice of Fellows and its choice of recipients for the Grant depended to a large degree on interscientific patronage. The nominations the Royal Society received for the Royal Medal were, for a long period, made by a small clique of the young guard of science.[93] Similarly, the State's patronage of science through problems in fiscal policy and control, or utilitarian legislation such as the Alkali Acts, could stimulate patronage among men of industry and commerce. The State's rather desperate attempts to patronize science education in the 1860s and 1870s did at least stimulate publishers like Macmillan to commission textbooks, while the textbook writers themselves came to publishers' attention through an informal system of mutual scientific patronage. It is this interaction, inosculation and diffusion, backwards and forwards among different categories, as well as the varying intensity of these catagories at different periods of history or at different periods of one man's career, that make the elucidation of the full extent of science patronage so difficult.

One final point requires emphasis. Although I have principally cited British examples, the categories of patronage enumerated were present in varying degrees during the careers of all nineteenth-century men of science. Taking a world-wide perspective, they were present as possibilities or actualities in varying degrees, in different countries, at different times.[94] It may well be that until the 1870s British science presents an overall impression of laissez-faire and self-help, and that Britain tended to rely on voluntaryist forms of patronage. But as I have tried to emphasize, other forms of patronage, including State support, were present to a degree, and were taken advantage of by men of science seeking their careers. In France, on the other hand, voluntaryist forms of patronage do not seem to have been present to anything like the same degree as in Britain. The French faith in a centralized educational administration patronized by the State was to be identified as a weakness by Bernard, Pasteur, Deville, Dumas, and others during the Franco-Prussian war. As for Germany itself, the country which first Britain, then France and finally America looked to as having solved the increasingly difficult problem of science patronage, the most intense lines in the spectrum were State and industrial patronage. The ignominy of Prussia's defeat by Napoleon helped to make State patronage of science and education seem a potent means of recovering prestige and strength. It was a successful formula, which, however, capitalized on the results won by Britain and France using other forms of patronism. It was a system which came to outweigh all other categories in the spectrum of patronage—though, undoubtedly, never to replace them entirely.

The Spectrum of Science Patronage

Notes and References

1. R. M. MacLeod, 'The Royal Society and the Government Grant: Notes on the Administration of Scientific Research, 1849–1914', *Historical Journal*, 14 (1971), 323–58 (p. 323); see also idem, 'Resources of science in Victorian England: the Endowment of Science Movement 1868–1900', in *Science and Society, 1600–1900*, edited by P. Mathias (Cambridge, 1972), pp. 111–66.
2. T. N. Clark, 'Institutionalization of Innovation in Higher Education: Four Models', *Administrative Science Quarterly*, 13 (1968), 1–25 (p. 16).
3. D. S. L. Cardwell, 'The Patronage of Science in Nineteenth-Century Manchester', above, pp. 95–113; R. Fox, 'Scientific Enterprise and the Patronage of Research in France 1800–1870', above, pp. 9–51; J. B. Morrell, 'The Patronage of Mid-Victorian Science in the University of Edinburgh', above, pp. 53–93. The original versions of these papers were read at a joint conference of the British Society for the History of Science and the Past and Present Society, held on 14 April 1972.
4. H. S. Miller, *Dollars for Research* (Seattle, Washington, and London, 1970).
5. Germany, through its pharmaceutical tradition, had a good working relationship between science and industry long before her states began to invest in higher education. Such pharmaceutical traditions also existed in Britain and France in the eighteenth century. The patronage of higher education was, no doubt, a necessary ingredient for 'success', but it does not seem to be a sufficient reason. My thanks to Mr B. Gustin, who intends to develop this point in his thesis on the professionalization of German Science. See also K. Hufbaur, 'The Formation of the German Chemical Community' (unpublished Ph.D. thesis, University of California, Berkeley, 1970).
6. For example, W. Gregory, *Letter to the Right Honourable George Earl of Aberdeen on the state of the schools of chemistry in the United Kingdom* (London, 1842).
7. H. Kolbe, *Das Neue Chemische Laboratorium der Universität Leipzig* (Brunswick, 1872), p. 1; idem, *Das chemische Laboratorium der Universität Marburg* (Brunswick, 1865), p. 16. The significance of a balance between State and individual subsidy for practical science teaching has been stressed by J. B. Morrell, 'The Chemist Breeders: The Research Schools of Liebig and Thomas Thomson', *Ambix*, 19 (1972), 1–46.
8. A. Zloczower, *Career opportunities and the growth of Scientific Discovery in Nineteenth-Century Germany, with special reference to Physiology* (Jerusalem, 1966), pp. 18–20.
9. S. E. Finer, 'Patronage and the Public Service: Jeffersonian Bureaucracy and the British Tradition', *Public Administration*, 30 (1952), 329–60 (p. 329).
10. R. Johnson, 'Administrators in Education before 1870: patronage, social position and role', in *Studies in the growth of Nineteenth-Century Government*, edited by Gillian Sutherland (London, 1972), pp. 110–38 (p. 115).

11. Of course, scientific and medical expertise was a factor in the transformation of the civil service. See below.
12. Evidence of A. W. H. Johnson, *Royal Commission on Scientific Instruction and the Advancement of Science* [hereafter Devonshire Commission]: *Minutes of Evidence (12 Feb. 1872 to 21 March 1873), Appendices, and Analyses of Evidence*, Parliamentary Papers [hereafter P.P.], 1874 [C. 958], XXII. 95 (p. 71).
13. See note 3 above.
14. C. Babbage, *The Exhibition of 1851* (London, 1851; reprinted 1969), pp. 191–2.
15. But see R. M. MacLeod, *Treasury Control and Social Administration: Establishment of the Local Government Board, 1871–1905*, Occasional Papers in Social Administration, London School of Economics, No. 23 (London, 1968), and his papers on the Alkali Acts, Alcoholism, Vaccination and the Salmon Acts cited therein.
16. However, see Gillian Sutherland's comments on 'zealots' and experts in her Introduction to Sutherland, ed., op. cit. note 10, pp. 1–10; also Maurice Wright, 'Treasury Control 1854–1914' in ibid., pp. 195–226; and especially, Royston Lambert, *Sir John Simon, 1816–1904* (London, 1963).
17. See Sir John Craig, *The Mint* (Cambridge, 1953), especially pp. 317–29, which, however, does not consider the Mint from the scientific employee's viewpoint.
18. Ibid., p. 323, suggests an average fee for these consultants of £1,278 per annum, i.e., approaching the salary of the Astronomer Royal, but I have not been able to corroborate this figure.
19. My thanks are due to G. P. Dyer of the Royal Mint for providing these details in a letter to my former student, J. Bentley, 16 June 1965; and to Mr Bentley himself I owe many details concerning Hofmann used in this essay.
20. J. B. Morrell, 'Thomas Thomson: Professor of Chemistry and University Reformer', *British Journal for the History of Science*, 4 (1968–9), 247–8.
21. *Reports of the successful application of Methylated Spirit, under the provisions of the Act 18 & 19 Victoria, cap. 38, to various purposes in the Arts, Manufactures, and Scientific Research*, P.P., 1854/5 (201), L.669.
22. Ibid., p. 681.
23. 'Report upon "Original Gravities", by Professors Graham, Hofmann, and Redwood', *Quarterly Journal of the Chemical Society of London*, 5 (1853), 229–56.
24. *Return of the Total Cost of the Excise Department, 1844 to 1885, Salaries of Officers and number selected for Chemical Education at University College, with cost; Sums paid to independent Chemists*, P.P., 1854/5 (509), XXX. 371. This gives details of courses followed.
25. 'If you go to the Government School, you will get government appointments, if not, not.' A. W. Williamson to Devonshire Commission: *First (Re-Printed), Supplementary, and Second Reports*, P.P., 1872 [C. 536], XXV. 1, qus 7,176–7,181. Compare the way students from Balliol College, Oxford, were patronized by the Education Department. See Johnson's article in Sutherland, ed., op. cit. note 10, pp. 119–21.

26. Sutherland, ed. op. cit. note 10, p. 6.
27. Wright, op. cit. note 16, p. 218.
28. Scientific aspects do not figure in the recent, excellent study by W. S. Hamer, *The British Army: Civil-Military Relations 1885–1905* (Oxford, 1970).
29. G. R. Paterson, 'Drug Entities in the Nineteenth Century', unpublished paper given to the British Society for the History of Pharmacy and the Pharmaceutical Society, 28 June, 1972. In the same period, the East India Company spent on average £21,582 with the Society.
30. W. T. Vincent, *Woolwich: Guide to the Royal Arsenal* (Woolwich, 1885). See also F. G. Guggisberg *'The Shop': The Story of the Royal Military Academy* (London, 1900). For some brief remarks on the notorious Abel-Nobel litigation, see R. W. Reid, *Tongues of Conscience* (London, 1969), pp. 1–9.
31. Devonshire Commission: *Third Report, with Appendix*, P.P., 1873 [C. 868], XXVIII. 637, qu. 11,923.
32. Wright, op. cit. note 10, p. 208.
33. 'Deodorization of Sewage', *Journal of the Society of Arts*, 7, no. 354 (2 Sept. 1859), 661–4.
34. *Report by the General Board of Health on the Supply of Water to the Metropolis*, P.P., 1850 [1218], XXII. 1.
35. 'Chemical Report on the Supply of Water to the Metropolis', *Quarterly Journal of the Chemical Society of London*, 4 (1852), 375–413.
36. *Reports to the President of the General Board of Health, on the Metropolis Water Supply*, P.P., 1856 [2137], LII. 251; *Report of the Chemical Investigations Relating to the Metropolitan Main Drainage Question*, P.P., 1857 (Sess. 2)(233), XXXVI, Appendix 1, pp. 1–60; *Report on the best means of utilising Sewage of Cities and Towns of England: First, with Evidence and Appendix*, P.P., 1862 (160), XIV. 321.
37. W. H. Brock, 'Edward Frankland', *Dictionary of Scientific Biography* (New York, 1970–), v (1972), pp. 124–7.
38. Such a listing has been started at the Victorian Studies Centre, University of Leicester, under the direction of Dr Shattock. See also Michael Argles, *British Government Publications during the Nineteenth Century*, History of Education Society (London, 1972).
39. *Athenaeum*, 6 December 1856, quoted in D. S. L. Cardwell, *The Organisation of Science in England*, second edition (London, 1972), p. 85.
40. Ibid., loc. cit.
41. *Report of Commissioners appointed by the Secretary of State for War on Re-organising the System for Training Officers for the Scientific Corps*, P.P., 1857 (Sess. 1) (57), VI. 1.
42. Something of the extent of the D.S.A.'s patronage of scientists can be gauged from H. Butterworth, 'The Science and Art Department 1853–1900' (unpublished Ph.D. thesis, University of Sheffield, 1968).
43. H. E. Armstrong, in *Chapters on the Aims and Practice of Teaching*, edited by F. Spencer (Cambridge, 1897), p. 223; reprinted in *H. E. Armstrong and the*

Teaching of Science 1880–1930, edited by W. H. Brock (Cambridge, 1973), p. 100.

44. Evidence of W. B. Carpenter, Devonshire Commission: *First* (*Re-Printed*), *Supplementary, and Second Reports*, P.P., 1872 [C. 536], XXV. 1, appendix 12, p. *41*.
45. Royal Institution Archives, The Journal of Thomas Archer Hirst, f. 1721. This Journal is to be published in four volumes by Mansell in 1976 under the editorship of W. H. Brock and R. M. MacLeod.
46. £12,494 between 1853 and 1860, and £39,339 from 1861 to 1870. Figures from *Seventeenth Report of the Department of Science and Art*, 1870, p. 534, See also Butterworth, op. cit. note 42, passim.
47. R. M. MacLeod, 'The X-Club. A Social Network of Science in late-Victorian England', *Notes and Records of the Royal Society of London*, 24 (1970), 305–22.
48. Babbage, op. cit. note 14, p. 193.
49. R. M. MacLeod, 'Science and the Civil List 1824–1914', *Technology and Society*, 6 (1970), 47–55.
50. J. M. Crook, 'Sir Robert Peel: Patron of the Arts', *History Today*, 16 (1966), 3–11.
51. MacLeod, op. cit., note 49.
52. For Frederick Penny (1816–69) see *Dictionary of National Biography* [hereafter *DNB*], and my notice in *Dictionary of Scientific Biography* (New York, 1970–), x (1974), 510–11.
53. *Chemical News*, 7 (1863), passim; ibid., 10 (1864), passim; ibid., 9 (1864), 249, 262; *Journal of the Photographic Society*, 2 (1856), 89.
54. For example, Simpson, Maule and Nicholson versus Wilson. See *Chemical News* 7 (1863) passim. Among those who gave evidence for one side or the other were Calvert, Frankland, Hofmann, Letherby, Medlock, Miller, Odling, Taylor, and Warrington.
55. Morrell, above, p. 84.
56. See, however, W. V. Farrar, 'Andrew Ure, F.R.S., and the Philosophy of Manufactures', *Notes and Records of the Royal Society of London*, 27 (1973), 229–324.
57. G. L'E. Turner, 'Microscopical Communication', *Journal of Microscopy*, 100 (1974), 3–20 (pp. 7 ff).
58. Referees' reports also shed much light on the *internal* development of science. For one such study, see S. G. Brush & C. W. F. Everitt, 'Maxwell, Osborne Reynolds, and the Radiometer', *Historical Studies in the Physical Sciences*, 1 (1969), 105–25.
59. Morrell, above, p. 84.
60. MacLeod, 'The Royal Society and the Government Grant', op. cit. note 1, p. 324.
61. R. M. MacLeod, 'Of Medals and Men: A Reward System in Victorian Science 1826–1914', *Notes and Records of the Royal Society of London*, 26 (1971), 81–105 (p. 83). Since 1736, the Royal Society had also awarded the Copley

Medal. A number of endowed lectureships, such as the Croonian and Bakerian, also dated from the seventeenth and eighteenth centuries. For a study of an American medal system, see R. Fox, 'The John Scott Medal', *Proceedings of the American Philosophical Society*, 112 (1968), 416–30. However, the John Scott Medal has only proved prestigious in this century; in the nineteenth century the patrons honoured gageteers alone.

62. Derek Hudson & Kenneth W. Luckhurst, *The Royal Society of Arts 1754–1954* (London, 1954), pp. 84, 117.
63. M. Berman, 'The Early Years of the Royal Institution 1799–1810: A Re-evaluation', *Science Studies*, 2 (1972), 205–40.
64. G. K. Roberts, 'The Royal College of Chemistry (1845–1853): A Social History of Chemistry in Early Victorian England' (unpublished Ph.D. thesis, John Hopkins University, 1973).
65. Letter A. W. Hofmann to J. Liebig, undated [1845]. An edition of this correspondence is being prepared by W. H. Brock and Elizabeth Wangermann.
66. *The Lancet*, 2 (1857), 318–19, gives A. Cayley as Professor of Mathematics and J. E. D. Rodgers as Professor of Chemistry. The College never opened. See *The Principles of Collegiate Education Discussed and Elucidated in a Description of Gnoll College, Vale of Neath, South Wales* (London, 1857).
67. John L. Thornton & R. I. J. Tully, *Scientific Books, Libraries and Collectors: A Study of Bibliography and the Book Trade in Relation to Science*, third edition (London 1971), pp. 360–1; R. M. MacLeod, 'Printing under the Golden Lamp: Taylor and Francis Ltd.', *Victorian Periodicals Newsletter*, 2 (1969), 11–12.
68. L. G. Wilson, *Charles Lyell. The Years to 1841* (New Haven, Connecticut, and London, 1972), pp. 270, 341–2, 375–6.
69. Frankland's Journal, 6 October 1848 (typescript copies in the Archives of the Royal Institution and the Royal Society).
70. *DNB*, s.v. Warren de la Rue.
71. *DNB*, s.v. J. P. Gassiot. See also F. G. Donnan, 'The Scientific Relief Fund and its Committee', *Notes and Records of the Royal Society of London*, 7, (1950), 158–71.
72. Cardwell, above, pp. 106f.
73. *DNB*, s.v. L. Mond.
74. T. S. Moore & J. C. Philip, *The Chemical Society 1841–1941* (London, 1947), pp. 38–40.
75. H. Wright, *Explorer of the Universe. A Biography of George Ellery Hale* (New York, 1966). For the general phenomenon of American science philanthropy, see Miller, op. cit. note 4 above.
76. M. P. Crosland, *The Society of Arcueil* (London, 1967).
77. R. M. MacLeod, op. cit. note 47; J. V. Jensen, 'The X-Club Fraternity of Victorian Scientists', *British Journal for the History of Science*, 5 (1970), 63–72.
78. M. Beach, 'Was there a scientific Lazzaroni?' in *Nineteenth-century American Science, A Reappraisal*, edited by G. H. Daniels (Evanson, Illinois, 1972), pp. 115–32.

79. W. Cannon, 'Scientists and Broad Churchmen: an Early Victorian Network, *Journal of British Studies*, 4 (1964), 65–88.
80. Hofmann to Liebig, 22 April 1850; see note 65 above.
81. Liebig to Hofmann, 12 April 1850. Liebig's and Hofmann's testimonials were reproduced by Frankland's daughters in *Sketches from the Life of Edward Frankland*, edited and concluded by M. N. W[est] & S. J. C[olenso] (London, 1902), pp. 122–3. Note that a confidential refereeing system does not seem to have come into practice much before the 1890s. Before then, the open printed testimonial system was used, and frequently abused by candidates. See J. B. Hope Simpson, *Rugby since Arnold*, (London, 1967), p. 134.
82. J. Bentley, 'Hofmann's Return to Germany from the Royal College of Chemistry', *Ambix*, 19 (1972), 197–203.
83. Hofmann to Liebig, 28 January 1855.
84. Idem, 26 June 1853.
85. Hirst, op. cit. note 45, 27 May 1855.
86. Liebig to Hofmann, 8 May 1852.
87. Hirst, op. cit. note 45.
88. R. J. E. Clausius, *The Mechanical Theory of Heat*, edited by T. A. Hirst, with introduction by J. Tyndall (London, 1867).
89. Hirst, op. cit. note 45, 29 November 1872. On the Naval College, see C. Lloyd, *Mariner's Mirror*, 52 (1966), 145–56.
90. J. Tyndall, obituary of Hirst, *Proceedings of the Royal Society*, 52 (1892–3), xviii.
91. W. Porter, *History of the Corps of Royal Engineers*, 2 vols (London, 1889). Both military and naval education underwent fundamental changes during the nineteenth century, while scientifically-trained engineers were increasingly required by both forces.
92. Clark, op. cit., note 2.
93. MacLeod, op. cit., note 61, p. 97.
94. Cf. the sociological perspective on patronage developed by Terence J. Johnson, *Professions and Power* (London and Basingstoke, 1972), especially pp. 41–7, 65–74.

II

Advancing Science: The British Association and the Professional Practice of Science

Introduction

For most historians the British Association is perceived in terms of crucial years – 1831 (foundation), 1847 (Thomson's discovery of Joule), 1860 (the confrontation between Wilberforce and Huxley), 1870 (Huxley on spontaneous generation), 1874 (Tyndall's Belfast Address), 1879 (Crookes on the fourth state of matter), 1898 (Crookes on wheat supplies), 1907 (Kelvin's swansong on atoms) and 1913 (Aston and the isotopes of neon) – years in which there were either important 'speeches from the throne of science',[1] or debates and controversies which captured the imaginations of both the scientific communities and lay public alike. At other times the historian's focus of interest is on the Association as 'Her Majesty's Parliament of Science';[2] or as a catalyst and container of a vigorously healthy provincial science; or as the chief promoter and sustainer of the popularization of science for a century or more. Behind all these images and uses to which historians can put the British Association in their writing about science during the past 150 years lies the assumption that the Association was of use to men of science in their careers. But how, in practice, did Victorian, Edwardian and inter-War professional scientists see and remember the Association? What was the role of the British Association in the lives of practising scientists compared with the other more specialized societies to which they might belong, or the institutions where they worked? Did membership of, and participation in, the annual peripatetic meetings of the Association contribute anything to the advancement of science? How, if at all, did the Association's function and status within the scientific community alter over time?

Like many 'obvious' questions, these are not easy for the historian to answer, despite the apparent wealth of data that the annual *Reports* of the British Association provide. It is clear, too, that although they were questions which repeatedly occurred to scientists participating in the Association's anniversaries, or when the Association was under criticism, the scientists did not find them easy to answer either, and were usually reduced to impressionistic, benevolent generalizations concerning the Association's 'usefulness' and indispensability. For example, at the Golden Jubilee in 1881, the President, John Lubbock, made the opportune comment that 'The history of the Association . . . is really the history of Science'[3] – a strategem which enabled him to avoid the difficult question whether Victorian science would have developed differently if the British Association had never existed and to replace it by a comfortable review of fifty years of scientific progress.

In this essay we shall look at the way the British Association helped to define science for both its full-time practitioners and the general public, examine the changing commitment and loyalty which practising scientists showed towards the Association during its first one hundred years, and survey some of the ways in which the Association actually promoted scientific research. Only close familiarity with a century of *Reports*, the published biographies and autobiographies of participating scientists (a sampling suggests that they are quite unrevealing of attitudes towards the Association), and above all, private correspondence, would allow confident generalizations to be drawn. What follows is based on only a small sampling of the materials.

An Association for Defining Science

Fundamental to any discussion of the role of the Association in the life of the scientific community during the last 150 years is the meaning of the word 'science' itself. In some sense, the term gained its modern meaning and import through the activities of the Association.[4] Although the word 'science' (from Latin, *scientia*), meaning any, or all, systematic knowledge, was in use before the nineteenth century, the more common term was 'philosophy'. With the seventeenth-century Scientific Revolution, and increasingly during the eighteenth-century Enlightenment, 'science' became used in the special sense of useful, empirically generated knowledge, and opposed to metaphysics, theology, and 'philosophy' whose subject matters were transcendent or revealed, and arguably without useful application. The distinction, which was also rendered by the terms 'natural philosophy' and 'experimental philosophy' was encapsulated in the British Association's title, even though 'science' in the wider sense of any 'systematic enquiry into the modes of action of the powers of nature'[5] continued to be implied in Victorian phrases such as 'Political Science', Cambridge University's Moral 'Science' Tripos and the short-lived sociological organization modelled on the British Association, the Social Science Association (1857-86).[6] To Carlyle, writing in 1829, it was a sign of the times 'that the Metaphysical and Moral Sciences are falling into decay, while the Physical are engrossing every day, more respect and attention'.[7] In this arrogation 'to itself the word previously used for all knowledge',[8] the British Association played a key role by its annual propagation of the term.

Furthermore, as Ross has pointed out, 'with the new meaning of *science* the need to designate a man of science became more pressing'.[9] 'Scientist', that is a person who cultivated or practised science, was the coinage of William Whewell at the third meeting of the Association at Cambridge in 1833.[10] Although use of the term scientist remained controversial for much of the century, often being avoided by circumlocutions like 'man of science', by the 1890s it had become firmly embedded in British and American speech.[11]

Most writers on the British Association have seen the coinage of the word 'scientist', as well as the creation of the Association itself, as symptomatic of the professionalization of science during the nineteenth century.[12] According to this viewpoint, the Association was a reflection of the rising self-consciousness of men of science during the nineteenth century – in J. F. W. Johnston's phrase, the drawing of 'public attention to science and scientific men' in order to 'make people inquire concerning both them and their pursuits'.[13] The goal, or end product, of such an awareness of the importance of the cultivation of science and of scientific expertise in an industrial society was – according to this historiographical tradition – the full-time, salaried, 'professional' scientist institutionalized in a university or industrial laboratory after following a rigorous and specialized education which had culminated in a certification of competence.

The difficulties with this interpretation are manifold, as the recent literature on professionalization attests.[14] If by a 'professional scientist' historians are referring to various attributes, trappings or traits of twentieth-century professions and white-collar occupations such as 'skills based on theoretical knowledge', 'special education', 'membership of a learned or qualifying society', etc.,[15] then the model accepts at the start the very features it exemplifies; it too naively and anachronistically accepts the definitions of twentieth-century professional groups; it is culture bound; and, from a sociologist's perspective, there is no theoretical principle dictating which of the dozens of possible traits of professionalism are necessary and sufficient.[16]

More useful in historical treatments are dynamical models of professionalization, most of which depart from Durkheim's analysis of the processes of the differentiation of labour in societies and concentrate on the functional relevance which full-time practitioners of science may have to the society in which they emerge.[17] Johnson's typology of Collegiate, Corporate and Mediative control patterns in professional-consumer relationships is, for example, a useful way of portraying the social development of medicine and engineering. By differentiating between 'science-professionals' who sell an esoteric service to knowledgeable colleagues, and 'client-professionals' who (like lawyers and accountants) live 'by selling an esoteric service to a client who is in a position to judge for himself the quality of the work he receives', Fisher has been able to draw a functional distinction between scientific 'amateurs' and 'professionals'.[18] The categorization of 'science-professionals' undoubtedly overcomes Cannon's appropriately severe condemnation of the historiography of professionalization[19] by meeting her intuitive evaluative definition:

> A professional scientist was characterised by the facts that he was involved in an endeavour of recognized social status and that he recognized its own standards of merit (he did not appeal to a general 'public'; it was a 'self-reviewing circle'), but he was distinguished from the gentleman, the amateur, the patron, the dilettante, the teacher, the interested cleric, and the retired naval officer, by his long-term attention to science (not necessarily to *a* science) as his major activity, by his technical expertise and by the number and quality of his accomplishments.[20]

Such an evaluative definition of 'professional' is obviously useful in the context of the British Association. It avoids the nonsense of historians being forced to say, in the light of a twentieth-century model, that many of the Association's leading figures were either 'amateurs' or not scientists in any modern sense. On the other hand, Cannon's definition offers little purchase on those members of the general public, industrial scientists or engineers who supported the British Association, not for self-advancement, but because they believed such an organization offered them a way of cultivating their scientific interests or of making their own contribution to the scientific enterprise. Here Reingold's threefold heuristic typology of 'researchers', 'practitioners' and 'cultivators' proves convenient. Reingold's researchers are Cannon's self-reviewing circle and Fisher's science-professionals; his practitioners

> are individuals wholly or largely employed in scientific or science-related occupations. Those who publish are less prolific and less significant in terms of accomplishments than researchers.[21]

Like the traditional professional groups of clergy, lawyers and physicians, they are engaged in employment in which their scientific knowledge is used. The 'cultivators of science' – a phrase significantly used by Brewster and others when speaking of the British Association's membership and audience – are the important residuum of cultured patrons and amateurs of science (Cannon's gentlemen, dilettanti, clerics and retired naval officers) whose cultural horizons embraced science and from whom the careers of practitioners and researchers were separated. From this perspective the British Association can be viewed, not only as an organization reflecting the process whereby practitioners and the more powerful elite of researchers increased in numbers and status at the expense of cultivators,[22] but also, paradoxically, as an organization in which all three groups have tried, with varying degrees of success at different periods, to find a forum of mutual interests.

As recent research on learned societies has illustrated, the specialization so often considered to be a hallmark of fully-fledged professional science was an early phenomenon of scientific societies and clubs.[23] 'The rise of Learned Societies [cannot] be seen as simply a consequence of the increasing complexity of individual specialities'.[24] Late eighteenth- and early nineteenth-century societies, such as the Linnean (founded 1788), Geological (founded 1807) and Astronomical (founded 1820), were all founded to deal with a particular project whose execution demanded co-operative participation and patronage. Many, if not all, were developments of eighteenth-century dining clubs attended by cultivators of science who included Fellows of the Royal Society. Despite Alexander von Humboldt's division of the Gesellschaft deutscher Naturforscher und Ärzte into sections in 1828, the sectionalization of the British Association must be seen in this light, namely as a continuation of the strategy whereby cultivators, practitioners and research experts could pool resources to report on, and advance, particular areas by means of specialized projects.

From its beginning the British Association met in convenient sections. These grew in number from the six sub-committees loosely grouped by Harcourt as Mathematical and Physical Science, Chemistry, Mineralogy, Geology and Geography, Zoology and Botany, and Mechanical Arts, into, by 1901, after various tinkerings, eleven alphabetically-defined Sections (see Appendix II). Agriculture was added in 1912 (separating it from either Botany or Zoology), and Psychology, in view of its growing prominence in the early twentieth century, was admitted separately from Physiology in 1921. Botanists, zoologists, physiologists and anthropologists had all met as Section D 'Biology' until 1884 when Anthropology separated as Section H. Physiology followed as Section I in 1893, partly in response to the virulent public treatment of J. S. Burdon-Sanderson, the President of the Association at the Nottingham meeting in 1893.[25] Two years later Botany became Section K, leaving the former Section D as Zoology alone. Attempts to re-establish a Medical Section (which had existed in the 1830s and 1840s before the foundation of the British Medical Association) were firmly resisted in the 1890s on grounds which would otherwise have led to the abolition of both the Engineering Section and the Section for Economic Science and Statistics – namely, 'medicine is not a science in itself, but merely an application of several sciences'.[26]

The position of engineering is not without interest. Throughout the nineteenth century, Section G was called 'Mechanical Science', probably deliberately echoing the Institution of Civil Engineers' (founded 1818, Charter, 1828) objective of 'the advancement of mechanical science'.[27] As 'a mediator between the philosopher and the working mechanic, and, like an interpreter between two foreigners',[28] the civil engineer was to be cultivated as an exemplar of an important aspect of the British Association's mission. According to the great Scots engineer, Rankine, Section G was to Section A what the classes of engineering and mechanics were to those of physics and mathematics in a university; that is, the Section which considered 'the practical application of those branches of science to whose theoretical advancement Section A was . . . devoted'.[29] Nevertheless, academics predominated. Between 1836 and 1881, 76 per cent of the Section's Presidents were Fellows of the Royal Society, and an engineer, William Fairbairn – an early member – was first elected to the Presidency of the Association only in 1861. It is small wonder that most writers have not seen the Association as playing any essential role in Victorian engineering affairs – though this view ignores the important stimulus it gave to the young J. S. Russell and the lead it took with engineering standards.[30]

It is clear that the principal expansion of Sections and their specialization took place in the decade 1884-1895,[31] though it is interesting to notice that Huxley, as President of Section D at Nottingham in 1866, was already lamenting 'that the sub-divisions of the Section [Botany and Zoology] had to meet separately as a result of specialization' in schools and universities.[32] This is the decade coinciding with the growth of the civic universities, the expansion of educational facilities for science, and a period of maximum activity for the Association in its publication of Reports on the State of Science and in its expenditure on co-operative research. These forces were clearly more significant in the establishment of new Sections than any accelerating tendency towards further specialization in new scientific societies; indeed, the period of maximum growth of specialist societies preceeded that of the Association's Sections by fifty years.

The fissiparous tendency of the Association's Sections was rigorously resisted in Section A. Although an Astronomy Division was permitted between 1900 and 1904, Council resisted a move by Schuster and Turner in 1908 to create a Section for 'Astronomy and Cosmical Physics',[33] though in view of the ease of formation of the later Agricultural and Psychological Sections, it is clear that there must have been opposition for Council not to have agreed. *Nature* once vividly compared the Association's sectionalism to 'a great industrial works in which each shop produces what it pleases, and no one has the duty of building up a noble structure from the various parts'.[34] However, the arrangements made for inter-sectional discussions and co-ordination during the 1900s, and for the popularization of science which the British Association introduced in the 1920s, appears to have dampened further budding, until the social relations of science debate, and the absorption of the British Science Guild in 1936, forced the creation of the Division for the Social and International Relations of Science in 1938.[35]

The British Association helped to define the scope, methods, and applications of science and its component specialisms for its own practitioners and researchers, and for the listening and reading public. But once defined and articulated, did the Association continue to be a useful instrument for achieving the aims and ambitions of practising scientists and researchers?

An Association for the Attendance of Scientists?

Attendance at meetings of the British Association has never, in practice, demanded a technical qualification.[36] The original intention to confine participation at meetings to male members of philosophical societies rapidly became a dead letter and, until 1919, participants, who received the printed *Reports* and who might serve on the British Association's Councils, could attend either as Life Members (who had subscribed between £5 and £10) or as Annual Members (who subscribed £1 annually). After 1845 it was also possible to join a particular meeting as an Associate, this being the universal form of membership after 1920. Inevitably the social character of the meetings encouraged the attendance of wives and other female relations, and special 'ladies tickets' were circulated between 1843 and 1919. This participation raised the question whether women were eligible for office[37] but, unlike the Social Science Association[38] and despite the arrival of women career scientists like Hertha Ayrton,[39] the British Association did not depart from 'the established custom' of male office holders until 1900.

Although there have been some direct and indirect scientific incentives for researchers and practitioners to attend meetings of the British Association, such motives as are recorded by participants are almost entirely social or careerist. It is well known that because of the ideology of 'declinism' in the 1830s the provincial founders of the British Association had a difficult time ensuring that the establishment figures of British science, who were chiefly based in London or at the Universities of Oxford and Cambridge, attended the early meetings.[40] Even when they approved the Association's ideology of giving 'a stronger impulse and more systematic direction to scientific inquiry', they saw it as a potential time-waster.[41] Charles Darwin seems to have found meetings only socially enjoyable. After the 1846 meeting at Southampton he wrote:

> We enjoyed (wife and myself) our week beyond measure: the papers were all dull, but I met so many friends and made so many new acquaintances (especially some of the Irish Naturalists), and took so many pleasant excursions.[42]

Dullness also repelled the chemist Edward Frankland. In 1855 he wrote to John Tyndall:

> The more I think of it the more I am convinced of the deep injury which it annually inflicts upon science. With the exception of the Red Lions dinner and Loch Katrine, which were the two bright gems of the [Glasgow meeting], I look back upon the whole matter with unmitigated regret.[43]

Six years later, he explained to Nicholson, his colleague at Owens College:

> I make a point of not going to the British Association meetings when they occur in the vacation, as the proceedings are merely a hash of what one has been listening to during the previous year at the meetings of the London Scientific Societies and consequently they

break upon one's vacation to no purpose whatever.[44]

However, Joseph Henry, touring Europe in 1837, had already astutely seen that the point of the British Association was not necessarily to hear of new nuggets of scientific information, but 'to form a kind of festival for men of science for mutual stimulous *[sic]* and to afford an opportunity of those engaged in the same researches of comparing notes and making explanations to each other which cannot so well be done by other than personal communication'.[45] The promotion of closer contacts between active research scientists was made easier in the 1860s as the 'old order of supreme management by a few magnates'[46] (like Sabine and Murchison), of individualism, and of science as useful knowledge and the pillar of natural religion, gave way to a 'Young Guard of Science', epitomized by the X-Club composed of, among others, Huxley, Hooker, Tyndall and Frankland himself.[47] Science, in becoming a system of knowledge institutionalized in universities, colleges, schools and museums, was to challenge other elites by claiming unique expertise to solve the problems of a nineteenth-century urban society;[48] and the British Association, together with Victorian periodicals, provided a great forum and lobby for the proclamation of scientific naturalism. This, together with the increasingly expensive, and laboratory, nature of research and the new seriousness with which clergymen undertook their parochial duties, was no doubt responsible for the dramatic decline in clerical office-holders and members between 1862 and 1919.[49]

Huxley and Tyndall had first met at the Ipswich meeting in 1851 where both had gone

> as scientific 'items' not, indeed, wholly unknown to the 'pillars' of that scientific congregation; and perhaps already regarded as young men whose disposition to keep their proper places could not, under all circumstances be relied upon. Being young, with any amount of energy, no particular prospects, and no disposition to set about the ordinary methods of acquiring them, we could conduct ourselves with perfect freedom.[50]

In fact Huxley's appearance was successful on two counts: that of 'doing a little necessary trumpeting, and . . . to get the assistance of the Association in influencing the Government' to pay for the publication of his Australian researches.[51] The following year Huxley went to Belfast in order to make himself known to Irish men of science in the hope of landing a job. As Huxley's friends told him, 'personal influence went for so much, and while men's reputations were better than themselves, he might flatter himself that he was better than his reputation'.[52]

Here we see common motives for attendance by younger men – explicit, for example, in the career of Oliver Lodge: the search for patrons, the demonstration of ability by a sectional paper or in the critical discussion of others' work. Huxley, for example, attended his fourth meeting at Liverpool in 1854, not to listen to papers, but to meet and seek the advice of Edward

Forbes,[53] the leading British naturalist and founder of the convivial dining club, the Red Lions. Similarly, Lodge, who had been inspired to take up a scientific career after hearing H. J. S. Smith's British Association address in 1873, and who became a secretary of Section A in 1878, set out deliberately to use the opportunity of the first overseas meeting of the British Association at Montreal in 1884 to advance his reputation by a brilliantly Tyndallesque lecture-demonstration on dust and smoke, and by exploiting his talents in Section A as an extempore discussant. The plan worked and Lodge was asked to organize one of the first inter-sectional discussions on the theory of solutions in 1885.[54]

Such inter-sectional meetings did not begin until the 1880s, and did not form part of the British Association's official programme policy until the 1900s. Before then, as Lodge tells us, the Organising Committee of the Association met at 10.00 a.m. on the first Wednesday to map out a programme, which it then reported to the individual Sectional Committees. The latter met every morning to list the papers to be read the following day, the list being dispatched to the printer immediately and proof-read that evening. Sections began their independent business at 11.00 a.m. and ran through without a break until 3.00 p.m. After a paper had been read, discussion was invited,

> and if it showed signs of life and interest, was allowed to continue, without much regard to the clock; there was no time-table, and the next paper was called on when the discussion ceased. Occasionally these discussions lasted quite a long time and they were sometimes the best part of the meeting. Lord Kelvin in particular had plenty to say, felt that he had the time available, and was occasionally discursive, but most of the audience were delighted, and in this informal manner, knowledge was advanced.[55]

In fact, because Section A comprised a loose confederation of sciences linked together only by their potential or actual mathematical character, some sketchy timetabling was practised, Thursdays being given to papers from the famous and overseas visitors, Fridays to promising novelties, Saturdays to pure mathematics, Mondays to meteorology and astronomy, Tuesdays to pure physics, and the final Wednesday to clearing any backlog.

Although in the 1920s and 1930s Lodge looked back upon such informality with nostalgia, he was certainly a catalyst of the change which took place in the 1890s. Following the Newcastle meeting in 1889, which had been held during the week of 11 September, Lodge complained in *Nature* that

> the British Association week undoes the benefit of the previous holiday, and that the conditions under which the work of the Sections is carried on are prejudicial to health, and have this time resulted in a considerable list of sick and wounded.[56]

Behind the apparently trivial complaint that Section A now worked from 10.00 a.m. until 5.00 p.m. without any refreshment breaks (Kelvin always 'sedulously attended' with sandwiches[57]) lay an urgent recommendation for change. H. E. Armstrong (never far behind Lodge in providing *Nature's* editors with copy) made this explicit:

> Machinery devised 50 odd years ago is no longer capable of satisfactorily coping with modern requirements. We do not go to British Association meetings to sit for hours together to hear papers read such as we have listened to *ad nauseam* during the previous sessions – our main object is to meet and exchange views; but everything seems to be done to prevent rather than promote this.[58]

Armstrong's remedy, which was supported by other correspondents, was to have fewer papers in each Section, greater selectivity of papers, more intersectional meetings, more encouragement of discussion, and a meeting date in August rather than September when 'most of those who are engaged in advancing science are then in the very middle of their holiday, and can only attend at grave personal inconvenience'. (Meetings were transferred to August for 1891-92, but with a few exceptions returned to September after 1893.)

Armstrong returned to the subject in 1892, urging the creation of sectional committees to select suitable thematic subjects:

> and that on each of these some one open a discussion by means of a carefully-prepared paper, printed and circulated at least a month beforehand, among those likely to take part in the debate. Such discussions should be carefully reported, and the edited report should be subsequently published, those who had taken part in the discussion having full liberty to give expression to their carefully considered opinion.[59]

Such a formula was used very successfully by Armstrong's Irish colleague, John Perry, in the influential joint discussion he stage-managed in 1901 between Section A and the newly created Section L for Educational Science.[60] In 1893 Section B (Chemistry) followed Armstrong's plan with discussions of interest to biologists and engineers, with debates on bacteriology and coalmine explosions.

Whether such amendments made any serious alteration in the attitude of clear indifference to the Association's meetings which the younger generation of 1890s scientists displayed is unclear, though there is some evidence that attendance was still a way of gaining attention. As early as 1878, Spottiswoode, a member of the X-Club's Young Guard of the 1860s, devoted part of his Presidential Address to the plea that the British Association could only continue to function given

> the personal scientific support of our best man . . . without their

> presence our meetings would fail in their chief and most important element, and had best be discontinued altogether. We make, it must be admitted, a demand of sensible magnitude in calling upon men who have been actively engaged during a great portion of the year, at a season when they fairly look for relaxation, to attend a busy meeting, and to contribute to its meetings; but unless a quota at least of our veterans, and a good muster of our younger men, put in their appearance, our gatherings will be to little purpose.[61]

Despite this plea the numbers of participants at Sheffield and Swansea in 1879 and 1880 plunged dramatically and only began to recover after the Jubilee in 1881. Most Presidential Addresses, during this period of a 'rising tide of opposition to science and the scientific outlook',[62] contain passages which clearly express the fear that the British Association was finished. C. W. Siemens, for example, in 1882, noted how the development of many more specialized scientific societies and institutes, the tremendous growth of scientific journalism,[63] and the emergence of new kinds of research facility had replaced most, if not all, of the British Association's original functions. Should it therefore 'yield its place to those special Societies it has served to call into existence'?[64]

Siemens's answer, as expected, was a resounding 'No', interestingly on the grounds that 'the necessities for scientific inquiry have increased in a greater ratio' to the facilities for research. Arguments for the Association's role as a medium for the popularization of science and scientific method still lay in the future.

Nevertheless, the decline in interest in the British Association by the scientific research community continued ineluctably. Although, for example, the numbers of Fellows of the Royal Society who held membership of the British Association remained constant at about 300 (9 per cent) between 1881 and 1892, over 60 per cent had gained the FRS after joining the British Association (15 per cent were elected within a year, 43 per cent within two to five years).[65] By 1919, though the effects of the First World War should be borne in mind, the number of FRSs had declined to 224 (7 per cent of the total membership). In this light, although not universally welcomed by those physical scientists who remained committed to the Association, the opening of the Educational and Psychological Sections undoubtedly helped the Association's survival by bringing in new kinds of practitioners and researchers.

On the other hand, since the public rewards of science remained few and far, the British Association's honours were gladly accepted. From the beginning the office of President became a distinctive honour which few research scientists wished to decline.[66] Presidents commonly laboured for several weeks to compose their Addresses – what Hooker termed 'Presidential Martyrdom'[67] – knowing that it would be 'printed more or less in full, and commented on in the leading newspapers'.[68] On only one occasion, in 1867, did an embarrassed Council not publish a Presidential Address.[69] At first Presidents were selected from the meeting place (e.g. Sedgwick from

Cambridge in 1833), but after about 1840 they were chosen in rotation to represent the different sectional interests.[70] Similarly, as a pattern of delivering Presidential Addresses to the individual Sections slowly emerged during the 1860s – arguably begun by Tyndall and Frankland as respective Presidents of Sections A and B in 1868 – these, too, became prestigous, much sought-after honours at the disposal of the scientific community. This is plain from the extraordinary amount of time which the X-Club spent on discussing possible candidates.[71] Thus, paradoxically, while a large proportion of the practising and researching scientific communities saw little point in attending meetings of the British Association, when approached, they were quite prepared to accept its patronage as office holders. This prestige can only be explained by the fact that among influential sections of the lay public – including the government – the British Association, together with the Royal Society, had become regarded as the principal mouthpieces of British science.[72]

Meetings of the British Association were suspended in 1917 and 1918, and although a hurriedly arranged meeting was held in Bournmouth in 1919, the first 'normal' post-War meeting was at Cardiff in 1920. Although scientifically a success, despite a tram drivers strike in the city, it was immediately noticed that attendance was down to 1378 from 1497 at the previous Cardiff meeting in 1891 – a negative trend not discernible in post-1900 attendance figures. This decrease was commented on in *Nature* by R. V. Stanford, Secretary of the Press and Publicity Sub-Committee at the Cardiff meeting – ironically not himself a member of the Association! The principal cause, said Stanford, was

> the apathy of local people of the educated classes to the presence of the Association. The plain fact remains that it is the exception to find anyone who has ever heard of the Association. Is this regrettable state of affairs due to something lacking in these individuals themselves, or is the blame to be laid at the door of the Association?[73]

Gregory, in a long critical editorial in *Nature*, and all the many correspondents who joined in three months of debate, were unanimous that the fault lay with the British Association.[74] This debate bore on the popularization of science, but it also questioned the scientific arrangements of the meetings. Both Gregory and Lodge, for example, wanted more extensive joint meetings of several sections, choosing broad themes connected with science and society which the general public could understand.

But unless men of science were actually paid to attend and to mix with the general public or came as elected office holders, we can see, in retrospect, that there was no way of persuading distinguished men and women researchers to attend the British Association's meetings – unless, like Lodge, they were convinced of the importance of such a function.

As a result of alterations to the programme consequent upon this debate, attendance at meetings did increase, the most favoured venues being academic towns. For by 1920, because of social alterations in holiday patterns and the scale with which the middle classes either wanted to, or could afford to, entertain, the 'social' side of the British Association, whereby men of science obtained free board and lodgings as a house guest of a local family interested in science, had gone forever.[75] Thus, unless the Association visited centres of thriving scientific research, industry or teaching, only the most dedicated or publicity-conscious researchers and practitioners attended meetings. By 1920, then, Spottiswoode's warning of 1878 had come true. As H. E. Armstrong, that ancient mariner of Victorian science who was always ready to hit out at all and sundry,[76] concluded in obituary mood in 1935:

> The meetings are but poorly attended by the profession and very few foreigners are present. In fact, in these locomotory times, scientific workers indulge in so much joy-riding to conferences on every possible and impossible subject that no gilt is left for the Association gingerbread, which can now only be wrapped in tin-foil like poor cheese and chocolate. There are no real 'lions' to show. The social value of the meetings is reduced to a minimum. The discussions are staged, a few performing companies being trotted about for the purpose.[77]

To Armstrong it was symptomatic of the Association's ailments that in 1935 it had elected an economist, Josiah Stamp, Chairman of the London, Midland and Scottish Railway, as President.

Nevertheless, by capitalizing on the captive communities of researchers, practitioners and natural history societies which it visited (Armstrong's 'performing companies'), and by awarding student visitors grants to attend meetings after 1922, the Association survived into the 1930s, to be piloted into a new harbour by Richard Gregory. It seems clear that, with obvious exceptions such as the creation of the Corresponding Societies in 1885, during most of the nineteenth century the British Association was controlled by, and therefore aimed at, those groups of researchers who became university academics. Such groups, rather than practising engineers, industrial scientists and laymen, were able to use the British Association for career advancement, to lobby for particular theories and to engage in broad, often polemical, discussions which served both to advance disciplines amongst peers and to proclaim the public significance of science. In this manipulation of the British 'popular participatory scientific movement' in the provinces,[78] the research functions of the British Association, to which we now turn, played significant roles.

An Association for the Advancement of Research

Nineteenth-century Presidents of Sections and of the Association were unanimous in their view that the principal ways the British Association advanced science were by means of research grants and the commission of reports on the state and progress of science. It will be useful in this section, therefore, to see how research grants were allocated, and with what effect; to examine the appointment, and effectiveness, of research committees; to suggest how the Association could function as a market place for the announcement of new results; and to raise the question of the kind and quality of the science which was permitted a public airing at annual meetings. Generally speaking, as with the attendance of academic researchers at meetings, the Association's research profile fell into two distinctive phases: the period before the 1890s, when the Association played a vital role in the research output of the scientific community; and the period following when the Association played less and less a role in either subsidizing research or providing a venue for the announcement of research findings.

When Joseph Henry visited England in 1837, it struck him forcibly that

> an important advantage of the Association is that from the great number of members a large fund of money is collected and applied to scientific purposes under the direction only of the principal men. Thus Sir David Brewster received a gr[ant] of £100 for carrying on his experiments on the lines of the solar spectrum produced by the intervension *[sic]* of some gaseous medium between the eye and the prism.[79]

Although it was 'a thoroughly British view of the subject . . . to appreciate the value of an investigation by the money it costs',[80] there can be no doubt that the Association's system of research grants advanced science in the nineteenth century and that it continued to act as a 'topping up' source of revenue well into the twentieth century.

The system of 'grants for scientific purposes' had begun in 1833, the brainchild of Harcourt. The Association's support for research rose steadily after 1850, from about 25 per cent of its income in 1851-55, to 70 per cent in 1866-70, then settling at about 50 per cent until 1911-15 when, as other Trust income (such as the Caird Fund, 1912-13) became available, it settled to about 45 per cent. Up to 1862, for example, the British Association had awarded £20,000, two-thirds of which had gone to Section A (£1800 on the Star Catalogue of 1837-45; £1300 on tidal research; £2000 on meteorological and magnetic observations; and nearly £6000 on the Kew Observatory which the Association sponsored from 1842 until 1872).[81] In 1894 the British Association agreed that £1000 should be regarded as the *normal* annual grant for research, more being permitted (as it usually was) if uninvested income allowed.[82]

As *Nature* pointed out in 1872, alone among other scientific societies which occupied themselves in giving publicity to results and speculations of the scientific community,

> and in rewarding with their medals successful labours, the British Association . . . systematically undertakes to distribute the greater

> part of its income, about £2000 per annum, in grants to enable men of science to conduct scientific investigations, and to institute inquiries with a view to possible future action.[83]

Although, as one might have expected, the total grant allocated to Section A between 1834 and 1931 was the largest of all the Sections, between 1880 and 1914, Section D (Zoology) was allocated £11,685 compared with Section A's £8562 during the same period. This came about from the high grants which the exploration of the mountain districts of eastern equatorial Africa attracted between 1880 and 1884. Subsequently, expeditions were always financed under Section E (Geography), though only minor sums were ever awarded compared to the principal private patron of exploration, the Royal Geographical Society.

High sums were also allocated to Section H (Anthropology) following its separate existence in 1884. Between then and 1914 its members received £5477, and another £5000 between 1914 and 1931. Although these periods coincided with the British Association's flirtation with Galton's eugenicist programme of anthropomorphic measurements (which attracted about £1000), the largest proportion was spent on archaeological investigations, including £1810 on the exploration of Kent's Cavern. Section B (Chemistry) received £1747 (41 per cent) from 1834 to 1879, and £2298 (53 per cent) between 1880 and 1914. £4764 (62 per cent) of the total given to geology over a hundred-year period was awarded after 1880. Indeed, the bulk of the Association's expenditure on research took place during the 1880s – despite and perhaps because of the public criticisms of science during the decade – when the British Association's attendances, income and numbers of collaborative research reports on the state of science all reached their peaks. Thus, the number of research committees allocated funds rose from an average of 13 in 1850-54 to 53 in 1885-89. Howarth, the Association's centenary historian, rightly drew attention to the fact that the research grant figures hid the considerable support which the British Association gave to investigations through 'expenditure upon printing the results of scientific investigations, whether by committees or individuals, which occupy so large a proportion of the annual reports issued by the Association'.[84] As printing costs rose, especially after the First World War, this hidden subsidy of research became an increasing drain on the Association's resources. Despite pleas to research committees to abbreviate their results, the comprehensive publication of all reports came to an end in 1938 with the launching of the quarterly, *The Advancement of Science.*[85]

Although, partly through the efforts of the British Association, a Government Grant of £1000 per annum, administered by the Royal Society, had become available for sponsoring research after 1850, the bulk of this money went to Fellows of the Royal Society who lived close to London and who were therefore able to lobby the administering committee.[86] Hence, though Fellows of the Royal Society also continued to enjoy the prospect of the small sums available from the British Association, the latter's small grants were often crucial to provincial non-FRSs who could no longer afford to finance projects from their own pockets. As J. B. Morrell has pointed out, the Edinburgh physicist, P. G. Tait (1831-1901), who was much disliked by the members of the X-Club, was indebted to the British Association for £345 between 1868 and 1876 – a sum which he spent on experiments on conductivity and thermoelectricity. He was always refused aid from the Royal Society's Government Grant.[87]

Spottiswoode was, therefore, quite right to argue in 1878 that the British Association had anticipated, or even inspired, the movement for the Endowment of Research[88] which, converging with the Association's commitment to the reform of science teaching in public schools, led to the appointment of the Devonshire Commission in 1870. The Association's grants, Spottiswoode suggested, were like the German university system of the *Privatdocenten:*

> [they] afforded to special individuals an opportunity of following out the special work for which they have previously proved themselves competent. And in this respect the British Association may be regarded as supplying, to the extent of its means, an elasticity which is wanting in our own Universities.[89]

Grant applications were scrutinized with considerable care by both a Sectional Committee and a special Committee of Recommendations formed from representatives of all the Sections, before being formally approved by the General Committee of the Association. After 1865, a sum of money was rarely given to an individual worker.[90] Both before and after the applicant was usually awarded the money as a member of a Committee composed of at least one, and sometimes as many as fifteen, co-workers. For example, the Jubilee meeting in 1881 disposed of £1280, the largest sums of £100 each being allocated to G. C. Foster for research on electrical standards, J. Evans for compiling a roster of geological research, Lieutenant-Colonel Godwin-Austen for a study of the natural history of Socotra and the ancient highlands of Arabia and Somaliland, W. T. Thistleton-Dyer for a natural history of Timor (Indonesia) and H. T. Stainton for compiling a record of zoological literature. Each of these men was already a distinguished Fellow of the Royal Society, but in every case the money was awarded to them as a member of a committee – Foster with fourteen other men, Evans with ten others, Godwin-Austen with fifteen others, and both Thistleton-Dyer and Stainton with two others.

Francis Galton thought the Committee system, reminiscent of the French Academy of Sciences in the eighteenth century, was admirable:

> The Association affords what is often the most appropriate means of ventillating new ideas. It can create a Committee with or without a grant of money, giving to its proposer the title of either Chairman or Secretary, which clothes him with an authority that an unknown individual would lack, when making inquiries of public bodies at home or abroad. It also provided him with colleagues to discuss and

criticise results before they are published.[91]

And he cited the case of the large Anthropomorphic Committee created by him in 1875, but for which Charles Roberts did all the hard work.[92] By following a Baconian ideal of collaborative research, the Association not only gained a responsible system of peer refereeing, but ensured plain dealing with the sums of money involved.

Successive Presidents of the British Association mentioned a number of areas of scientific research in which the Association had arguably advanced the topic concerned. The first President to justify the Association's existence in this way was the Irish astronomer, T. R. Robinson (1792-1882), who opened the proceedings of Section A at Dublin in 1857 by listing its achievements as:[93]

(1) Obtaining a government grant of £500 to enable the Astronomer Royal, G. B. Airy, to reduce all the planetary and lunar observations made at the Royal Observatory in Greenwich between 1750 and 1830.[94] This work was published in 1846.

(2) Obtaining a government grant of £1000 in 1843 for the publication of the catalogue of stars in J. J. Lalande, *Histoire Céleste Française* and N. L. Lacaille, *Astronomioe Fundamenta*, on which the Association had spent £753 between 1839 and 1842. With a further primer of £1374 the British Association was able to complete the project to map all known stars which had been abandoned by the Royal Astronomical Society for want of financial means. The British Association's *Catalogue of Stars*, which remained in standard use in the world's observatories throughout the nineteenth century, was published in 1845.[95]

(3) Awarding £973 towards research on tides and their analysis between 1834 and 1857 (a further £1180 was awarded between 1862 and 1887) which had been of great benefit to the British Admiralty, as well as to other maritime nations.[96]

(4) Setting on foot (with the effective support of the Royal Society) the establishment of magnetic observatories,[97] an enterprise which Whewell described as 'by far the greatest scientific undertaking the world has ever seen'.[98] Although, as Cawood has shown, little of theoretical or practical benefit resulted from the 'magnetic crusade', observations did lead to a correlation between magnetic activity and the sunspot cycle.

(5) Originating the research of Eaton Hodgkinson (1789-1861) and William Fairbairn (1789-1874) on the strength of iron in 1836 – research which enabled Stephenson to construct the tubular Menai Straits Bridge in 1848.[99]

(6) Encouraging the engineer John Scott Russell (1808-82) to work on ship design as a result of the support given to his first experiments on the hydrodynamics of waves and floating bodies which he discussed at the Edinburgh meeting in 1836. In 1857 Russell referred to his 'wave line' steamship *The Great Eastern* as 'a child of the British Association'.[100]

These six examples were repeated substantially by Robert Willis (1800-75) at Cambridge in 1862, and by John Phillips (1800-74) at Birmingham in 1865. Phillips also argued that it had been Airy's report on the present state of physical astronomy (1832) which first inspired J. C. Adams (1819-92) to search for, and to discover, the planet Neptune in 1845.[101] This, like Lodge's inspiration to publish a paper on chemical formulae after hearing Cayley lecture on graph theory ('trees'), is a demonstration that financial offerings by the British Association were not the sole incentives it gave to fundamental research.[102]

Four of these contemporary views on the British Association's contributions to the advancement of science have to do with what Susan F. Cannon has called 'Humboldtian Science', that is, the collection of vast storehouses of observations and measurements from which analytic laws can, or may, be deduced.[103] With the exception of the two commercially-significant engineering projects, they also strongly reflect university/academic interests in collaboration with other scientific societies such as the Astronomical and Royal Societies. The significance of men of science as government advisors and experts is also stressed. The two other investigations are concerned with Victorian engineering and technology and are a reminder that the British Association itself was dependent upon the heroic skills of Victorian railway engineers for its peripatetic functioning. In 1888, Bramwell took pains to argue that 'for the Advancement of science it is absolutely necessary that there should be *Applications* of Science',[104] thus justifying the presence of a Section for the mechanical arts.

Through its maintenance of Kew Observatory between 1842 and 1872, and its financial support of John Milne's seismological laboratory at Shide (on the Isle of Wight) from 1895 until 1915, the British Association provided a forum for large-scale Humboldtian research in the environmental and geophysical sciences. Moreover, as new areas of laboratory-based research, such as spectroscopy and electricity, began to be cultivated in the 1860s, the Association proved a valuable theatre for standardizing the results of research. It also provided an ideal platform for propagating the three great co-ordinating theoretical ideas of Victorian science: atomism, energy and evolution.[105]

The spectroscope had been invented by Bunsen and Kirchhoff in 1859. It was of fertile use in both chemistry and astronomy, and both parties, whether measuring the absorption spectra of chemical compounds or investigating what Lockyer termed 'solar physics', benefited considerably from the Association during the 1890s. The much consulted 'Tables of Wavelengths of the Spectra of Elements' appeared in the *Reports* between 1882 and 1909.[106]

Even more important nationally and internationally was the establishment of the Electrical Standards Committee by William Thomson in 1861. Beginning with its quest for a measurable and standardized unit of electrical resistance, the British Association 'Unit' or 'ohmad' (soon shortened to 'ohm'), and the Committee's firm adoption of the fundamental units of centimetre, gramme and second, did much to ensure the establishment of the 'c.g.s.' system in all branches of scientific measurement, and to prevent the development of a dual system of 'British' and 'metric' units in electrical engineering that was to plague mechanical engineering.[107] The original committee was dissolved in 1870, but the standards were redetermined by a new committee with a still greater accuracy after 1881, culminating in the adoption of British Association standards internationally in 1908.

It was the meticulous work involved in these determinations which convinced Oliver Lodge in 1891 that such work should be the permanent responsibility of a government laboratory.[108] Following discussions at Edinburgh in 1892, and powerful support which capitalized on fears of growing German supremacy,[109] the Prime Minister, Lord Salisbury, accepted the Association's arguments in 1897. The National Physical Laboratory, under the management of the Royal Society, was opened at Bushey Park, Teddington, in 1900, after the old Kew Observatory proved inadequate. Thus, the Association's action and expenditure on the Kew Observatory between 1842 and 1872

> not merely ensured for the future the maintenance of an institution of prime interest in the history of scientific research, but also led in direct line of succession to the establishment of a most powerful instrument for the advancement of science in the service of the State. The establishment of the National Physical Laboratory was an important achievement by Science in the education of the Government and the public, and the use of the British Association as a mouthpiece has never been more clearly demonstrated.[110]

In chemistry, the British Association provided Liebig with his first platform in England in 1837, and through the enthusiasm of pupils of the Scots chemist, Thomas Thomson, in Section B, helped in the establishment of the Chemical Society in 1841.[111] This, together with the tenuous connection which could be made between the British Association's Electrical Standards Committee and the Society of Telegraph Engineers (founded 1871) – the Institution of Electrical Engineers as it became in 1877 – is one of the few examples of a specialist society formed in consequence of sectional activity in the Association rather than the converse.[112] In the 1880s, Lodge and Armstrong used the discussion functions of Section B to express their doubts about Arrhenius's theories of ionization and of solution, making the Association a publicity machine for physical chemistry. However, after Rayleigh's dramatic announcement of the discovery of argon[113] at the meeting in 1894, and apart from Armstrong's use of the British Association to publicize his belief in heuristic laboratory teaching of chemistry, [114] chemists in the twentieth century seem to have found other venues for their activities.

Buckland, Sedgwick, Murchison, and the other giants of the 'heroic age' of British geology, were intimately involved in the early years of the Association which, in mid-century, did much to support the investigation of evidence for the antiquity of man. Between 1866 and 1931 some £2900 was spent on cave research, including £1810 upon a thorough investigation of Kent's Cavern in Devon by William Pengelly (1812-94).[115] Geology was one of the principal sciences which the amateur, despite the rival attractions of the Geologists' Association (founded 1858), was able both to benefit, and benefit from. A good example of amateur-science professional collaboration within Section C was the project to collect photographic evidence of a geological character. This was catalogued in the *Reports* from 1890 until 1937.[116]

It is easy enough to supply evidence that the British Association assisted in its primary function of advancing scientific knowledge in the nineteenth century; it is less easy to place a date on when this ceased to be a primary function, though the First World War, as in so many other areas of national life, forms a convenient watershed for the historian. Although the American physicist, C. J. Davisson (1881-1958) was inspired by discussions heard in Section A at the meeting in 1926 to search for, and to demonstrate, the wave nature of the electron by crystal diffraction (1927),[117] no scientist after Francis Aston announced the probable isotopic character of neon in 1913,[118] ever used the Association to reveal a new discovery. For British research scientists, *Nature's* weekly letters column had become the fastest and most favoured medium for scientific announcements. For example, all of Aston's post-War discoveries of isotopes were announced in *Nature*. Similarly, although sectional discussions have continued, there has been no dramatic confrontation since Kelvin's 'swansong' in the debate on radioactivity at Leicester in 1907.[119] Such controversies – for example, the confrontation between J. J. Thomson and his pupil Aston over the validity of the isotope concept[120] – have tended to occur at the Royal Society.

Even in retrospect it is difficult to define occasions on which the British Association failed to seize opportunities to advance science by supplying impulse and direction. Like the seventeenth-century Royal Society, however, it occasionally admitted deviancy to its proceedings by allowing a local worthy to air his views. As Lodge admitted, this permitted

> the crank to disport himself, and occasionally he was difficult to repress, but that didn't happen often, and the risk was worth running. The worst kind of crank was . . . one who attended the meetings year after year, and was anxious to display the bee in his bonnet . . .[121]

Despite 'an excited discussion' which lasted nearly a whole day, the British Association refused to print William Barrett's 'On Certain Abnormal Conditions of Mind' which he gave, with the approval of A. R. Wallace, W. Crookes and Lord Rayleigh, to the Anthropological Section in 1876. *Nature* thought it was 'doubtful whether the interests of science are best served by the introduction of subjects [i.e. telepathy] which are sure to provoke heated and unscientific discussion at a mixed meeting like that of the Association'.[122] This censorship of deviancy was a far cry from the Association's short-lived fêting of the poetical electrician, Andrew Crosse, in 1837, when he claimed to have spontaneously generated *acaria* from electrolysed solutions.[123] Despite a brave reference to his own researches into psychological phenomena in Crookes's Presidential Address in 1898,[124] the British Association firmly rejected such knowledge and closed its doors to the discussion of paranormal phenomena.[125]

On the other hand, it enthusiastically embraced what would now be considered a rather dubious research programme on anthropomorphic measurements – with its implicit racialist, imperialist and eugenic undertones. Similarly, the Educational Section, to its founder's disgust, and until the establishment of Section J in 1921, became a forum for psychological investigations of 'intelligence'. During the 1920s, Section J was much occupied with the issue of 'vocational tests'.[126]

Conclusion

In a discussion paper, circulated among delegates attending a conference on the dissemination of scientific information to the public in 1946,[127] four phases of scientific popularization were noted:

(1) the high noon between 1800 and the 1870s when Victorian 'stars' such as Davy, Huxley, Tyndall and Robert Ball used lecture demonstrations, didactic information, polemic and rhetoric to radiate an optimistic positivist message about science and its applications;
(2) the period from the 1880s until the First World War when scientific research had become 'bigger', more complex, and less easily disseminated by popular lectures, and when expositors like H. G. Wells propounded a more pessimistic message;
(3) the period from 1919 until the early 1930s when popularizers like Eddington lapsed into instrumentalism and mysticism;
(4) the period after 1936 when the presentation of the social, political and cultural implications of science and technology became the principal aim of popularization.

Although these phases articulate quite well with the history, aims and achievements of the British Association examined in this essay, from the researcher's and practitioner's perspective, three phases will suffice: the period from 1831 until 1880; 1880 until 1930; and from 1930 to the present day.

The first period from 1831 until 1880 was one in which the Association offered means to an end. It helped powerfully to lubricate the formation of the lobbies and power blocks, such as the X-Club, who defined Victorian science and who acted as gatekeepers to 'non-science'; it provided a stage for ambitious young men, such as Tyndall, Huxley and Lodge, to get noticed by their established peers,[128] or for older men (like Lord Kelvin) to inspire the younger generation to investigate certain subjects or to switch their allegiance to a particular paradigm such as the elastic atom, structural organic chemistry, or evolution; it gave clout to the sectional interests of metropolitan scientific societies, as well as peripatetically stimulating provincial enthusiasm; it encouraged the form of committee, 'team' or collaborative research which was to become the hallmark of twentieth-century science; it took up those causes, particularly educational and financial ones, which were to lead to the securer establishment of scientific careers; and it provided research funds and priming grants at a time when the state's endowment of research was small and narrow in its interest.

By the 1880s, however, the Association was becoming a victim of its own success, and less relevant to the growing army of researchers and practitioners. Its wheels had become caught in a rut from which it was only to be extracted in the 1930s.[129] *The Times,* in 1882, was in no doubt that 'the unprecendented scale and pervasiveness' of Victorian periodicals[130] had rendered annual general meetings such as those of the British Association and the Social Science Association out of date:

> For the general papers, the papers which deal with principles and their application, or with groups of facts that every one cares about, there are the monthly magazines in enormous numbers. For the technical papers there are the technical journals.[131]

Despite the contemporary (and later) reference value of the published reports of the British Association's research committees, there was much truth in this view. For with the development of the refereed, referenced and specialized scientific paper and abstracting system after 1850,[132] the written word came to play a far more important role in the advancement of science than it had done in the Association's early years.

However, this can only be part of the reason for the decline in interest in the British Association by the majority of researchers and practising scientists since the 1880s. Researchers, using the British Association and other platforms, successfully gained increasing government patronage and the security of academic careers in universities, colleges and state enterprises (such as the National Physical Laboratory). The need to invoke an annual public display of social cohesion in order to gain the attention and patronage of cultivators of science diminished. When the lobbying of government was required – as during the First World War – the British Association could still be usefully invoked, especially since its peak of activity and 'massed opinion' usually coincided with the summer trough of inactivity of the learned societies.[133] But the running, in new scientific ideas, was more often made by the Royal Society's *Proceedings*, or the weekly *Nature.*[134]

Similarly, although the numbers of science practitioners increased phenomenally during the Association's first century, the self-employed practitioner was being replaced by the full-time, academically-trained and certificated industrial and government scientist and technician. Consequently, except through their employer's permission or policy, practioners were less able to spare the time to attend meetings unless the Association visited their town. Moreover, the important social function of the British Association, whereby contact between researchers in different disciplines was promoted, was replaced by the regular activities of specialized societies and by the development of specialist national and international congresses. Furthermore, with the decline of the country house and the keeping of servants, and the advent of motorized summer holidays, there were no longer wealthy cultivators of science willing, or at home, to offer free hospitality to the touring matinée idols of science.[135]

Fragmentation of the common scientific culture[136] made it inevitable that the twentieth-century Association took on a more educational function, and sought new local audiences and constituencies, including the social sciences, in an effort to recapture, in Tyndall's phrase, its role as presenting a 'sphere of knowledge'.[137] Between 1919 and 1936, the Association dramatically shifted its posture towards being a forum for the discussion and analysis of interdisciplinary problems which had wide and significant social and cultural implications; but until then its achievement as far as the working and publishing scientist is concerned, was fairly summed up by the engineer Sir Charles Parsons, himself a liberal donor of funds to the post-War Association:

> Although the British Association cannot claim any paramount prerogative in . . . good work, yet it can certainly claim to [have provided] a free arena for discussion where in the past new theories of science, new propositions for beneficial change, new suggestions for casting aside fetters to the advancement of science, art and economics have first seen the light of publication and discussion.[138]

Acknowledgements

I would like to express my thanks to Roy MacLeod and Peter Collins for their comments on earlier drafts of this essay.

Notes

1. G. Basalla, W. Coleman and R. H. Kargon (eds.), *Victorian Science: A Self-portrait from the Presidential Addresses to the B.A.A.S.* (New York: Anchor, 1970), 4.
2. The phrase was first used by T. S. Traill, *B.A.R.* Liverpool, 1837, xlii; repeated by J. Phillips, *B.A.R.* Birmingham, 1865, lxvii. Cp. the satirical portrayal of the Social Science Association as 'The Parliament of Hobbies'.
3. J. Lubbock, *B.A.R.* York, 1881, 2.
4. W. J. M. Mackenzie, 'Science in the Study of Administration', *The Manchester School of Economic and Social Studies, 20* (1952), 4-5; S. Ross, 'Scientist: The Story of a Word', *Annals of Science, 18* (1962), 65-85; J. B. Morrell, 'London Institutions and Lyell's Career: 1820-41', *British Journal for the History of Science, 9* (1975), 133-4; S. F. Cannon, *Science in Culture: The Early Victorian Period* (New York: Dawson and Science History Publications, 1978), ch. 6, 'Professionalization and the B.A.A.S.', esp. 195-6.
5. J. S. Mill *The Principles of Political Economy* (London: Longmans, 1898), 1.

6. B. Rodgers, 'The Social Science Association, 1857-1886', *The Manchester School of Economic and Social Studies, 20* (1952), 283-310.

7. *Edinburgh Review, 49* (1829), 444-7, quoted by Ross, *op. cit.* note 4, 69.

8. Ross, *op. cit.* note 4, 70.

9. *Ibid.*, 71.

10. [W. Whewell], 'Mrs. Somerville on the "Connexion of the Sciences", *Quarterly Review, 51* (1834), 58-61; W. Whewell, *Philosophy of the Inductive Sciences* (London: Parker, 1840), Vol. 1, cxiii, where he also coins 'physicist'. See Ross, *op. cit.* note 4, for full discussion.

11. Ross, *op. cit.* note 4, 81. Cf. Francis Galton's choice of title, *English Men of Science* (London: Macmillan, 1874).

12. E. Mendelsohn, 'The Emergence of Science as a Profession in Nineteenth-century Europe', in K. Hill (ed.), *The Management of Scientists* (Boston: Beacon Press, 1963), 3-48; Basalla, Coleman, and Kargon, *op. cit.* note 1, Introduction; J. Ben-David, *The Scientist's Role in Society* (Englewood Cliffs, New Jersey: Prentice Hall, 1971); R. von Gizycki, 'The Associations for the Advancement of Science: An International Comparative Study', *Zeitschrift für Soziologie, 8* (1979), 28-49.

13. J. F. W. Johnston, 'Meeting of the Cultivators of Natural Science and Medicine', *Edinburgh Journal of Science, 4* (1831), 190, quoted by Mendelsohn, *op. cit.* note 12, 26.

14. T. J. Johnson, *Professions and Power* (London: Macmillan, 1972); A. Oleson and S. C. Brown (eds.), *The Pursuit of Knowledge in the Early American Republic* (Baltimore and London: John Hopkins, 1976). (Note esp. N. Reingold, 'Definitions and Speculations: The Professionalization of Science in America in the Nineteenth Century', 33-69.) A helpful orientation is provided by the Monopolies Commission, *Report on Professional Services*, Part II, Appendices 5-12, *Parliamentary Papers*, 1970-71, xxxii, Cmnd. 4463, 29-171. For a critical review of historical professionalization studies, see Cannon, *op. cit.* note 4, chs. 5-6.

15. 'Professionalization by traits' is the subject of G. Millerson, *The Qualifying Associations* (London: Routledge & Kegan Paul, 1964).

16. Johnson, *op. cit.* note 14, 23-31.

17. H. L. Wilensky, 'The Professionalization of Everyone?', *American Journal of Sociology, 70* (1964), 137-58.

18. T. Fisher, *The Role of the Amateur in Science* (unpublished Ph. D. dissertation, University of Leicester, 1980), 14-29.

19. Cannon, *op. cit.* note 4, ch. 5.

20. *Ibid.*, p. 150.

21. Reingold, *op. cit.* note 14, 38.

22. Cf. M. Berman, 'Hegemony and the Amateur Tradition in British Science', *Journal of Social History, 1* (1975), 30-50.

23. Oleson and Brown, *op. cit.* note 14; I. Inkster, 'The Askesian Society, 1796-1807', *Annals of Science, 34* (1977) 1-32; P. J. Weindling, 'Geological Controversy and its Historiography: The Prehistory of the Geological Society of London', in L. J. Jordanova and R. Porter (eds.), *Images of the Earth* (Chalfont St. Giles, Bucks.: B.S.H.S., 1979), 248-71; D. Sloan, 'Science in New York City, 1867-1907', *Isis, 71* (1980) 35-76; I. Inkster and J. B. Morrell (eds.), *Metropolis and Province: British Science, 1780-1850* (London: Hutchinson, in press).

24. R.Bud, *The Origins and Early Years of the Chemical Society of London* (unpublished Ph.D. dissertation, University of Philadelphia, 1980) ch. 1; P. Weindling, 'The British Mineralogical Society: A Case-study in Science and Social Improvement', in Inkster and Morrell, *op. cit.* note 23. Both Bud and Weindling stress the technical occupational (*i.e.* practictioner) element among the founders of specialized societies.

25. See comments of E. A. Schäfer, *B.A.R.* Oxford, 1894, 795-6. For the anti-vivisection movement, see R. D. French, *Anti-vivisection and Medical Science in Victorian Society* (Princeton: Princeton University Press, 1974). For the Association's Hippocratic Oath for physiologists using experimental animals, see *B.A.R.* Edinburgh, 1871, 144.

26. Quoted by P. Collins, 'The Origins of the British Association's Education Section', *British Journal of Educational Studies, 27* (1979), 238.

27. R. Wynne-Edwards, 'A New Look at the Old', in T. L. Dennis (ed.), *Engineering Societies in the Life of a Country* (London: Institution of Civil Engineers, 1968), 290.

28. W. F. G. Armytage, *A Social History of Engineering* (London: Faber & Faber, 4th edn., 1961), 122.

29. W. J. M. Rankine, *A Manual of Applied Mechanics* (London: Griffin, Bohn, 1858), 4.

30. A. F. Burstall, *A History of Mechanical Engineering* (London: Faber & Faber, 1963), 203; A. R. J. P. Ubbelohde, 'The Beginnings of the Change from Craft Mystery to Science as a Basis for Technology', in C. Singer *et al.* (eds.), *History of Technology* (Oxford: Oxford University Press, 1958), Vol. 3, 663-81; C. D. Woodward, *B.S.I.: The Story of Standards* (London: British Standards Institution, 1972).

31. Von Gizycki, *op cit.* note 12, 41-7.

32. Huxley's remarks were not printed. See L. Huxley, *Life and Letters of Thomas Henry Huxley* (London: Macmillan, 1900), Vol. 1, 277.

33. *B.A.R.* Dublin, 1908, cxxi. The appeal to Council was signed by Schuster, Turner, Shaw and Eliot, but not by Lockyer. The Division for Astronomy had been discontinued after 1904.

34. *Nature, 106* (1920-21), 69.

35. The Division folded in 1959, being absorbed into a Section X for General Science which had been established in 1954. A Section X for Popular Science was adumbrated by F. A. Bather, *Nature, 106* (23 September 1920), 112. Section N for Sociology began in 1960.

36. O. J. R. Howarth, *The British Association for the Advancement of Science: A Retrospect 1831-1931* (London: B.A.A.S. 1922, 2nd edn., 1931), 98-9.

37. *Ibid.* 100.

38. Anne C. Taylor, *The Role of the National Association for the Promotion of Social Science in the Women's Movement. 1857-1886* (unpublished M.A. dissertation, University of Leicester, 1976); L. Ritt, *The Victorian Social Conscience in Action: The National Association for the Promotion of Social Science, 1857-1886* (unpublished Ph.D. dissertation, Columbia University, 1959); Rodgers, *op. cit.* note 6.

39. E. Sharp, *Hertha Ayrton: A Memoir* (London: Arnold, 1926).

40. A. D. Orange, 'The Origins of the British Association for the Advancement of Science', *British Journal for the History of Science, 6* (1972), 152-76.

41. E.g. Charles Lyell. See Mrs. Lyell (ed.), *Life, Letters and Journals of Sir Charles Lyell, Bart.* (London: Murray, 1881), Vol. 2, 42, 45. For Lyell's attitude to holding office in scientific societies, see Morrell, *op. cit.* note 4, 138.

42. F. Darwin, *Life and Letters of Charles Darwin* (London: John Murray, 1887), Vol. 1, 351. C. Darwin to J. D. Hooker, October 1846.

43. Quoted by Tyndall in his *Journal*, Vol. 3 (1855-72), f.797, 27 October 1855 (Royal Institution archives).

44. M. N. W[est] and S. J. C[olenso], *Sketches from the Life of Sir Edward Frankland* (London: Spottiswoode, 1902), 286. Frankland to Nicholson, 13 October 1861.

45. N. Reingold (ed.), *The Papers of Joseph Henry* (Washington: Smithsonian Institution, 1979), Vol. 3, 511. Henry to Charles Coquerel, 1 October 1837.

46. F. Galton, *Memoirs of My Life* (London: Methuen, 1908; repr. New York: A.M.S. Press, 1974), 215.

47. R. M. MacLeod, 'The X-Club: A Social Network of Science in Late-Victorian England', *Notes and Records of the Royal Society of London, 24* (1970), 305-22; J. Vernon Jensen, 'The X-Club: Fraternity of Victorian Scientists', *British Journal for the History of Science, 3* (1970), 63-72.

48. F. M. Turner, 'Science and Citizenship: The Face of Public Science 1880-1919', *Isis, 71* (1980), 589-608.

49. F. M. Turner, 'The Victorian Conflict between Science and Religion: A Professional Dimension', *Isis, 69* (1978), 367. The address lists of members in the *B.A.*

Reports show that by 1919 the number of clergymen-cultivators holding membership was only 20, compared with 135 in 1881.

50. T. H. Huxley, *Nineteenth Century*, 35 (1894), 5-6, quoted by H. I. Sharlin, *Lord Kelvin: The Dynamic Victorian* (University Park, Pennsylvania: Pennsylvania University Press, 1979), 137.

51. Huxley, *op. cit.* note 32, Vol. 1, 88.

52. *Ibid.* 80.

53. *Ibid.* 113-14.

54. R. G. A. Dolby, 'Debates over the Theory of Solution', *Historical Studies in the Physical Sciences, 7* (1976), 297-404.

55. O. Lodge, *Advancing Science* (London: Ernest Benn, 1931), 148.

56. *Nature, 40* (17 October, 1889), 593.

57. J. J. Thomson, *Recollections and Reflections* (London: Macmillan, 1936), 424.

58. H. E. Armstrong, *Nature, 42* (28 August 1890), 414.

59. H. E. Armstrong, *Nature, 46* (28 July 1892), 291.

60. J. Perry (ed.), *British Association Meeting at Glasgow 1901: Discussion on the Teaching of Mathematics* (London: Macmillan, 1901). For Section L, see Collins, *op. cit.* note 26.

61. *B.A.R.* Dublin, 1878, 1-32, esp. 6-7.

62. R. M. MacLeod, 'The Support of Victorian Science: The Endowment of Research Movement in Great Britain 1868-1900', *Minerva, 4* (1972), 221.

63. W. H. Brock, 'The Development of Commercial Science Journals in Victorian Britain', in A. J. Meadows (ed.), *Development of Science Publishing in Europe* (Amsterdam, New York, Oxford: Elsevier Science Publishers, 1980), 95-122.

64. *B.A.R.* Southampton, 1882, 1-33, esp. 2.

65. Only 22 (9 per cent) of the 270 FRSs who were members of the Association in 1881 were elected FRS before the reforms of 1848. My figures are derived by comparing the British Association membership lists published in the *Reports* for 1881, 1892 and 1919 with the dates of election of Fellows of the Royal Society, recorded in *The Record of the Royal Society* (London: Harrison, 4th ed., 1940).

66. The office was declined by, among others, Edward Frankland; Prestwich (G. A. Prestwich, *Life and Letters of Sir Joseph Prestwich* (Edinburgh: Blackwood, 1899), 278); and Galton (Galton, *op. cit.* note 46, 218).

67. L. Huxley, *Life and Letters of Sir Joseph Hooker* (London: John Murray, 1918), Vol. 2, 116.

68. Galton, *op. cit.* note 46, 219.

69. 'The president [Duke of Buccleuch] on this occasion evidently considered it a condescension on his part to be there at all. He began by telling us that he had never written a speech in his life, and never intended to; that he knew very little about science . . .' A. R. Wallace, *My Life* (London: Chapman & Hall, 1905), Vol. 2, 48. Hooker was elected in 1868 to prevent a repetition of this embarrassing episode.

70. F. Bramwell, *B.A.R.* Bath, 1888, 4; Howarth, *op. cit.* note 36, 107-12.

71. Frankland, *op. cit.* note 44, 148-63. See also MacLeod and Vernon Jensen, *op. cit.* note 47.

72. *Nature, 134* (1934), 274.

73. *Nature, 106* (2 September 1920), 13. For fuller details of this debate, see P. Collins, 'The British Association as Public Apologist for Science, 1919-1946' in this volume.

74. R. Gregory, editorial, *Nature, 106* (2 September 1920), 69-72. For the correspondence of Lodge, Armstrong, *et al.*, see *ibid.*, 107-13, 144-7, 178-9, 211-13, 277-80, 373-4, 467-8, 565.

75. This explanation was canvassed by Napier Shaw, *Nature, 106* (7 October 1920), 178-9.

76. *Nature, 106* (28 October 1920), 277. Letter from J. L. Myres and H. H. Turner, the Secretaries under attack.

77. H. E. Armstrong, 'The British Association Meeting, 1935', *Chemical Age, 33*

(1935), 251.

78. Sloan, *op. cit.* note 23, 76.

79. Reingold, *op. cit.* note 45, Vol. 3, 511. For Brewster's grant, which led to publications, see *B.A.R.* Liverpool, 1837, xvii.

80. Rev. R. Willis, Professor of Applied Mechanics at Cambridge, *B.A.R.* Cambridge, 1862, liii.

81. *Ibid.* liv.

82. *B.A.R.* Oxford, 1894, lxxviii.

83. *Nature, 6* (15 August 1872), 297.

84. Howarth, *op. cit.* note 36, Appendix 1, 266-92 and discussion, 146-7. Howarth, 170-211, provides many more examples of the Association's researches than can be considered here.

85. *The Advancement of Science* was the brainchild of the then Secretaries, Boswell and Ferguson, who saw it as a mouthpiece for the new Division for the Social and International Relations of Science which would dispel the Association's 'once-a-year-only' image. A publication of the same title was also issued from 1920 at the time of the meeting. This contained all the Presidential Addresses. See Howarth *op. cit.* note 36, 257.

86. R. M. MacLeod, 'The Royal Society and the Government Grant: Notes on the Administration of Scientific Research, 1849-1914', *The Historical Journal, 14* (1971), 323-58.

87. J. B. Morrell, 'The Patronage of Mid-Victorian Science in the University of Edinburgh', in G. L' E. Turner (ed.), *The Patronage of Science in the Nineteenth Century* (Leyden: Noordhoff, 1976), 84; MacLeod, *op. cit.* note 86, 334-5.

88. MacLeod, *op. cit.* note 62, 207-13.

89. *B.A.R.* Dublin, 1878, 5.

90. E.g. J. A. Wanklyn, *B.A.R.* Bath, 1864, xlix.

91. Galton, *op. cit.* note 46, 213.

92. The Committee first consisted of a dozen members with powers to co-opt, *B.A.R.* Bristol, 1875, liv-lv.

93. *B.A.R.* Dublin, 1857, 1-2.

94. A. J. Meadows, *Greenwich Observatory*, Vol. 2, *Recent History (1836-1975)* (London: Taylor & Francis, 1975), 21.

95. *The Catalogue of Stars of the British Association* (London: 1845). The contributing editors were W. S. Stratford, T. R. Robinson, J. Challis and R. Farley.

96. Howarth, *op. cit.* note 36, 174-5.

97. J. Cawood, 'The Magnetic Crusade: Science and Politics in Early Victorian Britain', *Isis, 70* (1979), 493-518.

98. W. Whewell, *A History of the Inductive Sciences* (London: Parker, 3rd ed., 1858), Vol. 3, 55, quoted in Cawood *op. cit.* note 97, 493.

99. See N. Rosenburg and W. G. Vincenti, *The Britannia Bridge: The Generation and Diffusion of Technological Knowledge* (Cambridge, Massachusetts: M.I.T., 1978).

100. *B.A.R.* Dublin, 1857, 196; G. S. Emmerson, *John Scott Russell* (London: John Murray, 1977).

101. *B.A.R.* Birmingham, 1865, lxiv; confirmed by M. Grosser, *The Discovery of Neptune* (Cambridge, Massachusetts: Harvard University Press, 1962), 75.

102. Lodge, *op. cit.* note 55, 40-1. For Cayley, see *B.A.R.* Bristol, 1875, 257-303.

103. 'a star catalogue, tides, meteorology, terrestrial magnetism, fossil zoology, early marine biology, and the re-establishment of the Kew observatory', Cannon, *op. cit.* note 4, 191.

104. *B.A.R.* Bath, 1888, 4.

105. E.g. A. W. Williamson, 'The Atomic Theory', *B.A.R.* Bradford, 1873, lxx-xci; for Maxwell's 'Evening Lecture' on molecules at the same meeting, see *Nature, 8* (1873), 437-41. For thermodynamics, see S. P. Thompson, *The Life of William Thomson,* 2 vols. (London: Macmillan, 1910), and J. D. Burchfield, *Lord Kelvin and the Age of the Earth* (London: Science History Publications, 1975). For evolution, see E. B. Poulton, 'A History of Evolutionary Thought at the British Association', *B.A.R.* Nottingham,

1937, 1-23.

106. E.g. *B.A.R.* Southampton, 1882, 120-43.

107. The electrical reports were republished by Sir R. T. Glazebrook and F. E. Smith, *Reports of the Committee on Electrical Studies* (Cambridge: B.A., 1913). See I. B. Hopley, 'Maxwell's Work on Electrical Resistance', *Annals of Science, 13* (1957), 265-72; P. Dunsheath, *A History of Electrical Engineering* (London: Faber & Faber, 1962), ch. 18. For the adoption of the 'c.g.s.' system, see *B.A.R.* Bradford, 1873, 222-5.

108. Lodge, *op. cit.* note 55, passim; *B.A.R.* Cardiff, 1891, 549-51. The need for a national laboratory was first raised by W. Thomson, *B.A.R.* Edinburgh, 1871, lxxxvii. See also H. J. S. Smith, *B.A.R.* Bradford, 1873, 1-8.

109. See speech by D. Galton, *B.A.R.* Ipswich, 1895, 1-35.

110. Howarth, *op. cit.* note 36, 169; R. Moseley, 'The Origins and Early Years of the National Physical Laboratory: A Chapter in the Pre-history of British Science Policy', *Minerva, 16* (1978), 222-50.

111. Bud, *op. cit.* note 24, chs. 2-3.

112. R. Appleyard, *History of the Institution of Electrical Engineers 1871-1931* (London: Institution of Electrical Engineers , 1939). Earlier, in 1834, the Statistical Society had been founded as a consequence of the statistician Adolphe Quetelet's attendance at the Cambridge meeting of the Association in 1833. See V. L. Hilts, 'Aliis exterendum, or, the Origin of the Statistical Society of London', *Isis, 69* (1978), 21-43. Philip Lowe has also drawn my attention to the formation of the British Ecological Society in 1913 consequent upon Arthur Tansley's appeal in 1904 for the formation of a British Vegetation Committee, to map and survey vegetation. See *B.A.R.* Cambridge, 1904, 797-8. For the British Association's stimulation of new provincial societies see P. Lowe, 'The British Association and the Provincial Public', in this volume.

113. *B.A.R.* Oxford, 1894, 614 (not printed). See M. W. Travers, *A Life of Sir William Ramsay* (London: Edward Arnold, 1956), ch. 7.

114. W. H. Brock (ed.), *H. E. Armstrong and the Teaching of Science 1880-1930* (Cambridge: Cambridge University Press, 1973).

115. Howarth, *op. cit.* note 36, 191-2; L. K. Clark, *Pioneers of Prehistory in England* (London and New York: Sheed and Ward, 1961).

116. E.g. *B.A.R.* Leeds, 1890, 429-44; *B.A.R.* Nottingham, 1937, 274-80.

117. K. Koizumi, 'Davisson, Clinton Joseph', *Dictionary of Scientific Biography* (New York: Charles Scribner's Sons, 1971), Vol. 3, 597-8.

118. *B.A.R.* Birmingham, 1913, 403.

119. Thompson, *op. cit.* note 105, Vol. 2, 1200-2. Philip Lowe has also drawn my attention to F. W. Oliver's vehement attack on the botanical work of Kew Gardens and the Natural History Museum in 1906. See *B.A.R.* York, 1906, 733-8.

120. 'Discussion on Isotopes', *Proceedings of the Royal Society, 99A* (1921), 87-104.

121. Lodge, *op.cit.* note 55, 144.

122. *Nature, 14* (21 September 1876), 451. Barrett subsequently published the article in *Proceedings of the Society for Psychical Research, 1* (1883), 238-44.

123. *B.A.R.* Bristol, 1836, 47-8; Mrs. Cornelia A. H. Crosse, *Memorials Scientific and Literary of Andrew Crosse the Electrician* (London: Longmans, 1857). See Henry's acidic comments, Reingold, *op. cit.* note 45, Vol. 3, 320-1, and O. Stallybrass, 'How Faraday Produced Living Animalculae: Andrew Crosse and the Story of a Myth', *Proceedings of the Royal Institution, 41* (1967), 597-619.

124. *B.A.R.* Bristol, 1898, 3-38, esp. 29-32.

125. Roy Wallis (ed.), *On the Margins of Science: The Social Construction of Rejected Knowledge*, Sociological Review Monographs, *27*, (University of Keele, 1979).

126. E.g. *B.A.R.* York, 1931, 276-86.

127. *Advancement of Science, 4* (1946-48), 33-6.

128. This still happened in the 1890s. Cf. the case of D. T. Gwynne-Vaughan (1871-1915), who gained a lectureship in Botany at the University of Glasgow in 1897 on the strength of a British Association paper. See *Dictionary of Scientific Biography* (New York: Charles Scribner's Sons, 1972), Vol. 5, 604-5.

129. O. J. R. Howarth, 'The Future Pattern of the British Association', *Advancement*

of Science, *4* (1946-68), 68-74, esp. 68.

130. M. Wolff, *Urbanity and Journalism: The Victorian Connection* (Leicester: Victorian Studies Centre, 1980), 8.

131. *The Times*, 22 September 1882, 7, quoted in Ritt, *op. cit.* note 38, 294-5.

132. A. J. Meadows, *Communication in Science* (London: Butterworth, 1974), ch. 3, esp. 79-86; also Meadows, *op. cit.* note 63.

133. Spottiswoode, *B.A.R.* Dublin, 1878, 8.

134. A. J. Meadows, *Science and Controversy: A Biography of Sir Norman Lockyer* (London: Macmillan, 1972); R. M. MacLeod, 'The Genesis of *Nature*', centenary issue, *Nature, 224* (1969), 423-40; 'The Social Framework of *Nature* in its First Fifty Years', *ibid.*, 441-6. See also, O. J. R. Howarth, 'The British Association', *Endeavour, 3* (1944), 57-61.

135. A. D. Orange, 'The Idols of the Theatre: The British Association and its Early Critics', *Annals of Science, 32* (1975), 294.

136. Cannon, *op. cit.* note 4, ch. 9.

137. *B.A.R.* Norwich, 1868, 1.

138. *B.A.R.* Bournemouth, 1919, 3.

III

The Society for the Perpetuation of Gmelin: The Cavendish Society, 1846–1872

Summary

The Cavendish Society, which lasted from 1846 to 1872, was one of a large number of Victorian subscription printing clubs which published translations, re-issued historical works or commissioned original books which were too specialized for commercial publication. The Society's book production was limited, being principally devoted to a translation of L. Gmelin, *Handbook of chemistry*. Reasons for its limited success are sought in the institutionalization of chemistry during the 1840s and in a divergence of interests between academic and practising chemists.

Contents

1. Printing clubs

The Cavendish Society was a chemical book club which lasted from 1846 until about 1872. In its twenty-odd years' lifetime, financed by the annual subscriptions of its members, it managed to publish only seven titles in thirty volumes (eighteen of these being a single encyclopaedic work, Leopold Gmelin's *Handbook of chemistry*). Compared with its contemporary, the medical subscription library, the Sydenham Society, which was counted a failure even though it published thirty titles in forty volumes in a lifetime of only fourteen years (1843–57), the Cavendish Society was a flop.[1]

What is the explanation for this failure? Answers may be sought in the general development of Victorian printing clubs, in the organization of the Cavendish Society itself, and in the institutionalization of chemistry during the 1840s and 1850s.

Publication of books by subscription, which dates back to the seventeenth century, ' was a way of assessing the possible demand and of sharing the financial responsibility '.[2] In the eighteenth century the method became well-established for the production of works of natural philosophy, topography and local history, and was extended to include science subscription lectures by itinerant lecturers,

[1] On the Sydenham Society, and for some useful parallels with my treatment of the Cavendish Society, see William B. Ober, ' The Sydenham Society (1843–1857): rise and fall ', *Mount Sinai journal of medicine*, **41** (1974), 294–305.

[2] F. J. G. Robinson and P. J. Wallis, 'A preliminary guide to book subscription lists ', *History of Education Society bulletin* (Spring 1972), 23–54 (p. 27). Compare S. L. C. Clapp, ' The beginnings of subscription publishing in the seventeenth century ', *Modern philology*, **29** (1931), 199–224.

the support of schools by the Society for the Promotion of Christian Knowledge, or the maintenance of circulating libraries. Although an important British work of natural history was produced by subscription as late as 1849, the emergence of publishing houses, like John Murray, William Longman, Daniel Macmillan, Richard Taylor and his son William Francis, and John Van Voorst, who were prepared to invest in science, diminished the need, while the results of research were increasingly published in the form of papers in journals which were either supported by specialized scientific societies or launched as commercial ventures by these same publishers.[3] Where the financial risk remained great, however, the alternative to the production of a single volume by subscription patronage was the printing club which, like the Cavendish Club to be discussed, aimed to promote the sciences or some other area of knowledge, by the translation and/or publication of works and papers of a strictly limited appeal. Although some eighteenth-century societies, like the Dublin Society and the Antiquarian Society, had occasionally printed books from their members' subscription fees as an incidental part of their activities, the printing club proper developed in the nineteenth century.

Between 1812 and 1846 over twenty societies were founded with the declared intention of printing and publishing rare works otherwise inaccessible to their members. Such societies were often extremely specialized in their terms of reference, and exclusive in membership.[4] The earliest societies tended to be literary or historical in interest, although as the number of societies increased in the 1840s so did the range of subjects represented.

All the printing societies followed the example of the Roxburghe Club, which was founded in 1812 as a result of the sale of the third Duke of Roxburghe's library.[5] The aim of this society was that 'each member in turn should print some rare old tract, or composition chiefly of poetry' at his own expense and present other members with a copy. The result was that in the early years the society's publications were frequently amateurish, and the society itself gained the reputation of being an exclusive club of bookish dilettantes whose excessive gourmandizing did little to command respect. The quality of the Roxburghe Club's publications was considerably enhanced when it began to commission work from scholars.

[3] A late subscription work is J. D. Hooker, *The rhododendrons of Sikkim-Himalaya* (1849, London). Other examples are: George Green, *An essay on the application of mathematical analysis to the theories of electricity and magnetism* (1828, Nottingham); and R. Murchison, *The Silurian system* (1839, London). Note that Richard Taylor's *Scientific memoirs selected from the transactions of foreign academies* (5 vols., 1837–52, London) were published by subscription, together with a British Association subsidy.

[4] For the details given below I have used Rev. A. Hume, *The learned societies and printing clubs of the United Kingdom* (2nd ed., 1853, London); The British Library, *Catalogue of printed books*; Harrison Ross Steeves, *Learned societies and English literary scholarship* (1913, New York); H. Williams, *Book clubs and printing societies* (1929, London); and information contained in the works of individual societies. I have not seen Bernard Quaritch, *Account of the great learned societies and associations and of the chief printing clubs of Great Britain and Ireland* (1886, London).

[5] John Ker (1740–1804), who owned a library in St. James's Square which contained many items from Caxton's press (see *D.N.B.*). The Roxburghe Club was composed of 24 bibliophiles, including Lord George John Spencer (1758–1834), the President whose library became the nucleus of the Rylands Library at Manchester, and the Secretary, T. F. Dibdin (1776–1847). Each member was (and is) required to pay for the reprinting of a rare book. See Clive Bigham, *The Roxburghe Club: its history and its members 1812–1927* (1928, Oxford).

By 1828 the Roxburghe Club had been joined by two other societies: The Bannatyne Club (Edinburgh, 1823–75),[6] and the Maitland Club (Glasgow, 1828–59).[7] Although they were formed independently, many members were common to both societies and they were both concerned with printing works illustrative of the history and antiquities of Scotland. Both clubs printed valuable works often at the society's expense rather than that of individual members and the chief distinguishing feature between the two was that the Maitland was more concerned with the west of Scotland. Members of these two societies were often also members of the Abbotsford Club (1833–66) with its more general interest in miscellaneous pieces, illustrative of history, literature and antiquities.[8]

The formation of the Roxburghe, Bannatyne and Maitland Clubs paved the way for a mushrooming of such clubs in the next two decades. A distinctive feature of the majority of these societies was their extremely limited subject interest. Imitators of the early Scottish clubs became more and more obscure and specialized. Many societies exhibited strong local regionalism, following the example of the Bannatyne and Maitland Clubs, with their emphasis on the history and literature of Scotland. This is clearly demonstrated in the short-lived Iona Club (Edinburgh, 1833–*c.* 1838) which was founded for the purpose of investigating the history, antiquities and literature of the Highlands and Islands of Scotland. Likewise the Spalding Club (1839–71) was founded at Aberdeen for the purpose of printing the historical, ecclesiastical, genealogical, topographical and literary remains of the north-east of Scotland.[9] English clubs also demonstrated regionalism. The Surtees Society[10] was formed in 1834 for the purpose of publishing unedited manuscripts illustrative of the historical, intellectual and social conditions of the ancient kingdom of Northumbria, with particular emphasis on Durham. The Counties Palatine of Lancashire and Cheshire had their interests represented by the Chetham Society, founded at Manchester in 1842 to publish local records, cartularies and memorials. Ireland had its own societies: The Irish Archaeological Society (1840–53); and the Celtic Society, founded in 1845, which amalgamated with the Irish Archaeological Society in 1853.

Just as a strong local flavour could be detected in several printing clubs, so also could a tendency to represent esoteric subjects. Into the latter category fall the Percy Society (1840–52),[11] with its emphasis on reprinting obscure

[6] Sir Walter Scott was its first President. The aim of the Club, which was named after George Bannatyne (1545–1608), a compiler of a collection of Scots poetry, was to publish old Scottish documents (see H. R. Steeves (footnote 4), 108–112).

[7] Founded for the publication of Scottish literature and antiquities (see H. R. Steeves (footnote 4), 112–113).

[8] Founded to commemorate Sir Walter Scott, who died in 1832. Its aim was to publish materials on the history or literature of any country mentioned in Scott's novels (see H. R. Steeves (footnote 4), 113–114).

[9] Named after John Spalding (*fl.* 1650), the Aberdonian historian (see D. J. Withrington, Aberdeen antiquaries: the founding of the Spalding Club in 1839 ', *Aberdeen University review*, **44** (1971), 42–55).

[10] Commemorating Robert Surtees (1779–1834), antiquarian and topographer (see H. R. Steeves (footnote 4), 121–122).

[11] Named after Thomas Percy (1729–1811), the Bishop of Dromore, whose publications stimulated an interest in early English poetry. The founders included Thomas Wright, Thomas Croftob Croker, Alexander Dyce, J. O. Halliwell-Phillipps and John Payne Collier (see H. R. Steeves (footnote 4), 141–143).

ballad poetry, and the Aelfric Society (London, 1842–56), which aimed to publish Anglo-Saxon documents with an English translation. Slightly wider in appeal was the Shakespeare Society (1840–53), with its aim to reprint rare works and books illustrative of the history of Shakespeare's time. An awakening interest in the East was reflected in the foundation in 1828 of the Oriental Translation Fund of Great Britain and Ireland, which aimed to translate Eastern manuscripts into European languages. Its example was followed in 1840 by the Society for the Publication of Oriental Texts, which printed standard works in Syriac, Arabic, Persian, Sanskrit, Turkish and Chinese.

Another sphere in which the early printing societies were important was in the reprinting of ecclesiastical documents. Most prosperous and important among these societies was the Parker Society (1840–53), which aimed to print the best works of the divines of the English Reformed Church.[12] Scottish ecclesiastical history was well represented by the Woodrow Society (Edinburgh, 1841–64)[13], which concentrated on the early Presbyterian Church; and the Spottiswoode Society (Edinburgh, 1843–51), which was interested in the Episcopalian Church of Scotland. A short-lived English club was the Wycliffe Society (London, 1843–46), which only produced one volume before its dissolution.[14]

Not all the printing clubs established during this period were so specialized in their fields of interest. Indeed, several important general societies were established. The most noteworthy of these was the Camden Society (London, 1838–97), which aimed 'to perpetuate and render accessible whatever is valuable but little known amongst the materials for civil, ecclesiastical or literary history of the United Kingdom'.[15] Its particular significance was its production of antiquarian and historical works each as a cheap volume in its own right, thus making them accessible to a much wider market. Similarly, its low membership subscription provided a marked contrast to the aristocratic exclusiveness of the Roxburghe Club. Similar in aim, but considerably less successful than the Camden Society, was the English Historical Society (London, 1836–56). Another more general society was the Caxton Society (London, 1844–54) which proposed to print chronicles and other documents hitherto unpublished, illustrative of the history and miscellaneous literature of the British Isles during the Middle Ages.

Historical, ecclesiastical and literary works clearly provided the focus for most of the societies founded up to 1843. After this date it became apparent that there was scope for reprinting scientific and geographical works, both ancient and modern, and that there was an awakening market for such material. The establishment of the Hakluyt Society in 1846 a few years after the foundation of the Royal Geographical Society reflected the growing interest in geographical exploration of the period. Its aim was to print the most rare and

[12] Commemorating Archbishop Mathew Parker (1504–75) (see H. R. Steeves (footnote 4), 127–128).

[13] Named after Robert Woodrow (1679–1734), librarian of Glasgow University (see H. R. Steeves (footnote 4), 128).

[14] A more successful Wyclif Society, founded in 1882, completed its programme in the early 1920s.

[15] See F. J. Levy, 'The founding of the Camden Society', *Victorian studies*, **7** (1966), 295–305. Its work was taken over by the Royal Historical Society, which perpetuates the series.

valuable voyages and travels and geographical records from an early period of circumnavigatory exploration. It was not purely concerned with English explorers and provided translations of works in foreign languages.

It is interesting to note that, following the great proliferation of printing societies in the 1820s and 1830s, there was a marked nadir until the 1860s. After this some of the finest and most important English printing societies developed, as for example, the Early English Text Society (1864),[16] the Harleian Society (1869) for the publication of heraldic visitations and pedigrees, the Folk Lore Society (1878), and the Pipe Roll Society (1883), to name but a few of those societies which have rescued and made accessible to the research worker important works long neglected.

Undoubtedly the flood of publications of ancient, if obscure, works from the diverse printing clubs inspired men of science to investigate their own resources of forgotten literature. The result was that between 1843 and 1846 five societies emerged which were concerned with reprinting and printing scientific works. All modelled on the plan of the Camden Society, the first of these was the Sydenham Society (1843–57), whose aim was to meet certain deficiencies in existing means for diffusing medical literature, which were not likely to be supplied by the efforts of individuals.[17] Its elaborate terms of reference covered six categories: reprints of standard, rare or expensive works; miscellaneous selections from ancient and earlier modern authors reprinted or translated; digests of old and voluminous authors (English or foreign) with occasional biographical and bibliographical notes; translations of works of Greek and Latin medical authors, and of works of Arabic and other eastern languages; translations of recent works of foreign merit; and the publication of reference works which might never otherwise find a publisher. During its lifetime the Sydenham Society published thirty titles in 40 volumes. It was succeeded by the New Sydenham Society, which produced a further 74 titles in 185 volumes between 1859 and 1907.[18] The one guinea per annum subscription allowed members to receive all the books published in one year. This system was copied by other science book clubs, as was the appointment of local canvassing agents.[19] By 1844 the Sydenham Society had 112 local agents, a membership of 1700 (roughly 10% of licensed practitioners in Great Britain[20]) and an eminent masthead of President, 16 Vice-Presidents and 24 Council Members.

The Ray Society grew out of Friday evening botanical meetings at the home of John Stevens Henslow, which ended in 1836.[21] Despite the worthy efforts

[16] Founded by the scholar and philologist Frederick James Furnivall (1825–1910), an inveterate founder of clubs, reprint and commemorative societies, including the Chaucer Society (1868), the Ballad Society (1868), the New Shakespeare Society (1873), the Wiclif Society (1881), the Browning Society (1881), the Shelley Society (1886), and the National Amateur Rowing Society (1891) (see *D.N.B.*).

[17] William B. Ober (footnote 1).

[18] William B. Ober (footnote 1), 294. Compare Sir Jonathan Hutchinson, *The New Sydenham Society: retrospective memoranda* (1911, London).

[19] Local agents, or provincial representatives, had been introduced early into the organization of British geology and botany by the Geological and Botanical Societies (see D. E. Allen, *The naturalist in Britain—a social history* (1976, London), 110–111).

[20] William B. Ober (footnote 1), 295–297.

[21] [Anne M. Babington], *Memorials, journals and botanical correspondence of Charles Cardale Babington* (1897, Cambridge), 60; C. M. Yonge, ‘George Johnson and the Ray Society’, *Endeavour*, **14** (1955), 136–138; R. T. Gunther, ‘The Ray Society’, *Nature*, **133** (1934), 127–128; and Richard Curle, *The Ray Society. A bibliographical history* (1954, London).

of specialist publishers like John Van Voorst and Lovell Reeve, commercial considerations forced them to publish only naturalists' monographs which were ' best-sellers '. There remained ' a considerable residue of unmonographed classes of animals for which neither the British Association, nor the scientific societies nor publishers would have the funds necessary for publication '.[22] In 1843 the marine naturalist George Johnston of Berwick-on-Tweed, circulated naturalists on the need for an association for publishing zoological and botanical monographs and translations. On 2 February 1844, the Ray Society was formally constituted; and by October, with 400 members, it began its still flourishing programme of printing original works in botany and zoology.[23]

Also founded in 1844 was the Wernerian Club, which aimed to reprint and publish scientific works of antiquity in modern standard editions, miscellaneous essays, and certain modern works presented to and approved by the society. Unlike other scientific clubs, however, the Wernerian Club restricted its membership to 25 members in London and 50 associate provincial members who paid half the annual subscription of a guinea. Although its President (C. J. B. Aldis), treasurer (Charles Moxon) and Secretary (B. W. Croker) were interested in mineralogy, the membership seems to have been principally devoted to natural theology. The club does not appear to have survived into the 1850s.[24]

Three years later, in 1847, the London Clay Club, which had been founded by a convivial group of London fossil collectors in 1836, established the still-flourishing Palaeontographical Society for the publication of monographs illustrating British fossils.[25]

2. The Cavendish Society

The success of the Camden, Sydenham and Ray Societies, together with the establishment of the Chemical Society of London in 1841 and the Royal College of Chemistry in 1845, led the physician and lecturer in chemistry at St. Thomas's Hospital, Henry Beaumont Leeson (1803–72),[26] to suggest in 1846 that the fortunes of chemistry in Britain would be improved if a printing club were established for ' the translation and publication of works and papers on chemistry and its application to agriculture, mineralogy, manufactures, etc.' that were unlikely to be undertaken by commercial publishers.[27] Leeson's

[22] R. T. Gunther (footnote 21), 127.

[23] Richard Curle (footnote 21) gives the Victorian membership figures as: *1845*, 650; *1847*, 868; *1849*, 805; *1850*, 775; *1855*, 720; *1856*, 630; and *1857*, 697.

[24] The club republished Ray's *Wisdom of God* (1844), translated Werner's *External characters* (1849–50), and published an original essay on mineralogical classification (1846–47).

[25] D. E. Allen (footnote 19), 117. For bibliography, see *Catalogue of the Library of the British Museum (Natural History)*.

[26] *J. chem. soc.*, **26** (1872), 778–780. Leeson retired to the Isle of Wight to become a J.P. and landowner in 1853. Until then he had been a keen and quite imaginative teacher of crystallography. See his elaborate ' Observations on the circular polarisation of light by transmission through fluids ', *Mem. and proc. Chem. Soc.*, **2** (1843–45), 26–45.

[27] *Report of the first anniversary meeting of the Cavendish Society held on Monday, July 17th, 1848* (n.d. (1848), London), 1. This and succeeding Reports will be referred to as ' *First report*, 1848 ', and so on. These Reports are commonly (but not always) to be found bound in copies of the Society's books issued in the relevant year. Thus *First report*, 1848, is to be found in T. Graham (ed.), *Chemical reports and memoirs* (1848, London). I have not been able to find the *Eighth report*, 1855 despite inspection of collections of the Society's publications at the libraries of the Universities of Aberdeen, Cambridge, Manchester and Toronto, University College London, and the Science Museum, Chemical and Pharmaceutical Societies (see *Chemistry in Britain*, **9** (1973), 81–82, 187).

suggestion was supported by other members of the Chemical Society, notably by its founding-father Robert Warington (1807–67), the chemical operator at the Apothecaries' Hall, and by the Society's President, Thomas Graham (1805–69).[28] Following now established precedent, such a printing club was founded in October or November 1846 (the date is nowhere given[29]) at a meeting chaired by Leeson at the Society of Arts, then the first home of the Chemical Society. A managing, or provisional, committee of twenty ' gentlemen, resident in London ' was appointed, with Warington and George Day (1815–72), lecturer on *materia medica* at Middlesex Hospital, acting as joint-secretaries, and Leeson as Treasurer.[30] Graham was elected President, in place of Richard Phillips (1778–1851), who declined the office.

It was this initial Council,[31] in which the metropolitan pharmacists were well represented, that chose the title ' Cavendish Society '. The Sydenham and Ray Societies had clearly established a precedent of choosing a figure from the history of their respective disciplines. But why Cavendish? Why not Dalton, whose death in 1844 had been dignified by a public funeral in Manchester? In 1839, a furore started by William Vernon Harcourt at the British Association meeting in York had developed concerning the discovery of the composition of water. The title ' Cavendish Society ' was probably a deliberate anglophilic vindication of Henry Cavendish's work, for his life, by George Wilson, together with documentary material on the water controversy, was to be published by the Society in 1851.[32]

Between 1846 and 1848 several prospectuses of the Society were circulated among chemists and manufacturers throughout Great Britain, and subscribers were solicited both directly and by the appointment of provincial secretaries; for example, John Percy in Birmingham, W. H. Miller in Cambridge, Walter Crum in Glasgow, and Thomas Richardson in Newcastle.[33] According to the Society's *Prospectus*:

> 1. The object of the Society is the promotion of Chemical Science, by the Translation, and Publication of Works and Papers on Chemistry and its applications to Agriculture, Mineralogy, Medicine, and Pharmacy.

[28] For Warington, see *J. Chem. Soc.*, **21** (1868), xxxi–xxxiv; on Graham, see J. R. Partington, *A history of chemistry* (1964, London), vol. 4, 265–270.

[29] *Pharm. j.*, **6** (1846–47), 237, the issue for 1 November 1846, announced the intended formation of the Society whose first prospectus (printed on pp. 339–340, issue for 1 January 1847) was dated 13 November 1846.

[30] C. F. A. Marmoy and John L. Thornton, 'The anatomical and physiological bibliography of George E. Day (1815–1872) ', *Ann. sci.*, **28** (1972). 285–291.

[31] A. Aikin, D. T. Ansted, J. Bell, W. T. Brande, J. T. Cooper, M. Faraday, G. Fownes, A. W. Hofmann, Percival Johnson, W. A. Miller, J. Pereira, R. Phillips, L. Playfair, T. Redwood, G. O. Rees and E. F. Teschemacher (*First prospectus* in *Pharm. j.*, **6** (1846–47), 339). Note that all except Ansted were elected Fellows of the Chemical Society before July 1848.

[32] G. Wilson, *Life and works of Cavendish* (1851, London). The water controversy was revived in 1838 by the publications of Arogo's éloge of Watt, in which the claims of Cavendish were replaced by those of Watt. Cavendish was defended by Harcourt in 1839; subsequently Brougham, Jeffrey, Dumas, Berzelius, Whewell, Weld and Muirhead participated in a literary battle. See A. D. Orange, ' The early history of the Yorkshire Philosophical Society: a chapter in the history of provincial science ' (unpublished Ph.D. thesis: University of London, 1970).

[33] In Philadelphia the agent was the pharmacist William Procter (1817–74) (see *Dictionary of scientific biography*). The *Third report*, 1850, lists 24 American subscribers, 7 of whom were Philadelphians.

2. Every Subscriber of one guinea annually shall be considered a Member of the Society, and shall be entitled to one copy of every book published by the Society during the year to which his subscription relates; and no Member shall incur any liability beyond the annual subscription.
3. The number of copies of the Society's publications shall, unless otherwise directed by the Council, be limited to the number of actual subscribers, who shall have been enrolled, and paid their subscriptions.
4. The Society will commence its publications as soon as, in the opinion of the Council, a sufficient number of subscribers shall have been obtained.
5. The number of volumes to be published by the Society must be regulated by their size, and by the number of the annual subscribers; but the Council contemplate at present the publication of not less than three volumes each year.[34]

This document calls for two comments. Whereas Leeson had included ' manufactures ' among his objects for the Society, this term is replaced by ' medicine and pharmacy ' in the *Prospectus*. Despite this, as we shall see, the Society succeeded initially in attracting chemists from the manufacturing areas of Great Britain. It is also noticeable that the subscription was lower than that for the Chemical Society, which charged £2 per annum. Moreover, the Cavendish Society promised books for money, whereas it is hard to say what a member of the Chemical Society obtained in return for his subscription, for its publications were irregular and comparatively trivial in the 1840s. On the other hand, unlike the Cavendish Society, the Chemical Society did provide metropolitan chemists the privilege of attending regular scientific meetings.

It is clear that the Cavendish Society took far longer to organize than either the Sydenham or Ray Societies. ' The accession of members ', though steady, ' was by no means rapid ',[35] and there was no question of publication until an optimum number of subscribers (probably 600) was reached in the summer of 1848. It was felt, reasonably, that the numbers of subscribers would increase as soon as a volume was published. By 17 July 1848, when the Society held its first General Meeting, there were 537 subscribers (including 7 institutions), representing a potential income of £526. Membership of the Chemical Society stood at 224 on 30 March 1848,[36] but only 114 of the members of the Chemical Society elected before July 1848 also joined the Cavendish Society at the beginning. Thus the proportion of Chemical Society to Cavendish Society members was 21%.[37] On the one hand, this compares favourably with the Sydenham Society which is estimated to have recruited from only 10% of medical practitioners; on the other hand, it shows that there was a wider

[34] *First report*, 1848, 2.

[35] *Ibid.*, 3.

[36] *Quart. j. Chem. Soc.*, **1** (1849), 149.

[37] I have derived the figure of 114 by comparing the names of members in the *First report*, 1848 with the names of ' members ' of the Chemical Society noted in rough hand in ' Minutes of meetings 13 April to 7 Dec. 1846 ' and ' Minutes of ordinary meetings 21 Dec. 1846 to 18 March 1852 ' (Chemical Society Library). Although of course one cannot equate membership of the Chemical Society in the 1840s with the practice of chemistry, it is interesting to note that the Cavendish Society was far more successful in recruiting members:

	London members	Provincial members
Chemical Society (1848)	102	106
Cavendish Society (1848)	222	316

For Chemical Society membership, see *The jubilee of the Chemical Society of London 1891* (1896, London), 186–187.

audience for chemistry than the Chemical Society represented, and that the latter since 1841 had failed to publicize its programme. As Moore and Philip noted in one of their rare interpretative asides, ‘ There was apparently a wider demand for systematic textbooks and summaries of knowledge (in the chemical community) than for strictly original matter ’.[38] However, of the 537 only 165 were paid up subscribers, producing a working income of only £173.[39] The Society's estimated membership, income and expenditure, as suggested by its annual reports, are shown in Table 1.

	Estimated membership	Estimated income (to nearest £)	Estimated expenditure (to nearest £)
1848	537	173	—
1849	577	605	384
1850	750	1196	917
1851	854	1351	944
1852	890	1307	966
1853	1036	1388	794
1854	?	?	?
1855	?	?	?
1856	957	1464	824
1857	600	1323	1129
1858	420	842	641
1859	295	700	535
1860	502	874	592
1861	483	961	678
1862	422	781	406
1863	?	137	251
1864	342	790	570[a]

Table 1.

Estimated membership, income and expenditure of the Cavendish Society, 1848–64.

[a]The balance in hand on 1 March 1864, after which no more subscriptions were taken, was £220.

Nevertheless, the Committee felt confident enough to proceed with the first book, *Chemical reports and memoirs*, edited by Graham, when Leeson ‘ offered to advance any sum of money that might be required ’. Graham's book, which actually consisted of contemporary continental review essays on various chemical subjects translated by Day, was distributed in July 1848 and the impression exhausted by 1851. The Committee also commissioned Henry Watts to begin a translation of Leopold Gmelin's *Handbuch der Chemie*, which had begun to appear in Germany in 1843,[40] even though income from 600 members would be needed to publish both Graham's *Memoirs* and the first volume of Gmelin.[41] It is worth noting that the Society's interest in the translation of foreign works provided employment for needy men of science.

[38] T. S. Moore and J. C. Philip, *The Chemical Society, 1841–1941* (1947, London), 22.

[39] *First report*, 1848, 4.

[40] *Pharm. j.*, **7** (1847–48), 356, issue of 1 January 1848, announced that the Gmelin translation was already begun. For Ure's reviews of L. Gmelin, *Handbuch der Chemie* (4th ed., 1843, Heidelberg), Bd. 1, see *Pharm. j.*, **3** (1843–44), 83–91; Bd. 2 (1844, Heidelberg), *ibid.*, **4** (1844–45), 523–529; and Bd. 3 (1845, Heidelberg), *ibid.*, **6** (1846–47), 545–551. In the latter review readers were recommended to join the Cavendish Society to obtain the English translation.

[41] That is, £630, or approximately £315 per volume. I estimate that Leeson loaned about £145 to the Society.

In 1848 Playfair promised to furnish Frankland with some translations for the Cavendish Society if Frankland returned, unemployed, from Germany. Although Frankland never needed to take up this offer, the printed reports of the Society suggest that useful incomes were to be had. Graham and Day shared £104 for the *Chemical reports,* and Henry Watts £195 for the first two volumes of his heroic translation.[42]

The formally constituted Society of 1848 consisted of a President (Graham); twelve Vice-Presidents (two of whom had to retire every second year), a prestigious roll-call of eminent chemists and Fellows of the Royal Society, including Brande, Faraday, Prout and Thomson; sixteen Members of Council (four of whom had to retire every year), including Jacob Bell, Brodie, Hofmann, Playfair and Wheatstone; a Treasurer (Leeson) and a Secretary (Warington). The latter retired in 1849 to be replaced by the energetic pharmacist, Theophilus Redwood (1806–92), who remained Secretary for the remainder of the Society's existence.[43] J. J. Griffin, the chemical instrument maker and chemical supplier, no doubt with an eye to business, offered to stock the volumes at his Portman Square office, and a professional subscription collector was appointed until 1857 when the Society's printers, Harrison, took over.[44]

In March 1849 the Council decided to issue two further volumes of the Gmelin translation each year (the first having been issued in 1849), together with ' at least one volume each year of some other work '. The first such miscellaneous book to have been so commissioned was George Wilson's *Life of Cavendish*, whose publication, however, was delayed until 1851. Consequently, in 1850, the Society commissioned George Day (at his own suggestion) to translate C. G. Lehmann's *Physiological chemistry* from the German. This also began to appear in 1851 and was out of print by 1856. The edition was completed by two further volumes in 1853 and 1854, together with a *Supplementary atlas* by Lehmann's pupil, Otto Funk, in 1853.[45] This generous gesture to its medical members was practically the Society's undoing. Nevertheless, although the Society's finances showed that it could not afford to issue three volumes a year, this was used as an argument for increasing the membership rather than for reducing the Society's commitment to two volumes per annum. The plain truth was that there were no more subscribers to be found. In the

[42] Frankland's Journal, 6 October 1848 (typescripts in archives of Royal Institution and Royal Society). Watts became Editor of the Chemical Society's publications in 1849 and its Librarian in 1859—offices he retained until his death in 1884. Because of his increasing workload he was forced to ' sub-let ' some of the Gmelin translations to Francis Rudall (vols. 2–4), E. Atkinson and A. Dupré (vol. 13), C. E. Long (vol. 14), G. Carey Foster (vols. 15–16), A. E. Fletcher (vol. 16) and J. Robinson (vols. 17–18) (see Preface to L. Gmelin, *Handbook of chemistry*, vol. 18 (1871, London)).

[43] On Redwood, see Joseph Ince's obituary notice, *Pharm. j.*, **22** (1891–92), 763–766, in which it is stated that ' the duties incident [to the Secretaryship] involved much anxiety, of which not the least was the distribution of the Society's publications ' (p. 765). See also Peter H. Thomas, ' Medical men of Glamorgan: (3) Professor Theophilus Redwood (1806–92) ', *Stewart Williams's Glamorgan historian*, **3** (1966), 89–106.

[44] *Tenth report*, 1857, 2.

[45] Carl Gotthelf Lehmann, *Physiological chemistry* (trans. from 2nd German ed. by George E. Day: 3 vols., 1851–54, London); Otto Funk, *Atlas of physiological chemistry, being a supplement to Lehmann's physiological chemistry* (1853, London). The Preface is dated Leipzig, October 1852. There were 15 plates, 21 × 27 cm., each containing 6 ' microscopic views '. The text of this Atlas was translated anonymously for the Society by Liebig's English pharmaceutical pupil, B. H. Paul (see *Sixth report*, 1853, 2). On Day, see C. F. A. Marmoy and J. L. Thornton (footnote 30). Day was also an active translator for the Sydenham Society.

event, only in 1852 and in 1854 did the Society again produce three volumes—though more by accident than by design.

Another German encyclopaedic work which was considered for translation was Bischof's huge *Chemical and physical geology*. However, this was rightly seen to be too large a work for the Society to produce at the same time as the ever-enlarging Gmelin encyclopaedia. In 1852 Bischof agreed to condense and to update his book for the Society, and the first volume, translated by B. H. Paul and J. Drummond, appeared in 1854.[46]

A dramatic fall in subscriptions during the financial year 1857–58 seems to have been due firstly to the difficulties with Bischof's health, which meant that no volume was issued between April 1857 and March 1858, so that subscribers for this year received only one volume instead of two; and secondly, that new subscribers (who were barely coming forward in sufficient numbers to replace deceased members) were deterred when they found they could not purchase complete sets of the Society's publications because Graham's volume was out of print, and the first, important, volume of Lehmann's set was unavailable. Essentially, then, a new subscriber joined only for the sake of Gmelin's *Handbook*—unless he was sufficiently an historian to want a good biography of Cavendish and a bad one of Dalton,[47] or a theoretical organic chemist who was inspired by Laurent's neologisms.[48] To encourage subscribers, therefore, from 1858, the first six volumes of Gmelin were offered at the bargain price of two guineas, and Cavendish, Dalton and Laurent together for a guinea. This ruse evidently became a useful source of revenue, for in 1861 Harrison offered to reprint the exhausted first volume of Gmelin at no expense to the Society.[49]

Despite these efforts there was little improvement in the Society's financial position. By 1859 Council decided reluctantly that its first duty was to complete the translation of Gmelin before commiting the Society to new ventures. Thus, despite the efforts of an academically-minded ginger group of the Society (which included R. A. Smith, Odling and Warington) who met at the British Association in 1858 to urge the Society to translate Liebig and Kopp's *Annual report on the progress of chemistry*,[50] or to produce a variety of volumes (for which, however, the realistically-minded Council was hard-pressed to think of suitable authors), from 1859 the Society became a ' Gmelin Society '.[51]

[46] Gustav Bischof, *Elements of chemical and physical geology*, vol. 1. The German edition had begun to appear in 1847, and by 1854 it had reached 3300 pages. The Cavendish Society commissioned Bischof to reduce the work to about 1500 pages and at the same time to update the text. The 3 volumes of the English text in fact occupy 1544 pages. Paul translated half of the first volume and left the rest to Drummond (probably because of his research for the doctorate at Giessen). Volumes 2 (1855) and 3 (1859) were translated entirely by Paul, who also edited the final volume directly from the German edition because of Bischof's illness (see the Prefaces to vols. 1 and 3). For evidence of chemists' interest in geology, see Gerrylynn K. Roberts (footnote 58 below), ' Establishment ', 451–453.

[47] W. C. Henry, *Memoirs of the life and scientific researches of John Dalton* (1854, London). On its unique badness, see A. Thackray, Introduction to the reprint of H. E. Roscoe and A. Harden, *A new view of the origin of Dalton's atomic theory* (Johnson Reprint Co., 1970, New York and London).

[48] A. Laurent, *Chemical method* (trans. by W. Odling: 1855, London).

[49] *Fourteenth report*, 1861, 3.

[50] *Jahresbericht über die Fortschritte der . . . Chemie* (1849–1910, Giessen), a continuation of Berzelius's annual reports. A. W. Hofmann had translated the first four volumes with W. De La Rue and (for 1849–53) H. Bence Jones.

[51] *Twelfth report*, 1859, 2–3. Note, however, that Henry Watts added considerable information from Liebig and Kopp's *Annual reports*.

Even so, the Society's problems were not simplified. Watts could not produce the thirteenth volume of Gmelin, which was due in 1859, until after March 1860, both because the new German edition from which he was translating was incomplete, and because of the amount of new matter that he wished to add.[52] Subscribers therefore went without a volume in 1859. Inevitably there were disgruntled comments. At the March meeting in 1860, W. R. Basham, a London doctor, ' complained of the slowness with which the Society's books were published ',[53] but the Treasurer (Longstaff) explained that there were no suitable short works to fill in the ' gaps ' between successive volumes of Gmelin. Basham's complaint was probably stimulated by the letter of an anonymous Manchester subscriber to *Chemical news*,[54] who wondered what had happened to all the books promised to subscribers in 1848 and why the ' Society should degenerate into a society for the publication of Gmelin's *Chemistry* '. If that was so, let the subscribers know, he urged; complete Gmelin, close the Society down, and reanimate it as the Sydenham Society had recently done. The Society's *Fourteenth report* in March 1861 revealed that this is precisely what Graham and his Council thought best, perhaps because again Watts had found the German text delayed, and the Society could produce no volume.[55] Table 2 lists the publications of the Society.

Events after March 1864 are unclear. The seventeenth volume of Gmelin (issued against the subscription for 1863) was not ready until 1866, and the final eighteenth volume (for 1864) appeared only in 1871. Watts then rapidly completed the index which was issued by Harrison as a separate volume in 1872. The last *Report* of the Society, which was issued in 1864, never made it clear whether the Society would continue after Gmelin had been completed.[56] One suspects not, since Council's final action in 1864 or 1865 was to hand over to Harrison the Society's stock of publications on the understanding that the firm would see the completion of the Gmelin project,[57] thus losing the opportunity of resurrecting a ' New Cavendish Society ' from the proceeds of an auction.

3. Discussion of the Society's development

Why was the Society an apparent failure? A specific answer will only be found in the context of the institutionalization of chemistry in the 1840s and 1850s, and not in the general history of the printing club movement.

The first point to be considered is a possible connection between the Cavendish Society and the Royal College of Chemistry (founded 1845) rather than the Chemical Society. Using published information, it is possible to compare the lists of Cavendish Society subscribers with the 74% sample of

[52] *Chem. news*, **1** (1860), 163; *Thirteenth report*, 1860, 1.

[53] *Chem. news*, **1** (1860), 164.

[54] *Ibid*, 142.

[55] *Fifteenth report*, 1862, 1. Watts was delayed by the fact that the German text was issued in parts and he had to collect a sufficient amount together to make up a volume.

[56] *Seventeenth report*, 1864, 3.

[57] Harrison, advertisement dated 20 October 1871, in the 18th volume of Gmelin (1871) and the Index (1872).

Subscription year	Short title[a]	Pages
1848	Graham, *Chemical reports*[27]	370
	Gmelin 1[40]	519
1849	Gmelin 2	504
	Gmelin 3	488
	Cavendish *Life* (appeared 1851)[32]	478
1850	Gmelin 4	450
	Gmelin 5 (appeared 1851)	499
1851	Lehmann 1[44]	455
	Gmelin 6	428
1852	Gmelin 7	501
	Lehmann 2	465
	Funk, *Atlas*[45]	—
1853	Gmelin 8	499
	Bischof 1[46]	455
1854	Henry, *Dalton*[47]	249
	Lehmann 3	579
	Laurent, *Method* (issued 1855)[48]	382
1855	Gmelin 9	523
	Bischof 2	523
1856	Gmelin 10	566
	Gmelin 11 (may have appeared early 1857)	528
1857	Gmelin 12	559
1858	Bischof 3	566
1859	Gmelin 13 (issued 1860)	588
1860	Gmelin 14	534
1861	Gmelin 15 (issued 1862)	544
1862	Gmelin 16 (issued 1864)	536
1863	Gmelin 17 (issued 1866)	618
1864	Gmelin 18 (issued 1871)	469
	Index to Gmelin (issued 1872)	331

Table 2.

Publications of the Cavendish Society.

[a]Superscripts are to footnotes giving the full citation.

supporters of the Royal College of Chemistry whom G. K. Roberts identified in her doctoral thesis.[58]

Of the 223 London subscribers of the Cavendish Society listed in 1848, 43 (19%) had also been financial supporters of the Royal College of Chemistry in 1845. Although this may under-represent, since Dr. Roberts was unable to make a complete identification of all the College's supporters, the figure does not appear particularly impressive. That there should be some correlation between the ' membership ' of the two institutions is scarcely surprising, since any publicly available lists of ' chemical supporters ' would have been obviously

[58] Gerrylynn K. Roberts, ' The Royal College of Chemistry (1845–1853): a social history of chemistry in early Victorian England ' (Ph.D. dissertation: The Johns Hopkins University, 1973); partly published as ' The establishment of the Royal College of Chemistry: an investigation of the social context of early Victorian chemistry ', *Historical studies physical sciences*, **7** (1976), 437–485.

useful to the promotors of the Cavendish Society. Indeed, it is perhaps more surprising that the figure is as low as 19%. Since a rather similar overlap of membership occurs for London Fellows of the Chemical Society, it is clearly not possible to make a case for a special relationship with the Royal College of Chemistry on the basis of the *ordinary* membership of the Cavendish Society. Moreover, only twelve Cavendish subscribers were former students of the Royal College of Chemistry, while the number of provincial members of the Society who had also supported the College was virtually nil—in fact, three out of a total of 314. Finally, of the 61 honorary provincial agents of the Cavendish Society, only nine (14%) had been College supporters.

It seems clear, therefore, that there is little reward in attempting to find particular institutional connections between the Cavendish Society and the chemical community. However, Dr. Roberts, by using prosopographical techniques to analyse the reasons for the promotion and establishment of the Royal College of Chemistry in 1845, was able to provide striking evidence that provincial support from agricultural interest groups was an important source of revenue for the College. Is it possible that the Cavendish Society succeeded in stimulating another kind of source of provincial support for chemistry? The surviving evidence does, indeed, suggest that a major source of the Cavendish Society's support was ' industrial ' as opposed to ' agricultural '.

Whereas there appears to have been little support for the Royal College of Chemistry from the major industrial centres of Great Britain, the converse seems to have been the case with the Cavendish Society. Of the seventeen centres in Dr. Roberts's list of towns with more than four subscribers to the College (see Table 3), only four appear in a Cavendish Society list of towns with four or more subscribers. On the other hand, major industrial towns like Manchester (43), Glasgow (15), Newcastle (11), Bolton (9), Derby (9), Nottingham (8) and Sheffield (8) were all towns with fewer or no subscribers to the Royal College of Chemistry. All but one of the Manchester subscribers have been identified; apart from J. P. Joule (who had brewery connections), one clergyman and four surgeons, the remaining 36 subscribers to the Cavendish Society were working in Manchester's chemical industries.[59] The very high figure for Manchester compared with (say) Liverpool or Newcastle suggests that it was in towns where there were substantial applications of chemistry to other

Manchester	43	Bath	8	York	6
Liverpool	15	Cambridge	8	Durham	5
Glasgow	15	Nottingham	8	Hull	5
Edinburgh	15	St. Helen's	8	Norwich	5
Newcastle	11	Sheffield	8	Bedford	4
Birmingham	9	Cheltenham	7	Brighton	4
Bolton	9	Southampton	6	Coventry	4
Derby	9	Dublin	6	Plymouth	4
Gloucester	9			Winchester	4

Table 3.

Provincial centres with 4 or more Cavendish Society members in 1848.

[59] The principal industry represented was calico printing. For some relevant remarks on the Manchester chemists, see C. A. Russell, Noel G. Coley and Gerrylynn K. Roberts, *Chemists by profession* (1977, Milton Keynes), 31–32; and R. S. Crossley, *Accrington captains of industry* (1930, Accrington), *passim*.

industries (such as calico printing, soap making and metallurgy), as well as the production of chemical substances, that support was strongest.

It would appear, then, that even allowing for the three-year gap between 1845 and 1848 either the Cavendish Society was successfully appealing to a different audience than the Royal College of Chemistry had done, or a provincial audience was appealing to metropolitan chemists for a particular programme which had hitherto been underplayed.

If the Society's metropolitan promoters are analysed, a rather different picture emerges. Of the dozen Vice-Presidents seven (58%) had supported the Royal College of Chemistry; six (38%) of the ordinary Council had also done so; of the whole Council of 31 (including President, Secretary and Treasurer) 15 (48%) had supported the College. Once again, not unexpectedly, it would appear that some of the same London group were involved in the formation of the Royal College of Chemistry and the Cavendish Society—and, for that matter, the Chemical Society in 1841.[60]

Now Dr. Roberts's brilliantly-argued thesis is that the supporters of the Royal College of Chemistry could be analysed prosopographically in terms of various ' chemical interest groups ': medical, agricultural, and a heterogeneous miscellany which included engineers, manufacturers and educationists. However, ' by the end of the Spring semester of 1852–53, at which time it [the College] was taken over by the government and incorporated into a general science school . . . the number of supporters had dwindled to a handful '[61] because under A. W. Hofmann's leadership it had not become an ' institution devoted to the exploitation of research in *applied* chemistry ',[62] as its subscribers had been promised. If this imaginative interpretation of the financial failure of the Royal College of Chemistry is correct, it follows that the chemical literature most likely to appeal to a wide spectrum of ' chemically-interested audiences ' in the 1840s would have been of a practical, informative nature, rather than one of an academic or research-oriented character.

With this point in mind, we can examine the Society's proposed, and actual, publications. Of the Society's ten original proposals in 1846, three made a direct appeal to ' practising chemists '—translations of Rammelsberg, *Dictionary of mineralogy* from the Berlin edition of 1841;[63] Otto, *Economic chemistry* from the German edition of 1837;[64] and Pierre Berthier, *Essay on assaying* from the French edition of 1834.[65] Arguably Gmelin's *Handbook*, which was also proposed, would also have appealed to pharmaceutical practitioners who gave it a good press as a reference book despite its scholarly style;[66]

[60] The origins and early years of the Chemical Society are the subject of current research by Mr. R. Bud of the University of Pennsylvania.

[61] Gerrylynn K. Roberts (footnote 58), thesis, 1.

[62] *Ibid.*, 10.

[63] Carl F. Rammelsberg, *Handwörterbuch des chemischen Theils der Mineralogie* (2 vols., 1841, Berlin); with four Supplements (1843–49).

[64] Friedrich J. Otto, *Lehrbuch der rationellen Praxis der landwirtschaftlichen Gewerbe* (4 vols., 1837–55, Brunswick). Otto (1809–70), who was Professor of Technical Chemistry and Pharmacy at the Collegium Carolinium, Brunswick, had translated Graham's *Elements of chemistry* into German in 1844. His memoir on atomic volumes was, in turn, translated by Graham for the *Chemical reports*.

[65] Pierre Berthier, *Traité des essais par la voie sèche* (2 vols., 1834, Paris). There was no English translation.

[66] See footnote 40.

like a Beilstein or a Partington, it is always difficult to know how useful Gmelin was to chemists, for it is the essence of a reference book that it should be rarely cited. Two historical works were also proposed in 1846—a life of Cavendish, and a translation of H. Kopp's large *History of chemistry*.[67] We may allow these to have been of 'cultural' interest to chemical practitioners; but the four remaining suggestions in 1846—Buff, *Outlines of experimental physics*,[68] Gustav Rose, *Crystallography*[69] (no doubt chosen by Leeson), and two volumes of miscellaneous essays on chemical philosophy and animal chemistry—would obviously have made a more direct appeal to academic chemists. It should be noted, however, that Graham's eventual one-volume edition of *Chemical reports* did include two memoirs on dyeing and alkaloids.

Clearly, then, the Society's initial proposals did attempt to appeal to both academic and practising chemists. The curriculum proposed by the metropolitan centre was to the taste of both a Germanophilic academic centre and the more practically-disposed provincial periphery;[70] the 'theatre bill' appealed to audiences of research-oriented peers and practical appliers of knowledge.

In practice however, only three of the titles mentioned above were published and diffused by the Society. Moreover, on examining the seven titles finally printed by the Society, *none* (with the partial exception of Graham's *Chemical reports* and Gmelin's *Handbook*) had any direct use to the industrial and manufacturing chemist. Lehmann's *Physiological chemistry* appealed directly only to medical, and conceivably to some pharmaceutical, interests. Bischof's *Chemical and physical geology* was far too specialized to appeal to landowner interests, and like Laurent's *Chemical method*, it would have been incomprehensible to any but the academically trained.

Thus, like the Royal College of Chemistry and, for that matter, the Sydenham Society, the Cavendish Society failed to satisfy all the chemical interest groups to whom it had appealed successfully in its initial publicity. Like them it paid a financial penalty. What the centre offered was not what the periphery required, and, as in many other areas of Victorian society where payment by results prevailed, subscriptions lapsed when audiences found that they were not getting what they wanted. It is noticeable that whenever suggestions for future publications were made at Annual Meetings, they were always for works of an applied or useful character: a reference work of constants, Plattner on smelting, Persoz on dyeing, essays on photosynthesis (of

[67] Hermann Kopp, *Geschichte der Chemie* (4 vols., 1843–47, Brunswick). No English translation has ever been made.

[68] Heinrich Buff, *Grundzüge der Experimental-Physik* (1843, Heidelberg). Although no translation of this work was made, A. W. Hofmann translated for Richard Taylor, H. Buff, *Zur Physik der Erde* (1850, Brunswick) as *Familiar letters on physics* (1851, London). Buff (1805–78), a pupil of Liebig's, was Professor of Physics at Giessen, and Hofmann's brother-in-law.

[69] Gustav Rose, ***Elemente der Krystallographie*** (2nd ed., 1838, Berlin). His brother, Heinrich Rose, was the author of *Handbuch der analytischen Chemie* (1829, Brunswick); translated by John Griffin as *A manual of analytical chemistry* (1831, London). Rose produced a fifth edition, *Ausführliches Handbuch der analytischen Chemie* (1851, Brunswick), which the Cavendish Society planned to translate by Thomas H. Henry of Spitalfields in 4 volumes until it learned that Rose had approved a 6th edition in French (1859–61, Paris) (see *Twelfth report*, 1859, 2; J. R. Partington (footnote 28), 185).

[70] For the centre–periphery model of cultural diffusion, see D. Schon, *Beyond the stable state. Public and private learning in a changing society* (1971, London). I do not wish to imply that there were no industrial chemists in London.

agricultural significance), Rammelsberg's *Dictionary of mineralogy*.[71] Unfortunately, because of financial restraints and the heavy investment in the multi-volumed Gmelin, the Society was never able to satisfy these demands and complaints, even when it agreed they were legitimate suggestions.

Consequently others stepped in to fill the need. Ure, Muspratt and other compilers, publishers' series (such as Bohn's Scientific Library), or commercial journals, like the chemically-oriented *English mechanic*, filled the lacunae handsomely.[72] There was also a revival of chemical activity at the Society of Arts which, in 1839, undertook ' the dissemination of information about the industrial arts and sciences, and the publication of new discoveries and inventions of an industrial nature '.[73]

A problem which much exercises educationists today is the diffusion of curriculum innovations. How does one persuade schools and colleges to adopt a new science curriculum, like the various schemes supported by the Nuffield Foundation in England, or CHEM Study in America? Various models have been proposed, most of which are variations on the ' centre–periphery ' model originally proposed by Schon in 1971. Such a system, in which an innovation is diffused outwards from a centre to its potential users, is particularly prone to failure from the exhaustion of resources, overloading and mismanagement: ' When the . . . system exceeds the resources or the energy at the centre, overloads the capacity of the radii, or mishandles feedback from the periphery, it fails. Failure takes the form of simple ineffectiveness in diffusion, distortion of the message, or disintegration of the system as a whole '.[74]

The assumption in such a model is that the agent at the centre knows what the peripheral consumer wants or what is best for him. On the other hand, the experience of science curriculum reformers in the 1950s and 1960s on both sides of the Atlantic has been that the offered curriculum was frequently adulterated in diffusion, being adapted rather than adopted. More significant for our purpose, it was also found that there was often considerable consumer resistance.

[71] *Fourth report*, 1851, 3; *Pharm. j.*, **6** (1846–47), 339. K. F. Plattner, *Beitrag zur Erweiterung der Probirkunst* (1849, Freiberg). His *Probirkunst mit dem Löthrohre* (1835, Leipzig) had been translated by J. S. Muspratt as *The use of the blowpipe* (1845, London) (see *Tenth report*, 1857, 2). Jean François Persoz's *Traité théorique et pratique de l'impression des tissus* (4 vols. and Atlas, 1846, Paris) was to have been translated by John Graham, the Manchester local secretary and brother of the President, who managed the Mayfield calico printing works. It was abandoned on the grounds that the plates and illustrations would be too expensive. For Rammelsberg, see footnote 63. A volume of essays by Saussure, Ingenhousz and others on plants and photosynthesis was translated and prepared for the Society by Edward Solly (1819–86), Professor of Chemistry at Addiscombe Military College, who had a copy prepared by March 1851. However, for reasons which are unclear, it was published neither by the Society nor another publisher. The manuscript has not been traced.

[72] A. Ure, *A dictionary of arts, manufactures and mines* (1839, London); 4th ed. (2 vols., 1853, London); J. S. Muspratt, *Chemistry theoretical, practical and analytical* (2 vols., 1853–61, Glasgow), with numerous editions; B. H. Paul, *Manual of technical analysis* (1857, Bohn's Scientific Library, London), a partial translation of P. A. Bolley, *Handbuch der technisch-chemischen Untersuchungen* (1853, Frauenfeld-Zurich). See Francesco Cordasco, *The Bohn Libraries. A history and a check list* (1951), New York. *The English mechanic* was published weekly from 1865 to 1926.

[73] D. Hudson and K. W. Luckhurst, *The Royal Society of Arts 1754–1954* (1954 London), 176–180. Arthur Aikin, Professor of Chemistry at Guy's Hospital, was a Vice-President of the Cavendish Society, and had been Secretary of the Society of Arts from 1817 to 1839. In the latter year he established a Committee on Chemistry.

[74] D. Schon (footnote 70), quoted by B. MacDonald and R. Walker, *Changing the curriculum* (1976, London, Open Books), 14.

MacDonald and Walker have recently made two general points concerning curriculum change. First, that ' projects engage in image manipulation in order to disguise discrepancies between their own educational convictions and the convictions held by others, particularly teachers on the one hand and academic critics on the other '.[75] Thus the organizers of the Cavendish Society, who hoped to reform the public awareness of chemistry through the cheap dissemination of its literature, looked at two audiences for validation—their own, mainly academic, peers, and that of practising chemists. Second, ' there is a consequent generation of two distinct and conflicting views held outside a project concerning what the project essentially " is "; one view is held by teachers, the other by critics '.[76] For the successful diffusion of the new curriculum, MacDonald and Walker argue, the ' gap ' between the two images of the project must be sufficiently small for negotiation and fruitful compromise; if the gap is too great, the product fails. In the Cavendish Society we have an analogous example of ' the gap between the intent of a curriculum project '—to educate chemical practisers by making known academic work which would also be useful for academic chemists to possess in English—' and what actually happens in the classroom ': in industrial and commercial chemistry, the use of *ad hoc* recipes, and the retrieval of information as and when required. Because the metropolitan academic centre did not respond to close this gap, or could not because it was saddled financially with Gmelin's *Handbook* which it would have been ungentlemanly to have abandoned, the whole system collapsed. New centres of diffusion emerged to close the gap—commercial publishers, the (Royal) Society of Arts and last, but by no means least, the Chemical Society whose revitalized publications policy under Gmelin's translator, Henry Watts, was eventually to lead, in 1871, to the regular supply of chemical abstracts in English.

But the disjunction of interests between academic, or research-oriented chemists, and practising chemists which is glimpsed in the story of the Cavendish Society was only satisfactorily resolved in 1877 with the formation of the Institute of Chemistry and its associated qualifications and publications.[77] Perhaps the Cavendish Society is best viewed as one of three attempts to transform British chemistry in the 1840s by a tiny core of metropolitan chemists. In consciously aiming to promote the subject of chemistry and to provide it with the respect that Liebig claimed for it in Germany, the London chemists tried first the Chemical Society, then the Royal College of Chemistry, and finally the Cavendish Society. All three attempts were partial failures. However, these failures did suggest that the best chance of success lay in the Chemical Society. Ironically, the gap between academic chemists and practising chemists would not be so easily closed, and success was achieved only by creating the Institute of Chemistry as an independent offshoot of the Chemical Society.

[75] MacDonald and Walker (footnote 74), 42.
[76] *Ibid.*
[77] C. A. Russell and others (footnote 59).

Acknowledgments

I am grateful to Diana Dixon for bibliographical assistance with printing clubs; the late Dr. W. V. Farrar for identifying the trades of Manchester subscribers; and Dr. Gerrylynn K. Roberts and members of Professor A. Thackray's seminar in Philadelphia for their comments on an earlier draft. Professor W. Baker, Mrs. Kate Hammond, Dr. G. Molland and John A. Sharp kindly answered queries concerning sets of Gmelin in their care.

IV

The Cavendish Society's Wonderful Repertory of Chemistry

Summary

Contemporary correspondence is invoked to provide further information concerning the decision of the Cavendish Society (1846–1872) to publish an English translation of Gmelin's *Handbuch der Chemie* in 1846, and the opposition this provoked.

In an earlier essay, 'The Society for the Perpetuation of Gmelin',[1] the Cavendish Society, which lasted from 1846 to 1872, was placed in the context of the large number of Victorian subscription printing clubs which published translations, re-issued historical works, or commissioned original works that were thought to be too specialized for commercial publication. Despite its initial plans for a varied publishing programme, the Society's book production during its lifetime of twenty-six years was limited to only eight titles, the chief of these being a translation by Henry Watts of Leopold Gmelin's *Handbuch der Chemie*. In my previous paper, it was argued that the reasons for the Society's limited success as a printing club compared with, say, the Sydenham Society,[2] must be sought in the circumstances of the institutionalization of chemistry during the 1840s, and, in particular, in the divergence of interests between academic and practising chemists during this decade.

The present note complements this interpretation with an examination of how the decision to print Gmelin was taken and how it proved immediately controversial.

Although there is no reason to doubt that the Cavendish Society was instigated by the physician and lecturer at St. Thomas's Hospital, London, Henry Beaumont Leeson (1803–72), like the Chemical Society itself,[3] the Society seems to have been principally shaped by Robert Warington (1807–67), the Chemical Operator at Apothecaries Hall. In particular, it seems clear from Warington's intimate friendship with Heinrich Will (1812–90), Liebig's chief assistant at Giessen, that it was Warington who committed the fledgling society to the translation of Gmelin's *Handbuch*. Will, who had been a contender for the Highland and Agricultural Society's position of Chemical Analyst in

[1] W. H. Brock, 'The Society for the Perpetuation of Gmelin: the Cavendish Society, 1846–1872', *Annals of Science*, 35 (1978), 599–617.

[2] G. G. Meynell, *The Two Sydenham Societies. A History and Bibliogrpahy of the Medical Classics Published by the Sydenham Society and the New Sydenham Society (1844–1911)* (Acrise, Kent, 1985).

[3] Robert Bud and Gerrylynn K. Roberts, *Science versus Practice. Chemisty in Victorian Britain* (Manchester, 1984), Chapter 2.

1844, and a possible Director of the Royal College of Chemistry in London in 1845,[4] had attended the meeting of the British Association for the Advancement of Science at Southampton in September 1846. He had been handsomely entertained in London by Warington, to whom he was probably introduced by Liebig's former pupils, John Lloyd Bullock (another chemical operator) and John Blyth.[5] It is clear that Will had discussed the significance of the organic part of Gmelin's encyclopaedic work, the enlarged fourth edition of which had begun to appear in Germany in 1843 and enjoyed favourable reviews in the English pharmaceutical press. Following the appearance of the third volume on inorganic chemistry at Heidelberg in 1845, Gmelin turned to the horrendous task of reviewing the whole of organic chemistry—by then the fastest-growing, most exciting, but most difficult and controversial, part of chemistry.[6] Will commented presciently to Warington that

> the whole of that organic part will certainly fill not less than four large volumes. I suspect rather five or six than four. I do not think that he can finish more than one volume every year, so that it will take *at least* four years until the whole is finished.

In the event, the German edition of the organic section appeared in six volumes and kept Gmelin busy until his death in 1853. And because of Henry Watts' conscientiousness as translator, and his addition of considerable amounts of extra material to the English Edition, the latter came to occupy nineteen volumes, the final index volume not being published until 1872.

It was Will who acted as an intermediary between Warington and Gmelin, and his press editor, Herr K. Wimber, in October 1846. Will was happy to do this because Gmelin himself wanted Will to supply a Supplement to the *Handbuch* that, on the model of the sixth and final volume of Berzelius' *Lehrbuch der Chemie* (1832), would contain 'an accurate description of chemical manipulations, apparatus, analyses of every kind, [and] an explanation of chemical expressions, etc'.[7] In the event, probably because Gmelin's project took so long to complete, Will never composed this supplement; on the other hand, its likelihood must have played a part in Warington's determination to see the Cavendish Society launch a translation since Will gave him

[4] Ibid., 57. The Scottish position was filled by James F. W. Johnston. Liebig arranged for Will to be given an Assistant Professorship at Giessen in order to ensure that he could not take the London post, while at the same time arranging for his pupil, A. W. Hofmann, to accept the Royal College of Chemistry Directorship. See Will to Warington, Giessen, undated (1845), Lawes-Gilbert papers, file G.108, Rothamsted Experimental Station, Harpenden, Hertfordshire, AL5 2LQ, U.K. There is an excellent calendar of these papers: Margaret Harcourt Williams, *Catalogue of the Rothamsted Archives* (Harpenden, 1987).

[5] Will to Warington, Giessen, 25 October 1846. Rothamsted archives, G.108. On Bullock, see Bud and Roberts, (footnote 3), pp. 51–3 and 201; and for Blyth, see J. R. Partington, *A History of Chemistry*, IV (London, 1964), 327. Both Bullock and Blyth had worked with Liebig and Will at Giessen.

[6] Gmelin's project began as *Handbuch der theoretischen Chemie*, 2 vols (Frankfurt, 1817–1819); second edition 1821–1822; third edition 1827–1829, The fourth, enlarged edition, entitled *Handbuch der Chemie*, was published at Heidelberg between 1843 and 1866, and occupied nine volumes. This was completed by a team of German editors who simultaneously worked on a fifth edition begun by Gmelin in the year of his death and completed in thirteen volumes in 1870. Interestingly, this edition incorporated material from the Cavendish Society's English edition.

[7] Will had previously published an account of the Liebig-Giessen method of analysis, *Anleitung zur qualitativen Analyse* (Heidelberg, 1846); translated as *Outlines of the Course of Qualitative Analysis Followed in the Giessen Laboratory* (London, 1846).

permission to use the information if he thought it useful. Will also secured Wimber's co-operation in sending sheets of the German edition to Watts as they came off the Heidelberg press. On the other hand, his suggestion that if the Cavendish Society had

> the intention to bring out two or three volumes [of Gmelin] every year, it will be perhaps of some importance if you publish immediately the organic part too, when leaving the German press, so that in every year there is appearing (in England) one volume (or two) on inorganic chemistry and one volume on organic chemistry

never proved practicable.

It is clear from Will's October letter than the idea of the Cavendish Society had already been actively discussed in London chemical circles during the summer of 1846. Indeed, in a letter to Warington dated 25 July 1846, George Wilson, the Regius Professor of Technology at the University of Edinburgh, expressed delight at the idea of a 'Chemistry Publication Society' and supposed that since Robert Boyle was recognized as 'our Father' of chemistry, the proposed society would be called the Boyle Society.[8] Ironically, it was Wilson's espousal of the claims of Henry Cavendish against those of Lavoisier and Watt in the controversy over the first synthesis of water in the 1780s which ultimately justified the new Society's choice of title.[9]

Another letter from George Fownes to Warington, which precisely dates the foundation meeting of the Society at the Society of Arts as 13 November 1846, shows that, from the start, its organisers' intention was the translation of Gmelin, and that this was stated in a *Prospectus* circulated to a selected number of chemists that has not survived. Fownes' letter[10] also confirms that there was an initial tension between those, like Warington, who wanted the Society to publish works which would be of *practical* use to members, and those (like Fownes) who saw the project as primarily bibliophilic, gentlemanly and 'academic'. (That this argument should come from someone who earned his living as a teacher of *practical* chemistry is a delicious irony.)

> It appears to me that a mistake has been made in the choice of a book to begin with, namely Gmelin's *Chemistry*. In the first place, we should publish our books for the use of the members and such books as are most likely to be valuable, and those I take to be careful reprints of old books and papers difficult to obtain under ordinary circumstances which serve to illustrate the history of chemistry, giving a preference to those in the English language. We should have nothing to do with chemistry books of any kind. The chemical writings of Cavendish, Boyle, Mayow, Priestley, Scheele and others who could be named might be republished in this manner, with lives, if persons could be found able to draw them up.
>
> The particular objections to Gmelin in my mind, in addition to that of its not being the right kind of book, are, its great size and bulk and the very heavy expense which will attend mere printing and correction, and the immense difficulties of translation and editing in a satisfactory manner. I do not for my own part see how these are to be accomplished without very great expense and loss of

[8] Wilson to Warington, Edingurgh, 25 July 1846. Rothamsted archives, G.108.

[9] George Wilson, *Life and Works of Cavendish* (London, 1851). This was the fifth volume published by the Society. See Brock (footnote 1), p. 605 (reference 32).

[10] Fownes to Warington, University College, London, 12 November 1846. Rothamstead archives, G.108. Fownes (1815–49) was Professor of Practical Chemistry at University College, London.

> time. It must be remembered that the editor will be responsible for the whole work and certainly for all errors of translation.
>
> The Council will now have to decide whether they will proceed according to the terms of the prospectus, which course I consider attended with great danger, or withdraw that prospectus and issue another stating that unexpected difficulties have occurred with Gmelin's book, and that at present it will not be attempted.
>
> You will perhaps do me the favour to read this note at the meeting.

The Society's founding Council—a committee of twenty 'gentlemen residing in London' which included Warington, Leeson, Thomas Graham (President of both the Cavendish and Chemical Societies) and George Day, evidently had little difficulty in combatting Fownes' argument when it was read out at the inaugural meeting. All the surviving evidence implies that unbreakable commitments had already been made to Gmelin and his translator, Watts. As it turned out, Fownes proved correct in his assessment of the risks involved in publishing Gmelin's enormous work.

Although historians of science might have been grateful if Fownes' suggestion of reprinting the classics of British history of chemistry had been adopted—a plan later adopted in part imaginatively by the Edinburgh Alembic Club[11]—arguably the Gmelin project was of greater use to the mid-nineteenth century chemist. However, that view has to be rephrased as a conditional *could* have been useful, for the truth of the matter is that the *Handbook's* sluggish publication over a period of twenty-four years must have negated its potential usefulness, besides monopolising the assets of the Cavendish Society. Nevertheless, when, in 1859, Angus Smith and Henry Roscoe voiced Mancunian complaints at the slowness with which the edition of Gmelin was being completed and the consequent lack of variety in the Society's publication plans, Graham reminded Roscoe that Gmelin was a 'wonderful repertory of Chemistry'.[12] He further assured Roscoe that by transferring many of the Society's secretarial duties to Harrison's (the bookseller and printer) and communicating its publication plans to provincial members more efficiently, a 'quicker regularity' of publication would be enforced and 'all grounds of dissatisfaction speedily removed'.

It was not to be. It took a further thirteen years to complete the edition of Gmelin and no other titles were published.[13]

[11] James Kendall, 'The Alembic Club', *Endeavour*, 13 (1954), 94–6. The Club was founded in 1889 and issued its first reprint in 1893.

[12] Graham to Roscoe, London, 2 March 1859. Royal Chemical Society archives. The phrase was apparently Charles Daubeny's. I thank Dr M. Stanley for this reference.

[13] Brock (footnote 1), p. 609.

V

THE SELECTION OF THE AUTHORS OF THE BRIDGEWATER TREATISES

THE argument for the existence of God from Design using the model of Newtonian scientific methodology formed the basis of the Natural Theology of the eighteenth and nineteenth centuries. Since the proof of God's existence rested on what God was believed to have accomplished, and His continual benevolent activity was regarded as the explanation of everything which science could not explain, the power of the argument decreased proportionally as the horizons of science expanded. Nevertheless, a steady stream of learned and more popular literature on natural theology was published during the one hundred years after Newton's death in 1727. The spate of publications continued despite Hume's contention at the end of the eighteenth century that the use of analogical reasoning in theological discourse had no logical validity (1). In theology, unlike science, hypotheses cannot be tested either directly or by the fulfilment of certain predictions based on the original hypotheses. Although the use of analogical reasoning in scientific discourse is an invaluable aid to the formulation of hypotheses, these always have to be tested against experience. This is impossible in theology since the design analogy does not produce (in Hume's requirement) a constant conjunction between the hypothetical first cause, God, and its effects, the materials of the observed world. We cannot confirm the existence of a World Designer by observing Him make one. Even so, as has been pointed out recently, the design argument continued to be used after Hume because 'previously acquired and emotionally grounded prejudices in its favor' were psychologically persuasive (2). In fact, this kind of natural theology remained a subject of considerable popularity, and it survived, in Great Britain at least, until the biological theory of evolution demolished the last vestiges of the design argument's empirical foundations.

The epitome of this continuing tradition was Archdeacon William Paley's very successful *Natural Theology* which appeared in 1802 and became a set-book at the Universities of Cambridge and Oxford during the remainder of the century. At Cambridge, Paley's arguments became indelibly impressed upon the mind of the young Charles Darwin (3); while at Oxford, Paley's chapters were used by the Duncan brothers as part of their scheme for the classification of the University Museum's Natural History collection in 1836 (4).

Paley (1743-1805), who was committed to the 'Sovereign Order of Nature' of the English Newtonian tradition, reacted strongly against the atheistic 'blind chance' of the French savants and of such English radicals as Erasmus Darwin. His design arguments were drawn principally from examples of organic contrivances; that is, he argued that the various species of plants and animals were all examples of beautifully designed artifacts. Paley paid little attention to the premises of the argument for design, and he completely ignored Hume's critique. This is not so surprising; for the Christian religion seemed to many people at that time to be the bulwark of an ordered, decent society, and Paley sincerely believed that the only alternative would be the blind chance, atheism and anarchy of the French materialists and revolutionaries.

There were some twenty editions of Paley's manual before 1820, and a further ten or more had appeared before the end of the century. The book became the Bible of Nature for many a scrupulously pious scientist, especially physicians who, despite their frequently barbaric education, usually became models of ethical and religious propriety (5). It is not a coincidence that four of the eight authors of the Bridgewater Treatises were practising physicians. For the clergyman, on the other hand, Paley's study was full of useful ammunition which could be fired against freethinkers and atheists.

The terms of the Bridgewater legacy have been interpreted by most historians as one sign of the worry and confusion over contemporary scientific developments, especially those which were taking place in geology (6). The dispute over 'Genesis and Geology' which took place between scientists of a uniformitarian or catastrophic persuasion, together with the clergy nourished on natural theology, was over both the interpretation of the geological record, and over whether this record did or did not reveal a providential deity who constantly intervened and actively participated in the development of nature by *supernaturally* adapting means to benevolent ends. It was certainly in this environment that the Bridgewater Treatises came to be written, and it is interesting to notice that all the authors believed in a catastrophic geology, a providential deity, a God who had created the universe specifically for man, and a God who continually supervised its operations with this purpose in view.

The Bridgewater Treatises were sponsored by the will (7), dated 1825, of the eighth and final Earl of Bridgewater, the Reverend Francis Henry Egerton (1756-1829). The son of a bishop, Egerton was a clergyman who 'assiduously neglected his parish' (8), a literary scholar rather than a scientist, and a family historian; but these were sufficient qualifications to gain him an 'F.R.S.' in the days when the Royal Society was to a large extent a Gentleman's Club, not a thriving scientific society. His rich store of 'Egerton Manuscripts', which deal with the history and literature of France and Italy, were left to the British Museum together with a sum of £12,000 for their provision. The Earl of Bridgewater, Viscount Brackley, Baron Ellesmore, Prince of the Holy Roman Empire, and Fellow of All Souls, Oxford, had both title and money. The *Dictionary of National Biography* notes politely that he also had his eccentricities. He filled his 'Hotel Egerton' in Paris with cats and dogs, many of which he dressed in men's and women's clothing, fed at his table, and took out with him in his carriage. An aristocrat to the end, 'in his last feeble days he stocked his garden with large numbers of rabbits, and with pigeons and partridges with clipped wings, in order to enjoy the *sport* of killing a few heads of game for his table' (9).

Egerton had written a Paley-inspired *Treatise on Natural Theology* for private circulation (10), and in his 'very long and very extraordinary' will (7) he bequeathed £8,000 to the Royal Society as a payment to the person or persons chosen by its President who would:

> . . . write, print and publish, one thousand copies of a work on the Power, Wisdom and Goodness of God, as manifested in the Creation: illustrating such work by all reasonable arguments, as for instance the variety and formation of God's creatures in the animal, vegetable and mineral kingdoms; the effect of digestion and thereby of conversion; the construction of the hand of man, and an infinite variety of other arguments; as also by discoveries ancient and modern, in arts, sciences, and the whole extent of literature (11).

Royalties were to be given to the author or authors.

The terms of this legacy placed a considerable burden of responsibility on Davies Gilbert, the personal friend of Sir Humphry Davy, who had succeeded to the Presidency of the Royal Society on Davy's resignation because of ill-health in 1827. Gilbert's responsibility was made difficult by the fact that throughout his Presidency he was the centre of the controversy over the power structure of the Royal Society between the Scientists and the Gentlemen Amateurs. Gilbert's position finally became intolerable and he resigned from the Presidency in November 1830 in favour of the non-scientific Duke of Sussex. Gilbert, therefore, had no direct hand in the final printing and publication of the Bridgewater Treatises which appeared between 1833 and 1836; this administrative responsibility was directed by the Society's capable Secretary, Peter Mark Roget (1779–1869), who was himself one of the chosen writers. However, before his resignation, Gilbert co-ordinated the search for suitable authors, and commissioned the writing of eight treatises. His apparently personal decision that the legacy, which had been quickly invested in 3% Bank Annuities, should be broken down into *eight* divisions, did not go uncriticized (12), and it became used as yet another excuse for dissatisfaction with Gilbert's Presidency. With some foresight, however, Gilbert had wisely aligned both the Archbishop of Canterbury, William Howley (1766–1848), and the Bishop of London, Charles Blomfield (1786–1857), on to his side, and invited them to suggest names and even to negotiate terms with writers of their own choice. The collection of correspondence connected with the Bridgewater authors that was published privately later in the century by Gilbert's nephew, John Enys, reveals how these eight men were chosen (13).

Gilbert noted in 1831 that he had been 'fully aware of the duty imposed on him to select persons amply qualified for discharging in an adequate manner the task they would have to perform; and he [had been] also impressed with the conviction, that however carefully a selection might be made, several gentlemen must be omitted possessing the requisite qualifications, equally, perhaps with those who received the appointment' (14). As soon as the terms of the bequest were made known to him, Gilbert contacted the elderly politician Baron Farnborough (Charles Long, 1761–1838), who had been related to Egerton by marriage, in order to make certain that neither Egerton's large family or his three executors intended to challenge the endowment (15). Secure with Farnborough's reassurances, early in 1829 the Reverend G. D'Oyley, chaplain to the Archbishop of Canterbury, informed Gilbert that both Howley and Blomfield were willing therefore to assist him in the choice of authors (16).

Meanwhile, one may read between the lines of the Enys correspondence that Gilbert received a number of letters which either solicited a commission to write the Bridgewater Treatise, or recommended their friends as excellent candidates. One such claimant was John F. (*sic*) Graves, who was described by his supporter, the Bishop of Cloyne (John Brinkley), as 'a young man of most excellent character and attainments' (17). This must have been John *Thomas* Graves (1806–1870), the Irish mathematician and lawyer who became Professor of Jurisprudence at University College London, in 1831 and F.R.S. in 1839 (18). Since Charles Babbage later felt obliged to write an unofficial mathematical Bridgewater Treatise, the fact that Graves was primarily a mathematician might explain why no call was made on his talents. However, Howley revealed in 1830 that 'Mr Grave's (*sic*) application does not tell in his favor . . ., and I conceive the intention of the Testator was not so much to give the prize to an ingenious young man, as to encourage the publication of a work or works which might be really useful to the public' (19).

Howley suggested, somewhat impractically, that a number of 'eminent persons' should be selected to plan a large work on natural theology whose parts 'might be filled up by them respectively, as they might agree amongst themselves' (19). For some time Gilbert held to this plan; for example, in June 1830, he still clearly envisaged eight separate essays in two octavo volumes (20). The eventual production of 8 large volumes in 12 seems to have been a purely practical decision taken by the authors themselves at a much later date when they found the essay form was too restrictive.

By June 1830 names were being examined and rejected by the three selectors. Howley remarked to Gilbert that he had been pleased with a book of Peter Roget's which Gilbert had sent him (21), and also by Herbert Mayo's *Outlines of Physiology* (22). He also noted suggestively that the prospectus of John Macculloch's *Proofs and Illustrations of the Attributes of God*, which had been sent to him recently for approval, appeared 'to be very well done' (23). Of these three men only Roget was eventually chosen to be an author. Mayo (1796-1852), an excellent anatomist and physiologist, was Professor of Anatomy at King's College, London, but despite his membership of that 'Godly Institution' he was turned down by the three selectors in favour of John Kidd (1775-1851), who eventually wrote the anthropomorphic treatise on the physical constitution of man (24). As for the geologist Macculloch, nothing more is mentioned of him in the correspondence. Ironically, Macculloch deliberately withheld the publication of his own large book until the way had been prepared by the publication of all the Bridgewater Treatises. The book had to be published posthumously in 1837 because of his tragic honeymoon death in 1835. Macculloch would have been a good choice, but he was ignored perhaps because he was personally disliked by most scientists, and because William Buckland, who was in holy orders, was such an obviously dignified choice for the controversial subject of geology.

The successes of eighteenth-century mathematical astronomy had also made the astronomical design argument rather controversial and difficult. Gilbert spent some time anxiously searching for a suitable writer on astronomy, and with perhaps a diplomatic thrust at John Herschel, who was one of the principal spokesmen for the reform of the Royal Society, he first offered the commission to the great astronomer. Herschel's dignified reply is worth quoting, for it was critical of the manner in which Gilbert and his advisers were directing this embarrassing legacy (25).

Slough, July 1st, 1830

My Dear Sir,

I am duly sensible of the honor you do me by proposing that I should write the treatise on Astronomy in fulfilment of the conditions of Earl Bridgewater's bequest for publishing a work 'on the Power, Goodness, and Wisdom of God as manifested in the Creation'; though I must decline the undertaking.

No one, as you well know, is more deeply impressed with the great truths intended to be inculcated in this work; but in precisely the same proportion is the repugnance I feel to weaken the weight of my testimony in their favour by promulgating them under the direct and avowed influence of pecuniary reward. To any one who considers how argument loses in persuasion, and praise in value (humanly speaking) by the bare mention of such an origin, it will I fear appear questionable whether much effective support is likely to accrue to Religion or Glory to God from the adoption of the means in question. To yourself, as President of a great scientific body, there is however another point of view in which it may very reasonably be expected to occur, and has very likely done so—viz: as an opportunity of calling forth to something like lucrative exertion the talents of men, who with *real science* and *irreproachable character*, have their zeal chilled and their sphere of utility contracted by the 'res augusta domi.' To such persons (men, for example, such as Ritchie, who I mention as a specimen of the class I allude to) who live, or rather starve on their science, but who prefer hunger in that good cause to competency in a less dignified calling, a thousand pounds (the sum you propose to devote to each of the eight sections of the work) would indeed be a more material and noble assistance. Such men, who deserve all encouragement, but who receive none, in the ordinary course of events, in this country, are the people who every lover of science would wish might receive at least the benefit of windfalls such as these.

William Ritchie (1790-1837), the physicist referred to by Herschel, had been educated for the Scottish Church and had only taken up physics when he was in his twenties (26). A born experimentalist, Herschel and others helped him to become successively Professor of Natural Philosophy at the Royal Institution and at the University of London. Although he was a Fellow of the Royal Society, Ritchie suffered from the same defect as Howley saw in

Edward Hawke Locker (1777-1849), a commissioner of Greenwich Hospital who was suggested by Baron Farnborough (27): both were ingenious men, and very useful as writers for the lower classes. 'But I question whether [their] depth of knowledge is such as to qualify [them] for the production of a book which would stand the test of professional criticism' (28). Herschel's point was never taken by the selectors.

Another candidate, the Reverend Samuel Wix (1771-1861), was suggested by one of Egerton's executors, the banker J. C. Clarmont. However, Blomfield dismissed Wix brusquely with the confidence that although he was 'a highly respectable clergyman, his turn of mind and habits of thought are not precisely those which would qualify him for a share in this undertaking' (29). Wix, a Fellow of the Royal Society, was of the High Church party, but his ecumenical ideas had involved him in much controversy. Complete theological orthodoxy was obviously a fundamental criterion of choice for the Bishop and Archbishop.

In mid-August 1830 Roget pointed out to Gilbert that all sorts of demarcation problems could arise later if a decision about the authors had not been reached before Gilbert officially resigned on the 30 November (30). Fortunately some measure of agreement between Gilbert, Howley and Blomfield had been reached already. They unanimously agreed that the anatomist Charles Bell, who had already published a work on natural theology, should tackle the subject of 'Human Anatomy including . . . Lord Bridgewater's favorite topic the Human Hand' (31); that Roget should be given 'Physiological or Comparative Anatomy'; that William Whewell (Blomfield's choice) should write the work on astronomy; and that William Buckland (Howley's choice) should be given geology. Four more authors were still to be found.

It is clear from the Enys correspondence that initially Gilbert did not think in terms of a separate chemical treatise, but of one which would include an incidental discussion of the imponderable elements of light and heat (32). The name of David Brewster (1781-1868), who was famous for his work on polarization, was proposed and seriously considered. Brewster, a member of the established Church of Scotland, had preached as a young man before nervousness had caused him to abandon the thought of a clerical profession. Prompted by the Bishop of London (33), by 2 September 1830 Gilbert instead suggested a treatise 'Chemistry chiefly in reference to the Etherial or Imponderable Fluids or especially to Light, including Optics with the recent discoveries of Polarization'. This enlarged treatise was to be written by *either* William Prout (1785-1850), *or* by Brewster. Blomfield then advised Gilbert that Prout was 'a first-rate chemist', and 'I should think would do whatever he undertook with great ability'; but he spoiled the effect by admitting in a postscript that 'I don't know what sort of *writer* Dr Prout is; never having seen his publications' (34). He concluded his advice by emphasizing that the literary merits of Prout and Brewster were an important factor in deciding between them. In fact, although Brewster was the more productive of the two men, there was little to choose between their respective literary gifts and scientific experience. Why Prout rather than Brewster was eventually chosen is not revealed by the Enys correspondence; for with Gilbert's resignation, the task of organization seems to have fallen entirely on to the shoulders of Roget. It may be conjectured, however, that Prout was successful because he was well-known personally to both Roget and Gilbert through his Council membership of the Royal Society, and above all because he was known to be interested in the subject of digestion which had been mentioned specifically in Bridgewater's bequest. The result was the delectably titled *Chemistry, Meteorology, and the Function of Digestion considered with reference to Natural Theology*. A sarcastic *Athenaeum* reviewer noted wittily:

> To dismember *Meteorology* from *Geology*—the one involving *causes* of which the other presents the *effects*—in order to make it the link between *Chemistry* and *Digestion*, was the work of no ordinary mind; and to separate *Digestion* from *Physiology*—a part from the whole—and again place *Physiology* deprived of the *Hand*, in opposition to the *Physical Nature of Man*, allowed to retain his hands, evinced most uncommon tact in classification. The ingenuity with which Dr. Prout has connected his subjects, does not render their combination a bit the less ridiculous (35).

There was also some difficulty over the choice of an author for natural history. Blomfield had immediately thought of John Leonard Knapp (1767-1845), 'the author of that very ingenious and pleasant work *The Journal of a Naturalist*' (36), and both Gilbert and Howley seem to have agreed. But in September of 1830 Blomfield began to make further enquiries and 'upon looking again into Mr Knapp's book' he doubted 'his competency to the task' (34). Whatever the reason, perhaps that Knapp had written a best-seller for the 'lower classes', he was rejected in favour of the orthodox Suffolk septuagenarian clergyman and entomologist, William Kirby (1759-1850).

The final two treatises were a curious example of demarcation. Howley noted from Gilbert's preliminary list of authors that none were Scotsmen. (He overlooked Bell who had lived in London for so long.) Both he and

Blomfield had immediately thought of the Scots reformer Thomas Chalmers (1780–1847) who had given scientific sermons from his Glasgow pulpit. Blomfield urged that:

> . . . there should be an Essay on the adaptation of the *Physical Constitution* of man to his intellectual and moral faculties, his social propensities, and the provision made for his wants in the works of nature. This part of the subject might perhaps be advantageously divided into two, and the *intellectual* and *moral* discussion proposed to Dr Chalmers, the physical to Dr Kidd of Oxford, or Dr Mayo whose Outlines of Physiology is a very ingenious book (37).

Apart from the substitition suggested by Archbishop Howley of Kidd for Mayo which was noted earlier, the task of the selectors was finished.

Blomfield's invitation to Chalmers was printed by the Reverend John Cumming in his posthumous edition of Chalmers' treatise (38). It implied that the expenses of publication would be defrayed out of each £1,000. But it is difficult to believe that this was done; in any case its legality would almost certainly have been challenged. Such an arrangement would have been especially unfair on Buckland, half of whose treatise on geology was made up by costly illustrations on which three artists had to be employed (39).

During the winter of 1831 scandalous rumours began to appear in the press concerning the 'competitors for the legacy' (40). Buckland was appalled (41) and appealed to Gilbert for a public statement from him which would give the names of the authors and explain that there had been no lobbying. Gilbert obliged with a suitable statement of the true facts in the *Philosophical Magazine* for March 1831 (42).

> The Reverend and Right Honorable Francis Henry Egerton Earl of Bridgewater died in the month of February 1829, at Paris, leaving his last will and testament bearing date on the 25 February 1825, in which he desired and directed his trustees to lay out and invest in their own names in some or one of the public stocks or Funds of Great Britain, the sum of eight thousand pounds sterling; the said sum with all accruing dividends thereon to be held at the disposal of the President for the time being, of the Royal Society of London, to be transferred, paid and applied, according to the order and direction of the said President of the Royal Society, in full, and without any diminution or abatement whatsoever, in such proportions and at such times, according to his direction and judgement, and without being subject to any control or responsibility whatsoever, to such person or persons the said President, for the time being, should or might nominate or appoint and employ. . . . And

he [Bridgewater] desired that the profits arising from and out of the circulation and sale of the aforesaid work should be paid to the said President of the Royal Society, as of right, as a further remuneration to such person or persons as the said President of the Royal Society should so nominate, appoint and employ, with a further power to advance the sums of £300 and of £500 during the writing and printing of the said work.

. . . Each [author had] been pledged to take a part, as designated by the testator, most adapted to his acquirements and to his pursuits; and thus it is confidently hoped and expected, that a work intrusted to such individuals will appear, as a whole, worthy of the age and of the country about to give it birth.

The names of the authors and the titles of their books were then listed.

Meanwhile Roget had great difficulty finding a publisher (43). Murray accepted the series, then backed down because of a decline of trade; Longmans refused to publish only the initial limited edition of a thousand copies which the will seemed to legally demand. (If interpreted to the letter, only 250 copies of each book could have been printed—a most uneconomic proposition.) At last, at a meeting of the nominees in London on 11 October 1832 (44), the final allocations were decided and contracts signed with the publisher William Pickering (45). Lord Bridgewater's desire for a professional *a posteriori* demonstration of the existence of a benevolent deity was finally satisfied by the publication of the eight 'strange and deadly' volumes intermittently between 1833 and 1836 (8).

There were still two further episodes before the Royal Society was able to finally close its file on the Bridgewater legacy. In February 1837 Buckland sent the following legal bombshell to Gilbert (46).

Ch.[rist] Ch.[urch]
Oxford
Feb. 28, 1837

My Dear Sir

You will no doubt recollect that in the appointment you were pleased to make of myself and others to write the Bridgewater Treatises, it was expressly stipulated as directed by the will of the Earl of Bridgewater that the sum of £8,000 should be vested in the Funds until the whole of the Treatises were finished to the satisfaction of the President of the Royal Society and that then the said sum with interest and accumulation of interest should be divided among the eight Authors as *Tenants in Common.*

On the completion of the work by the publication of my Treatise in September last the Authors conjointly applied for their respective shares of interest on the £8,000, and were informed that the whole of the money was sold out in 1833, and has not been kept at interest as the will directed until all the Treatises were finished. This communication from Mr Parkinson, the Solicitor to Lord Bridgewater's Trustees, was accompanied by an opinion of Mr Knight of the Chancery Bar that the Trustees were under the circumstances of the case, justified in the course they had taken. I have subsequently submitted this opinion of Mr Knight to the Solicitor General [Sir Robert Rolfe] and have obtained from him a decided opinion that interest is due to each of the Authors up to the time when he received his £1,000 on the completion of his Treatise.

Having these two conflicting opinions it has been agreed by the Authors to refer them both to the Attorney General [Sir John Campbell] as an Umpire, if Mr Parkinson on behalf of the Trustees will agree to such an Arbitration, and that all the parties concerned should abide by the Attorney General's decision. It is the wish of the Authors to settle the matter in question amicably and quietly, and we consider the course proposed of referring it for final decision to the Attorney General to be the most delicate that can be pursued, and to be that which we should be gratified to find approved of by yourself.

I was in hopes of seeing you in London in the course of two or three recent visits to the Geological Society when I might have explained to you in person the details of this matter as I hope soon to have an opportunity of doing.

Wm. Buckland

Presumably the authors had been contracted to produce their manuscripts by a certain date, otherwise, on the brief of Buckland and the Solicitor General, an author could have gained more interest merely by dallying with his manuscript. As it was, Buckland stood to gain most from the success of his argument; for his book took him six years to write and was the last of the Bridgewater Treatises to be published. However, no record of any legal action has been discovered and it must be assumed that an amicable quiet settlement was reached. The possibility of a civil action by the Bridgewater authors against the Bridgewater estate, or against the ex-President of the Royal Society would have led to an appalling scandal. The commonest criticism of the Bridgewater Treatises was that they had been written for money (47). Although the authors publicly and privately denied the credibility of this, Buckland's letter, and an earlier one from Whewell to Gilbert

(48), reveal that whatever their individual piety, the motives of the eight writers may not have been entirely altruistic when offered a thousand pounds.

Finally, as if to underline altruism, Charles Babbage, who had been unpleasantly critical of Gilbert's Presidency, published his extraordinary and unofficial *Ninth Bridgewater Treatise* in 1837 (49). Ostensibly Babbage's book was written in opposition to Whewell's denial in his Bridgewater Treatise that pure and applied mathematicians could marshal natural theological arguments from mathematics. Babbage proceeded to demonstrate that they could, noting caustically that his demonstration was voluntary and that no money was involved. This Tristram Shandy of a book—Chapter 6 is largely blank space, and Chapter 14 consists of a single sentence—was very much the 'Fragment' indicated by Babbage's sub-title. However, it is a most interesting document which contains dissertations on free will and miracles supported from arguments drawn from his calculating engine.

It is clear that the eight Bridgewater authors had to fulfil three conditions:

(i) They had to be good, and if possible, well-known scientists. (One suspects that they were expected to be Fellows of the Royal Society.)

(ii) They had to have literary ability and be able to write for an educated, if not, academic public.

(iii) They had to be pious individuals who were theologically orthodox, and who would not write solely through financial inspiration.

How successfully were these criteria achieved?

With the exception of Chalmers all the authors were Fellows of the Royal Society and very distinguished scientists. As Canon Raven remarked in his famous Gifford Lectures of 1951:

> A team which included Thomas Chalmers of Edinburgh, William Whewell, afterwards the great Master of Trinity, William Buckland the geologist, and William Kirby, the doyen of entomologists, could not easily be improved (50).

To this we need only stress that the team also included John Kidd, who with Buckland and Daubeny pioneered the teaching of science in nineteenth-century Oxford, Charles Bell the skilled artist and investigator of the nervous system, and William Prout, who is famous for his hypothesis about the nature of matter and for his analysis of gastric juice. All eight men wrote fluently, and, as may be judged from the number of editions recorded in the Bibliography to this paper, their Treatises sold exceptionally well. In general, too, the authors were well-treated by reviewers. To the modern reader, of course, all the treatises seem to suffer from the basic fault that evolutionary change appears to be perversely unrecognized. Each species of plant, animal and mineral is assumed to have been created specially, and the authors seem to us to labour under the impossible task of demonstrating the final intention of God for every structure which they describe. Frequently the authors are led into logical incongruities, for when final causes are assigned they have an unfortunate tendency to cancel one another out. On the other hand, in their own time, the Bridgewater Treatises were of high merit as scientific textbooks. For example, they stressed the advantages of Cuvier's classification of animals, and the neologism, *convection*, first came into scientific use through one of the Treatises (51a).

Theologically, they were less successful. Four of the authors were in holy orders, and in their Christian orthodoxy the eight men ranged from the quaint fundamentalism of Kirby to the High Anglicanism of Whewell. Although the Treatises were sincerely written—only Whewell and Buckland appear to have inquired the exact size of their emoluments—their logic was only as good as the Design Argument can ever be. None of them really tried to get to grips with the difficulties of their reasoning; the problems of pain, disease, disaster and death were either ignored, or disposed of in a hurried and faintly embarrassed fashion as the mysterious ways of the beneficent deity.

Despite the careful vetting of Howley and Blomfield, two of the Treatises were heterodox in tendency. Oddly, these two treatises, those of Buckland and Prout, were to my mind the best and most creative scientifically. As was to be expected, Buckland's *Geology and Mineralogy* was the most controversial book simply because he had to say something about Genesis and Geology. Dictated at night to his devoted wife (52), the book was awaited with eager anticipation by Buckland's followers. Would he attack Charles Lyell's uniformitarian *Principles of Geology*? How would he reconcile the biblical chronology with geological time? Although he said nothing about Lyell's thesis, his first chapter was devoted to the problem of reconciling Genesis with Geology. The Bible, said Buckland, was not a scientific textbook, even though it frequently harmonized with the scientific record; the Bible had been written to reveal God and instruct men of the Divine purpose, and not to teach men science. For these opinions, which echo those of Galileo, Buckland was attacked in the daily press which had evidently expected a defence of Mosaic geology from him. Instead, what Buckland gave his contemporaries was an extremely good geology textbook which even made favourable reference to Lyell.

Prout's heterodoxy was rather different and potentially more serious theologically. Because of his research interests in animal chemistry Prout believed that an animal or organic agent was necessary to organize the organic substances which went to make up living things. This agent was of a different order from the inorganic forces that held together the molecules of organic substances. In later editions of his Bridgewater Treatise Prout postulated a large number of organic agents which he described as 'intelligent' conscious *beings* 'possessing knowledge, will and power' (51b). 'We must confess', wrote one hostile critic *a propos* this 'ugly excresence' to an otherwise brilliant book, 'our inability to discern the essential difference between this figment, and the *mythical* notions of the ancients respecting their deities or semi-deities, which we presume that Dr Prout would condemn as absurd' (53). Prout's metaphysical ideas sounded dangerously like polytheism, a doctrine with which Archbishop Howley and Bishop Blomfield must have taken care to avoid when they vetted potential authors for the Bridgewater commissions. But to be fair, these tendencies of Prout's were not at all obvious in 1830, and they were only explicitly voiced by him in 1845.

Acknowledgements. The Librarian of the Royal Society, Mr I. Kaye, for answering my postal inquiries; Mrs D. Brinkler of the South Western Regional Library System, Miss W. M. Bennetts the County Librarian of Cornwall, and P. L. Hull the County Archivist of Cornwall for their intensive and successful search for a copy of the 'Enys Correspondence'. My paper could not have been written but for the great kindness of the Reverend C. R. S. Enys of Enys, Penryn, who allowed me to examine the copy of this pamphlet which is in his library.

Bibliography

William Paley, *Natural Theology, or Evidence of the Existence and Attributes of the Deity Collected from the Appearances of Nature*, London, 1802. (See British Museum Catalogue for the large number of later editions.)

The Bridgewater Treatises On The Power Wisdom And Goodness Of God As Manifested In The Creation, 8 vols. in 12, William Pickering, London, 1833-1837; also 12 vols., Bohn Library, London, 1852-1869. These and other editions and impressions are noted individually below.

Treatise I: The Rev. Thomas Chalmers, *On the Power, Wisdom, and Goodness of God as Manifested in the Adaptation of External Nature to the Moral and Intellectual Constitution of Man* (dedicated to Blomfield), 2 vols., 1833 (1833, 1834, 1840, 1853, 1884).

Treatise II: The Rev. Dr John Kidd, *On the Adaptation of External Nature to the Physical Condition of Man* (dedicated to Howley), 1833 (1833, 1834, 1837, 1852, 1887).

Treatise III: The Rev. William Whewell, *Astronomy and General Physics Considered with Reference to Natural Theology* (dedicated to Blomfield), 1833 (1833, 1834, 1836, 1837, 1838, 1842, 1847, 1852, 1864).

Treatise IV: Dr Sir Charles Bell, *The Hand: Its Mechanism and Vital Endowments as Evincing Design*, 1833 (1833, 1834, 1837, 1852, 1860, 1865, 1874, 1885).

Treatise V: Dr Peter Mark Roget, *On Animal and Vegetable Physiology Considered with Reference to Natural Theology* (dedicated to the Duke of Sussex), 2 vols., 1834 (1839, 1840, 1862, 1867, 1870).

Treatise VI: The Rev. William Buckland, *Geology and Mineralogy Considered with Reference to Natural Theology* (dedicated to Gilbert), 2 vols., 1836 (1837, 1856, 1860, 1869).

Treatise VII: The Rev. William Kirby, *On the History, Habits and Instincts of Animals*, 2 vols., 1835 (1835, 1852, 1853).

Treatise VIII: Dr William Prout, *Chemistry, Meteorology, and the Function of Digestion Considered with Reference to Natural Theology* (dedicated to Gilbert), 1834 (1834, 1845, 1855).

Charles Babbage, *The Ninth Bridgewater Treatise: A Fragment*, London, 1837; 2nd ed., 1838.

Rev. (now Canon) D. W. Gundry, 'The Bridgewater Treatises and their Authors,' *History*, **31**, 140 (1946); **32**, 122 (1947); **33**, 125 (1948).

A. C. Todd, *The Life of Davies Gilbert (1767–1839): A Study in Patronage and Political Responsibility*. Ph.D. thesis, London, 1958.

Notes

(1) See 'Dialogues concerning Natural Religion' in R. Wollheim (ed.), *Hume on Religion*, Collins Fontana paperback, London, 1963.

(2) R. H. Hurlbutt, III, *Hume, Newton, and the Design Argument*, Lincoln, U.S.A., 1965, p. 164.

(3) See F. Darwin (ed.), *Life and Letters of Charles Darwin*, 3 vols., London, 1887, vol. 1, p. 47; vol. 2, p. 219.

(4) H. M. and K. D. Vernon, *A History of the Oxford Museum*, Oxford, 1909, p. 36: 'Happily at this time a taste for the study of Natural History . . . [has] been executed in the university by Dr Paley's very interesting work on Natural Theology, and the very popular lectures of Dr Kidd on Comparative Anatomy and Dr Buckland on Geology.' The Museum's contents were arranged so that: 'The first division proposes to familiarize the eye to those relations of all natural objects which form the basis of the argument in Dr Paley's Natural Theology; to induce a mental habit of associating natural phenomena with the conviction that they are the *media* of Divine manifestation; and by such association to give proper dignity to every branch of natural science.'

(5) E.g. Charlotte Hall, the devoted wife of Dr Marshall Hall, recalled her husband's engrossment in Paley, *Memoirs of Marshall Hall*, London, 1861, p. 58.

(6) C. C. Gillispie, *Genesis and Geology*, Cambridge, U.S.A., 1951, pp. 209–216, 298; L. Eiseley, *Darwin's Century*, London, 1959, pp. 177–178.

(7) For a full description of the will see *Gents Mag.*, **99**, i, 559 (1829).

(8) C. C. Gillispie, *op. cit.*, p. 209.

(9) C. W. Sutton, 'Francis Henry Egerton', *Dictionary of National Biography* (cited as *D.N.B.*), **17**, 154. This account was based largely on the material of note (7) *supra*.

(10) Cited in refs. (7) and (9) *supra*; there is no copy in the British Museum.

(11) Quoted in the front papers of all eight Bridgewater Treatises. The Bridgewater legacy became the model for many other zealous benefactions. For instance, in 1838, a certain Mrs Hannah Acton, widow of Samuel Acton, 'from motives of respect and regard for the memory of her deceased husband, and in order to carry into effect his desire and intention, caused an investment to be made of the sum of £1,000 in 3%

Consol Bank Annuities, in the names of the Trustees of the Royal Institution of Great Britain, the interest of which was to be devoted to the formation of a fund out of which the sum of one hundred guineas was to be paid septenially, as a reward or prize to the person, who in the judgement of the committee or managers for the time being of the Institution, should have been the author of the best essay illustrative of the wisdom and beneficence of the Almighty, in such department of science, as the committee of managers should, in their discretion, have selected'. The first winner of this Actonian Prize was the Professor of Practical Chemistry at University College London, George Fownes. His *Chemistry As Exemplifying the Wisdom and Beneficence of God*, London, 1844, from which the quotation is taken, was inferior to the Chemical Bridgewater Treatise of William Prout. (See G. Wilson, *Religio Chemici*, London, 1862, pp. 1-50, for a joint criticism of Fownes and Prout.) The Royal Institution's Actonian Prize is now given to an invited lecturer, and is not competitive. Note also The University of North Carolina's 'John Calvin McNair' lecture bequest of 1857 to 'prove the existence and attributes . . . of God from Nature', and the terms of the Scottish Universities' Gifford Lectures of 1887.

(12) E.g. *Athenaeum*, **58**, 349-350 (1834); *Edinb. Rev.*, **58**, 422-426 (1833-1834).

(13) John D. Enys (ed.), *Correspondence Regarding the Appointment of the Writers of the Bridgewater Treatises Between Davies Gilbert and Others*, priv. printed, Penryn, Cornwall, J. Gill & Son, 1877, 32 pp. Hereafter cited as '*Enys*'.

(14) *Phil. Mag.* (2), **9**, 200 (March 1831); *Enys*, p. 23.

(15) 14 February 1829, *Enys*, p. 3. Farnborough inherited property worth £4,000 p.a. from the Bridgewater estate because Egerton was his wife's uncle. See 'Charles Long', *D.N.B.* **34**, 99.

(16) 14 May 1829, *Enys*, p. 4.

(17) 12 May 1829, *Enys*, p. 3.

(18) See 'John Thomas Graves', *D.N.B.* **22**, 430.

(19) 8 June 1830, *Enys*, p. 4.

(20) Letter to John Herschel, 29 June 1830, *Enys*, p. 5. (For original letter, see the Royal Society Herschel correspondence, HS. 8. 127.) Farnborough, on the other hand, was under the impression that the work was to be divided into only *three* parts, 7 August 1830, *Enys*, p. 7.

(21) 8 June 1830, *Enys*, p. 4. Roget's 'treatise' must have been either his *An Introductory Lecture on Human and Comparative Physiology*, London, 1826, or one of his essays for the Society for the Diffusion of Useful Knowledge: *Natural Philosophy*, vol. 2, *Electricity, Galvanism, Magnetism, Electromagnetism*, London, 1829, etc.

(22) H. Mayo, *Outlines of Physiology*, London, 1827, with several later editions. For Mayo, see *D.N.B.* **37**, 172.

(23) John Macculloch, M.D., F.R.S., F.L.S., F.G.S. (1773-1835), *Proofs and Illustrations of the Attributes of God from the Facts and Laws of the Physical Universe, being the Foundation of Natural and Revealed Religion*, 3 vols., London, 1837, posthumous. See *D.N.B.* **35**, 17.

(24) Note that John Kidd, Regius Professor of Medicine at Oxford, had previously published *An Introductory Lecture to a Course in Comparative Anatomy, illustrative of Paley's Natural Theology*, Oxford, 1824.

(25) *Enys*, pp. 6-7. See *supra*, note (20) for Gilbert's letter. Enys did not print the second halves of Gilbert's letter and Herschel's reply; these were concerned with comments on a paper on optics by John Herapath. For Herschel's original draft, see Roy. Soc. Herschel correspondence, HS. 25. 1. 5.

V

(26) See *D.N.B.* **48**, 326; *Phil Mag.* (3), **12**, 275-276 (1838).

(27) 7 August 1830, *Enys*, p. 7. Locker had edited *The Plain Englishman* with Charles Knight from 1820-1823. A Fellow of the Royal Society, he is now principally remembered for his creation of the gallery of naval portraits at Greenwich. See *D.N.B.* **34**, 41.

(28) 11 September 1830, *Enys*, p. 15.

(29) 28 August 1830, *Enys*, p. 12; note also *ibid.* pp. 8, 10. For Wix, see *D.N.B.* **62**, 275, and *Gents Mag.* **12**, i, 94 (1862).

(30) 17 August 1830, *Enys*, p. 8.

(31) 23 August 1830, *Enys*, p. 9. Bell's earlier volume was *Animal Mechanics, or Proofs of Design in the Animal Frame*, London, 1827 (Society for Diffusion of Useful Knowledge). Later he produced an annotated edition of Paley with Lord Brougham, *Paley's Natural Theology, with . . . notes by Henry Lord Brougham . . . and Sir Charles Bell; to which are added supplementary dissertations by Sir C. Bell*, London, 1836, 1842 and 1845.

(32) *Enys*, pp. 9, 10, 11.

(33) 'I agree with you in thinking that *Light* might be the subject of a distinct treatise, provided that it embraces *Optics*. Is chemistry to form a branch? If not *Heat* should be coupled with Light', 28 August 1830, *Enys*, p. 11.

(34) 1 October 1830, *Enys*, p. 16.

(35) *Athenaeum*, **58**, 349, col. 1 (1834); note also *Edinb. Rev.* **58**, 424 (1833-1834).

(36) 28 August 1830, *Enys*, pp. 11-12. Knapp, a botanist, was never made a Fellow of the Royal Society. His *Journal of a Naturalist* was issued anonymously in 1829, with further eds. in the same year and in 1831. See *D.N.B.* **31**, 235.

(37) 28 August 1830, *Enys*, p. 11.

(38) T. Chalmers, First Bridgewater Treatise, 5th ed., edited by J. Cumming, Bohn Scientific Library, London, 1853, p. xxvii. For Chalmers' reply to Blomfield, 15 December 1830, see *Enys*, p. 20.

(39) Buckland noted there were 87 plates and 705 figures, Sixth Bridgewater Treatise, London, 1836, vol. 2, p. vii. According to *Alliborne's Dictionary of English Literature*, London, 1885, **1**, 277, Buckland spent much of his £1,000 on the illustrations.

(40) E.g. *The Literary Gazette and Journal of Belles Lettres, Arts and Sciences*, no. 733, p. 88, 5 February 1831.

(41) 8 February 1831, *Enys*, pp. 20-21.

(42) *Phil. Mag.* (2), **9**, 200 (1831); *Enys*, pp. 21-24.

(43) 13 October 1832, *Enys*, p. 17.

(44) Chalmers and the infirm Kirby were unable to attend.

(45) Aldus' English disciple; see S. H. Steinberg, *Five Hundred Years of Printing*, Pelican Books, 2nd ed., Harmondsworth, 1961, pp. 308-309.

(46) *Enys*, pp. 31-32.

(47) Cf. Herschel's feelings *supra*; also note the later reaction of John Tyndall when asked to edit a posthumous edition of Prout's treatise in 1854, A. S. Eve and C. H. Creasey, *Life and Work of John Tyndall*, London, 1945, p. 56. (The date of Tyndall's remark is incorrectly given as 10 June 1854; it should be 19 November 1854.)

(48) 17 October 1830, *Enys*, p. 19.

(49) C. Babbage, *The Ninth Bridgewater Treatise: A Fragment*, London, 1837, 2nd ed., 1838. It was published by Murray who was perhaps somewhat vexed by the success of Pickering's eight treatises.

(50) C. E. Raven, *Natural Religion and Christian Theology*, Cambridge, 1953, p. 173.

(51) W. Prout, Eighth Bridgewater Treatise, London, 1834. (a) p. 65. (b) 3rd ed., 1845, p. 399.
(52) Mrs Gordon, *Life and Correspondence of William Buckland, D.D., F.R.S.*, London, 1894, p. 193.
(53) *Br. & For. Med. Rev.* **21**, 123 (1846).

VI

WILLIAM BOLLAERT, FARADAY AND THE ROYAL INSTITUTION

WHEN Faraday died in August 1867 contemporary scientific journals paid him every respect by way of formal obituaries, meditative editorials, and reviews of his life's work. Some of these offerings, such as those of John Tyndall, Henry Bence Jones, and John Hall Gladstone, grew into longer separately-published memoirs. On 31 August 1867, the short-lived chemical journal *Laboratory* carried an extensive essay on Faraday by its editor, John Cargill Brough, and reproduced Maclise's caricature of Faraday.[1] Unique to *Laboratory*, however, were three pages of anonymous "Recollections of Faraday and the Royal Institution."[2] While these pages add nothing of great significance to our knowledge of Faraday and the Royal Institution, they are not without interest; and, to my knowledge, they have not been noted previously by historians, or writers on Faraday. Consider the following reminiscence:

> Mrs. Greenwood, the housekeeper of the institution, a Yorkshire-woman, would often talk to the writer of past times, when crowds of great people attended Davy's lectures, and the lines of carriages extended all down Albemarle Street into Piccadilly. "Ah, William", she would say, "Sir Humphry *is* a great man, a wonderful man! He made our institution. Mr. Brande is the kindest man alive, so gentle and always so nicely dressed. He lectures beautifully, and any one can understand him. But there are not the strings of carriages in the street as in former times, and the institution is not so rich as it used to be." Mrs. Greenwood had a high opinion of Faraday, whom she affectionately called "Mickey", he was so steady and persevering. She was quite sure that he would be

> another great man, and was not surprised to hear that *some* people were a little jealous of him. This good woman was a Methodist, and could she have had Davy, Brande, and the writer, in the same fold with herself, her happiness would have been complete. "Ah!" she would exclaim, "what is the use of all this science, all this strange brainwork, all this to-do, if you are not sure of salvation? Sir Humphry and Mr. Brande are Episcopalians, and have chosen, I fear, but a sorry road to heaven. Mickey belongs to the Sandemanians, who have got a chapel in the City; and I hear that when they have done praying and discoursing they all dine together in the very chapel—a curious sort of religion that! And as for you, William, I don't think you get much good from the chapel opposite, for you spend your Sunday afternoon in drying plants and reading about botany".

Who wrote these words? According to Brough they were by "an eminent *savant*, who commenced his scientific career as 'Laboratory Assistant' under Brande and Faraday." This clue immediately identifies the writer as William Bollaert, whose adventurous career was worthy of any Victorian hero created by Kingston, Ballantyne, or Henty.[3] "From Laboratory Assistant to Eminent Savant" might well be a description of Faraday; but, curiously, Bollaert's name escaped a place in the *Dictionary of National Biography*. Although the *Royal Society Catalogue of Scientific Papers* lists fourteen of his enormous number of papers, he is probably better known—if at all—to historians and geographers of the Americas than to historians of science.

In his editorial allusion to Bollaert, Brough added, tantalisingly, that he had not had sufficient space to reproduce the whole of Bollaert's manuscript. On the basis of what he did print it is clear that the recovery of the manuscript *could* be of some significance. Here, for example, is an elaboration of the account of an accident to Faraday which is briefly recorded by Bence Jones.[4]

> About the end of 1823 Faraday was nearly deprived of his sight. Oxygen had been made in the usual manner from the black oxide of manganese, and as he was in the act of wrenching the delivery tube out of the iron bottle some water from the trough passed through the tube and fell upon the red-hot manganese in the bottle. The scalding steam burst forth, carrying with it particles of the heated manganese, some of which got into Faraday's eyes. The writer led the sufferer to the sink, and, leaving him with his face in a pail of water, he called Mrs. Greenwood, and despatched the porter for

> Dr. Paris, who lived in the next street. He then rushed off to Saville Row, the hotbed of medicine. Finding that the first doctor called upon was from home, he hurried to the house of a second, but before he reached it he came to grief himself, by running against a post in the street with such force that he fell down as if shot. Ultimately he found Dr. Kerrison, of Old Burlington Street, at home, and took him, as fast as possible, to Albemarle Street. Other medical men soon arrived, and Faraday was conveyed to his apartments and well cared for. His face was badly burnt, and some days elapsed before his eyes were pronounced to be out of danger. In the laboratory there were masks of wire and thick glass, also goggles for protecting the eyes when dangerous experiments were performed, but these were rarely used.

Would Bollaert have given scientific information, or fuller details in his original? Or was the whole manuscript filled out on a pleasant, but perhaps uncritical, anecdotal level? Perhaps he was writing his autobiography? There can be no answer, yet, or perhaps ever, to these questions; for although extensive collections of Bollaert's papers are scattered in libraries of England and America, no trace of this particular essay has come to light.[5]

There was, however, one person who was priviliged to read and utilise Bollaert's manuscript, namely J. H. Gladstone. In his gentle study of Faraday, published in 1872, he says:

> And conjurers, indeed, the two [Brande and Faraday] appear in the eyes of the youth on the left, who waits upon them, then the "laboratory assistant", now the well-known author, Mr.William Bollaert, from whom I have learned many details of this period.[6]

From Gladstone's narrative it is clear that although he made no direct quotations, he extracted from, and reworked material from nine of the eleven recollections published in *Laboratory*.

William Bollaert was born on 21 October 1807 at Lymington in Hampshire, the eldest son of an unqualified apothecary whose ancestors were Dutch.[7] In personal circumstances his background was probably similar to Faraday's. From the ages of fourteen until nineteen (1820-25), he assisted Brande and Faraday (or Davy whenever he put in an appearance), on a salary of twelve shillings a week. (Faraday had started at the Royal Institution with twenty-one shillings a week.) Under Faraday's guidance Bollaert published his first work, on the extraction of benzoic acid from oil of mace, in the Institution's journal in 1825.[8] Clearly, he was all set to follow in Faraday's footsteps and to become a chemist. But then, in 1825, Bollaert's father became blind, and the need to support his family forced him to leave the Royal Institution for a more lucrative employment.[9]

Bollaert accepted an offer to act as an assayer and chemist in the Peruvian silver mines of Guantajaya, a decision which was to launch him on a geographical, ethnological, antiquarian, literary, and business career.

> [These] celebrated silver mines are seven miles from Iquique. They are reached from the port by proceeding over a plain through which runs an immense ridge of sand. The track is then a winding one along the slope of the mountains to the *caracol* or steep zigzag road. At the summit the ground is loose and sandy, thickly covered with large angular pieces of rock, some in an advanced state of degradation, to which ages of solar heat has given a calcined appearance, some having crumbled into powder. Here is much salt, of a variety called clinkers, looking as if they had oozed out of the earth and crystallized by the sun's heat. They have at a distance the appearance of a collection of bones, and the scene is one of absolute sterility.[10]

Water to work the mines had to be transported from twenty-one miles away at a cost of four shillings per fourteen gallon drum (a fact also noted on another occasion by Darwin, who found the whole area depressing.) Peru, which had revoked Spanish domination between 1821 and 1824 to become a Republic, was in economic and political chaos; but it was ripe for exploitation by British business interests. Unfortunately for Bollaert, the Guantajaya mines, whose economic heydey had been the eighteenth century, were closed down soon after his arrival.

Not outdone, in 1827-28, together with another young Englishman, George Smith, Bollaert made an official survey of the province of Tarapaca for the Intendente Castilla—a later President. While searching for meteoric iron the two men crossed the Peruvian (now Chilean) desert of Atacama, and were reputedly the first white men to have done so. Their Tarapaca survey helped create the nitrate industry which replaced guano, and Smith was to remain in Peru to become a leading industrialist.

Bollaert, who had plenty of business sense when the need

arose, still had other ambitions. He returned to London in 1830 by way of Cape Horn (travelling part of the way with Captain King on the *Beagle's* sister ship *Adventure*), with good references, and money in his pocket. King's College in the Strand was appointing its first academic staff, and Bollaert hoped to get the Chemical Assistantship under J. F. Daniell. It was not to be:

> I was sorely and unexpectedly disappointed. This changed my views of life; my strong natural propensities had been thwarted.

In this depressed state Bollaert could have done anything; he therefore leaped at the opportunity of further adventure as a mercenary in Portugal during 1832-3.

> Sir John [Milly Doyle, the liberal who was illegally recruiting volunteers for the cause of the thirteen-year-old Donna Maria II] observed to me in glowing terms that my scientific and mining knowledge would be most available [*sic*] in connection with artillery and engineering. I had a liking for such studies, having the idea of entering later the Turkish service, where pay and preferment were satisfactory.

For nine months Bollaert lived amid the squalor of the besieged city of Oporto, serving first as quartermaster, then as a corporal, in a Rifle Corps; when this was disbanded, he joined the Artillery. We learn such personal details from the ramshakle book that he published in 1870, *Wars of Succession in Portugal and Spain.*[11] This dreadful compilation only springs to life when the author produces a personal observation. For instance:

> We gave a party at our new quarters in the convent, and a right merry night we had of it, with port-wine negus in abundance. We profanely wished that some of the former fair occupants had been there, then we might have had a dance.

He was wounded in the calf, survived a cholera epidemic, but had to go to hospital with inflamed lungs after prolonged starvation. There he taught himself Portuguese, and,

> I also had the opportunity of investigating the habits of no less than three species of hospital parasites—pediculi; one appeared to be peculiar to the hair of the head, another to the beard (for all those who could cultivate the hirsute appendage did so, *swearing* not to shave until Lisbon was attained by them, the Miguelites [the Pretenders to the throne] on the other side had *sworn* not to shave

> until the Pedroites [supporters of Maria] had been annihilated), the third those that tormented the body, which after luxuriating there, could be found in swarming colonies between the stitches of the shirt. I say nothing of the other *inconveniences* of the hospital, excepting fleas, and they were outrageous.

On 5 September 1833 Donna Maria II assumed her rightful throne. Bollaert expected to receive a commission as Lieutenant in the Artillery, but it was not forthcoming; though some twenty years later he was created a Knight of the Order of the Tower and Sword of Portugal. In ill-health he left for England again. He had never been interested in the politics of the Portuguese War. Indeed, in a fine, honest passage, he wrote in 1870:

> Long before daylight one morning I was stretched out on the bare boards at the picket house of the Rifles at Lordello, unable to sleep from hunger and cold, my comrades being in about the same plight; our pillows our cartridge-boxes, our only covering our threadbare and thin soldier's greatcoats; each man with his rifle by his side, ready to be up and doing at a moment's notice. As day dawned I look around at our miserable condition. Not many were much interested in the political circumstances which had brought us there; a few had come to the theatre of war to learn strategy, and on the field of battle tactics, many for employment, some for glory and promotion —mere soldiers of fortune, having all to win, nothing to lose, but their lives. As for myself I had been disappointed in not obtaining a scientific appointment in London, and in truth I was rather reckless about the present or the future, and unphilosophically left my fate to the chapter of accidents. . . .

Armed with such an ambivalent attitude towards war, it is not surprising to find Bollaert becoming involved in the Spanish Wars of Succession on the side of the Carlists—that is Don Carlos, who saw himself as the legitimate heir to the Spanish throne on the death of Ferdinand 8, in 1833. Carlos plotted for the throne with the aid of the Portugese pretender, Dom Miguel, Bollaert's former enemy. For six years, between 1834 and 1840 Bollaert worked for the Carlist cause as a spy and emissary. His initial contact with the Carlists was made through the financier, Baron Maurice de Haber; and Bollaert's principal task seems to have been to deliver messages between various European centres, including Spain, Holland, London, Rome and Germany. Later he may have become a personal assistant to Miguel. But he does not tell us a great deal about these intrigues—perhaps it would still have been dangerous to reveal everything when he wrote up the *Wars* between 1860-70.

On one occasion he fell into the hands of robbers who removed everything of value, including a prized silver pencil case "given me by Mrs. Faraday in 1825, on leaving the Royal Institution for South America".[12] He begged the thieves so hard for its return that they relented—failing altogether to detect the top-secret documents he carried in his false-soled shoes! By 1830 the energies of Don Carlos and Dom Miguel were spent, and the succession of Isabella to the Spanish throne was secured. Bollaert received the Order of Fidelity, but he does not seem to have made much money from his activities.

Ever since 1834 there had been brewing in his mind an audacious scheme to explore the African continent. By 1837 the plan had reached sufficient maturity to be laid before the Royal Geographical Society: Africa was to be crossed east to west from Zanzibar, and its interior explored and surveyed. Unfortunately, Bollaert and his collaborator, a Captain Turner, disagreed over the financing of the expedition. Bollaert thought 600 pounds adequate, with 160 pounds for presents, 100 pounds for instruments, 240 for travel expenses, and 100 pounds for sundries. Turner was less optimistic and said the Government should help them; whereupon Bollaert pressed the Royal Geographical Society to recommend someone to replace him.[13] The Society declined, probably because they were actively supporting other expensive expeditions at the time. However, it appears that, on 24 February 1838, Admiral Beaufort commended to the Council of the R.G.S. that Bollaert and Lt. Ormsby, R.N. should together explore Africa.[14] The matter does not seem to have gone any further than this, though in 1854, when reports appeared that Arab traders had traversed Africa from Zanzibar to Benguala, the Secretary of the R.G.S. recalled Bollaert's plan.[15] In the following year, of course, Livingstone became the white man who fulfilled Bollaert's ambition.

About 1840 the President of Texas, General Houston, gave a grant of 4½ million acres of land to the English traveller, William Kennedy. Kennedy's agent in London then interested the British Government in placing settlers in these lands of the Apaches and Comanches. Somehow Bollaert became interested in these colonial plans, and in February 1842 he visited Texas to survey and explore its fitness for European immigration. At the same time he officially surveyed the Texas coastline for the British Admiralty, joined the Texan Navy during a Mexican invasion threat which never materialised, and seemed all set on making Texas his permanent home. However, when Kennedy became British Consul at Galveston, he returned his land grant to the Republic. With an estate agency no longer possible, or at least made more difficult; with the threats of the U.S.A. to annex Texas which put British speculators in peril; and because of continual bouts of yellow fever and malaria, Bollaert decided to return to England in July 1844.

Besides publishing a large number of papers on Texan life, its botany, its natural history, and its Indian tribes, Bollaert kept a journal of his experiences which remained unpublished until 1956. Despite its unfinished state, it is in many ways, the most satisfactory and informative of Bollaert's books. At one point in his diary, in a style not unlike Jerome K. Jerome, he describes a hilarious concert promoted to raise Army funds. The evening was only saved from chaos by Bollaert's able rendering of Spanish songs. When all was over he was entertained by the ladies, and sought out by one particular beauty.

> She was very young, very pretty, interesting and intelligent. . . . In the course of our brief conversation, I mentioned that there were but two things worth living for.
> "Pray tell me."
> "But you must keep it to yourself."
> "That I will."
> "Then Lady, one is ambition."
> "And the other?"
> "No, I do not think I ought to mention the other. You are yet too young."
> "Pray tell me."
> "Then it is love."
> "You are quite right."

He sees the Lady home.

> Somehow I could not relinquish the well-rounded arm that was enfolded in mine. . . . Arrived, what think you awaited me in the shape of retribution? Said she, "I hope the dear boy is not awake." In mere astonishment and dumbfoundedness—trembling I ejaculated,

WILLIAM BOLLAERT, FARADAY AND THE ROYAL INSTITUTION

> "Whose boy?"
> "Why my dear little Donald."
> I thought I had been flirting with Mr. Mac's daughter, Charles, It was his wife. Gods! Ye Gods!

He adds drily (and scientifically): "I mention this anecdote to shew how young the American ladies marry."[16]

Marriage was just what awaited Bollaert when he returned to London. He took a job as a merchant clerk in the Spanish department of the City firm of Rickard, Little & Co., and in November 1845 he married Susannah McMorran from Stamford, a girl he had met in 1837. There were five children, a boy and four girls.

The 1840s were a period of depression, and frequently Mrs. Bollaert had to supplement her husband's salary by needlework, while he increased it with literary earnings. His business still took him frequently abroad; but his last adventures took place between 1853-54, when he returned to Peru, and found time to explore New Granada (Colombia), Equador, and Chile. He returned home through Argentina, Paraguay and Brazil, armed with material for another book.[17]

Bollaert's business in South America would appear to have been connected with the formation of the Equador Land Company—another immigration scheme rather like the earlier one in Texas. This too was to prove a failure, for Peru eventually declared war on Equador to prevent land being sold to Europeans. He may also have wanted to verify his previous Tarapaca survey in order to see whether the borax monopoloy of Italy and India could be broken by exploiting the boron minerals of the region. Smith, on the other hand, evidently suspected that there was insufficient boron to make an industry economically viable. In a paper read to the British Association in 1852 Bollaert discussed these problems, and also hinted that it might be worth exploiting South America's iodine.[18b]

While in Peru, Bollaert examined Inca tombs which had been disturbed by railway excavations; subsequently he presented a collection of Incarial antiquities to the British Museum. In Chile, as a skilled surveyor, he was asked by a railway company for advice on routing. Another Chilean firm employed him to inspect Chilean coal resources. Bollaert's report on coalmines was first read to the University of Santiago, which subsequently made him a Corresponding Member.

Bollaert was a very sociable man, and consequently an active member of many clubs and societies. He contributed several papers to the Royal Geographical Society before being elected a Fellow in 1850. He was a Corresponding Member of the Ethnological Societies of both London and New York, a founder-member of the Anthropological Society in 1863, and its Honorary Secretary from 1865-6; earlier in the 1830s he had been a founder member of the Medico-Botanical Society. He contributed papers to the Meteorological Society, the Society of Antiquaries, the Royal Society of Literature, and the British Association for the Advancement of Science. In 1853 Prince Albert awarded him a Bronze Medal of the Royal Society of Arts for a geochemical memoir on the origin of salt deposits.[18a]

Despite his business and literary activities Bollaert fell on hard times during the last sixteen years of his life. When he ruptured his right lung in 1860 he became an invalid, and could only earn his living by the pen. He gradually gave up attending meetings of the societies; he resigned from the R.G.S. in 1863, presumably because he could no longer afford the subscription. As he drifted into lodging houses, suffering from continuous ill-health, he disappeared from the public ken. He died overlooked and largely forgotten on 15 November 1876, from "paralysis of the functions of speech and swallowing, complicated with congestion of the right lung". The only obituary appeared in the *Transactions* of the R.G.S.

So ended, rather sadly, the career of an adventurous man who, despite his talents, does not seem to have been a worldly success. In his writings Bollaert always displayed sympathy for the indigenous populations of the countries he visited, and he is full of feeling for the fate of their cultures at the white man's hands. At the same time, however, he was a typical Victorian Empire Builder who sincerely believed that the *best* form of government was a constitutional monarchy. Since Faraday was a personal friend of Daniell, it may seem odd that Bollaert failed to obtain the chemical Assistantship at King's College in 1831. Is it possible that Faraday did not think much of Bollaert's scientific abilities? It is certainly curious that he was never invited back to the Royal Institution to lecture on his travels. It is suggestive that in Texas, as Bollaert admitted, he was familiarly known as "Blowhard".[19] This need not have referred to his lung condition, for in colloquial American a blowhard is a braggart.[20] Perhaps, then, Bollaert's personality was not admirable; and he was not the sort of man Faraday trusted or wanted as a friend.

Yet, if this were the case, Bollaert never seems to have borne any ill feelings towards either Faraday or the Royal Institution. At the age of sixty, in 1867, he had been able to recall with evident pleasure the days when he shared a laboratory bench with Michael Faraday; and remember with delight one particularly marvellous piece of intimacy.

> In the autumn of 1821 Faraday repeated with great care the experiments of Ampère on the mutual action of magnets and electric currents, and was eventually led to the discovery of electro-magnetic rotations. The experiments were mainly carried on in the battery-room, Hare's calorimotor being much used as the source of electricity; and many were the devices resorted to with the view of showing the rotation of a conducting wire round a magnet, and that of a magnet round a wire. The writer was by Faraday's side when the two rough arrangements began to work. He can see the face of the experimenter now, as it appeared forty-six years ago, beaming with joy—a joy not unmixed with thankful pride—when the magnet commenced to move round the wire and the wire round the magnet. Well does he remember the exclamation, "There they go! there they go! we have succeeded at last!" After this great discovery Faraday considered that he was entitled to a little recreation, and to please the writer he suggested a visit to Astley's [Royal Amphi-] Theatre. There was a crush at the pit entrance, and the conduct of a rough fellow who subjected the writer to some unnecessary squeezing excited Faraday's anger. He ordered the fellow to behave himself, unless he wished to get into trouble, and he appeared quite ready to support his words by thumps.[21]

Have the many other laboratory assistants of the Royal Institution had such remarkable careers, and such pleasant memories?

EXHIBITS IN THE LIBRARY

(*a*) Books and periodicals from the Library of the *Royal Institution*.
(*b*) Books lent by the *London Library*.
(*c*) Manuscript lent by the *Royal Geographical Society*.

REFERENCES

1. Bates, W. (Ed.) (1898), *The Maclise Portrait Gallery of Illustrious Literary Characters with Memoirs*, London, p. 357.

2. *The Laboratory: A Weekly Record of Scientific Research* (1867). **1**, 338-390.
3. Identification was first made in 1964 through the kind offices of Mr. K. D. C. Vernon, former Librarian of the Royal Institution. Neither Bollaert nor his "Recollections" is noted by Jeffreys, A. E. (1960), *Michael Faraday, A List of his Lectures and Published Writings*. London, Chapman & Hall, or Williams, L. P. (1965), *Michael Faraday*. London, Chapman & Hall.
4. Jones, H. B. (1870), *Life and Letters of Faraday*, 2 vols., vol. 1, p. 373.
5. The principal collections are: The Royal Geographical Society, the Newberry Library, Chicago, and the University of Oklahoma. Note also the bibliography, *William Bollaert's Researches from* 1823 *to* 1865*: principally on South American Subjects*. London 4 pp. 1865.
6. Gladstone, J. H. (1872), *Michael Faraday*. London p. 19.
7. This biography is an amplification of the anonymous obituary of Bollaert, (1877). *Proceedings of the Royal Geographical Society*. **21**, 422-424; and the additional facts given by Stanley Pargellis in Hollon, W. E. and Butler, R. L. (1956), *William Bollaert's Texas*. Norman, University of Oklahoma Press. The latter work is an edition of Bollaert's unpublished Texan journals and diaries, 1841-1844.
8. Bollaert, W. "Experiments on Oil of Mace", *Quarterly Journal of Science*. (1825). **18**, 317-319.
9. In July 1824 Bollaert's salary was raised to fifteen shillings a week, *Managers' Minutes*, vol. 6 (June 1814-December 1824), p. 441 (5 July 1824). In 1826, his successor, Thomas Griffiths, was given twenty-one shillings a week, *ibid*., vol. 7, pp. 11, 70.
10. Bollaert, W. "Observations on the Geography of Southern Peru", *Journal Royal Geographical Society*, (1851) **21**, 107. This paper, and others, were translated into Spanish by Bollaert's friend, Jorge Smith. (1853), *Vistas en la provincia de Parapaca y del puerto de Arica*.
11. Bollaert, W. (1870), *The Wars of Succession of Portugal and Spain, from* 1826-1840*: With Resumé of the Political History of Portugal and Spain to the Present Time*, 2 vols., vol. 1 (Portugal), vol. 2 (Spain).
12 *Ibid*., vol. 2, pp. 66-7.
13. Bollaert to Captain Washington, 24 June 1837, Royal Geographical Society, Bollaert Correspondence, Block, 1834-40.
14. I am indebted to Mrs. M. B. Hughes, Archivist of the R.G.S., for this information.
15. Shaw, N. (1854), *J. Roy. Geog. Soc*., **24**, 271. Note this dating removes the editorial uncertainty expressed in *Bollaert's Texas*, p. xiii.
16. *Bollaert's Texas*, pp. 60-1.
17. Bollaert, W. (1860), *Antiquarian, Ethnological and other Researches in New Granada, Equador, Peru, and Chile, with Observations on the Pre-Incarial, Incarial, and other Monuments of Peruvian Nations*. London. He also translated for the Hakluyt Society in 1861, Pedro Simon, *The Expedition of Pedro de Ursua and Lope de Aguirre in Search of El Dorado and Omagua in* 1560-61. London (reprinted 1964).
18. **a** "Of Common Salt—The Sources from whence obtained, and the Processes involved in its Manufacture. With observations on the Origin of Salt and other Saline Bodies", *Journal Royal Society Arts*. (1853). **i**, 451-54, 464-67, 478-80. **b** *British Association Reports* (*Belfast*) (1852), **21**, 100.
19. *Bollaert's Texas*, p. 45.
20. Craigie, W. A. and Hulbert, J. R. (1960), *A Dictionary of American English*, 4 vols, London, vol. 1, p. 249. I am grateful to Professor R. Siegfried for this interpretation.
21. *Laboratory*. (1867), **1**, 389. In Thompson, S. P. (1901), *Michael Faraday. His Life and Work*. p. 51, Bollaert's role in this story is attributed to George Barnard.

VII

GLAUCUS : KINGSLEY AND THE SEASIDE NATURALISTS

Fanny Kingsley and subsequent biographers have all noted Charles Kingsley's many contacts with natural history -- his mother's love of nature and his father's sporting knowledge of wild-life; the move to Clovelly (1830) where Kingsley studied conchology with William Truton (1762-1835), whose shell collection is now in the Smithsonian Institution; his absorption in natural history at school in Clifton; his friendship with the prolific field naturalist, Rev. C.J. Johns, at Helston School; the many letters to his parents which are full of allusions to botany and geology; his vow to become a clergyman in 1841 after meditating on the seashore: *before the sleeping earth and the sleepless sea and stars I have devoted myself to God.*[1] Indeed, for Kingsley, Church and Cathedral were but poor imitations of nature's glades and forests where all living things joined in proof and praise of God.

In 1849 we find Kingsley excitedly collecting marine specimens at Lynmouth with the extraordinary Charles Mansfield; again, at Torquay in 1853 following Fanny's miscarriage, he began serious shore collecting with Robert Battersby (one of the founders of the Torquay Natural History society in 1844), locating the sixty species of molluscs, annelids, crustacea and polyps he later listed in *Glaucus*[2], and earnestly sending specimens to his new friend in London, Philip Gosse. Kingsley was made a Fellow of the Linnean Society in 1851, and of the Geological Society in 1863 – not, of course, for original scientific work (which was confined to an unfinished classification of the water gnat, *Phryganae*, and a few published observations in *Nature*[3]), but for his popularisation of natural history in the essays variously collected in the Macmillan volumes *Health and Education* and *Sanitary and Social Essays*, and the splendid children's eye-opener to geology, *Madame How and Lady Why*. Kingsley's most enduring monuments in natural history are, however, firstly the Chester Natural History Society, which grew out of his educational activities in Chester in 1871 and for whose members he produced the Ruskinesque *Town Geology*; and secondly, *Glaucus*,

> *the fisherman, who, enamoured of the wonders of the sea, plunged into the blue abyss once and for all, and became himself «the old blue man of the sea».*[4]

In their beginnings, seaside holidays were health-cures, with sea water recommended for not only dipping in, but for imbibing. The Napoleonic scares promoted the development of many watering places in Great Britain by preventing the aristocratic Grand Tour of the Continent. Tor Bay, on the south Devonshire coast, first became a tourist attraction when Napoleon was captured and his ship-prison anchored there. From 1811 onwards Lyme Regis became well-known to savants like William Buckland from the paleontological activities of Mary Anning, a local vendor of natural curiosities who discovered the bones of the icthyosaurus, plesiosaurus and pterodactyl in local cliffs. Here, and at other resorts, the seaside was soon revealed to offer unrivalled possibilities for the observation, collection and classification of flora and fauna.[5]

The reasons seem straightforward. Bathing was a rapid affair, while seaside amusements and varietieswere few and far between in the '30s, '40s, and '50s. As Kingsley put it in the review which because *Glaucus:*

> *What... people do at watering places is a matter of perrenial wonder, or rather what they think of – for they do nothing; and every wharf to them is but a «wharf of Lethe», by which they «rot dull as the oozy weed»...* [a] *life-in-death in which thousands waste their summers.*[6]

Both Harvey in his *Seaside Book* (1849) – one of the subjects of Kingsley's review, and G.H. Lewes in his *Sea Side Studies* (1858) make similar points. Lewes, for example, comments on the youth who finds the seaside so dull *that he must read even when in the open air.* Such people are, as Kingsley put it in Madame How, *no eyes.*[7]
For Christopher Marsden, the potential boredom of the seaside explained

> *If only as an alternative to staring at the sea, the fever of botanising which seized the mid-Victorians.*

– a fever which seemed

> *to have been canalized into an endless stream of zoophytes and polyps, fucus and sea-sponge.*[8]

Conchology, or the collection of mollusc shells, the collection of pebbles (with a lapidary in every resort), of ferns or seaweeds, were all popular pastimes before the 1850s. These hobbies were served by a considerable literature ranging from specialized works like W. Turton, *Conchological Dictionary of the British Isles* (1819), J.G. Francis, *Beach Rambles in Search of Seaside Pebbles and Crystals,* (1859), Edward Newman, *History of British Ferns* (1840), Mrs. Gatty, *British Seaweeds* (1863) to more general works like Anne Pratt, *Common Things of the Sea-Coasts*, and W.H. Harvey, *The Sea-Side Book* (1849), both of which were reviewed by Kingsley. Professional collectors and suppliers soon emerged to cater for the enthusiasm. In 1836, Mrs. Mary Wyatt of Torquay, who sold sets of shells and seaweeds, «published» with Simpkin and Marshall a set of 50 dried seaweeds with accompanying texts – a pattern followed in the fern-books of the 1840s.

Generally speaking seaside literature emphasized the collection and individualization of objects. Thus, the conchologist ignored the mollusc itself which had exuded the shell[9], and the lapidary ignored the forces of erosion which produced the beach and coastline. There were facts, but there was no philosophy to serve as a torch, said Lewes correctly[10]. What is striking about the 1850s, however, is that collectors did begin to be concerned with living flora and fauna; to use a modernism, they became ecological. We can detect the new mood of naturalism in Thoreau's comment that *the proper herbarium, the true cabin of shells and the museum of skeletons...* [*is*] *in the meadow where the flower blossomed, by the seaside where the tide cast up the fish*[11]; or in Emerson's poem, *Each and All* (1847).

> *I wiped away the weeds and foam,*
> *I fetched my sea-born treasures home;*
> *But the poor, unsightly, noisome things*
> *Had left their beauty on the shore*
> *With the sun and sand and the wild uproar.*

The aquarium transformed collection, for *sea-born treasures* now survived by being brought from Torquay or Tenby to London by railway carriage. It was through Philip Gosse, as his son Edmund put it, that natural history ceased to be *a necrology* and became instead a science of living creatures.[12]

The aquarium craze swept through Great Britain between 1853 and 1860 as a result of Gosse's writings. The sociologist and botanist David Allen has amusingly described the Victorian Fern Graze (or what Kingsley sarcastically called pteridomania[13]) which seized the country in the 1840s and 1850s following Nathaniel Ward's invention of a case for transporting growing plants from one geographical site to another. Today, horticulturalists use polythene bags, but the equilibrium principle is the same – in the sunshine, transpired water evaporates and condenses on the leaves and soil as a dew in the night, while the input and output of carbon dioxide and oxygen remain in equilibrium by day and night. Incidentally, Ward was an ardent advocate of the public health aspects of his principle, for cased plants imbibed carbon dioxide, and thus purified the ambient air. The whole principle might therefore be used to purify the air of a home or a city through conservatories and parks. Hence he and John Lindley lobbied Parliament succesfully in 1845 to repeal the tax on glass in order to cheapen the cost of home conservatories. The Crystal Palace of 1851 was really a giant Wardian Case![14]

The aquarium was a logical development of Ward's device, though scientifically its credentials followed from the principle of photosynthesis by plants slowly unravelled by Priestley, Ingenhousz and Senebier. The Victorians were given to debating the name of the inventor of aquaria, for it seems to have occurred to the chemist Robert Warington, and the naturalist Philip Gosse, and others independently and simultaneously in the early 1850s.[15] But is was Gosse who captured the public's imagination. The principle was simple – establish an equilibrium between aquatic plants which give off oxygen in light to aerate the water, and a few fish and marine creatures, together with scavengers such as snails and anemones to remove decomposing matter, and the water never need be changed.

> *It has now become one trade to supply tanks and vases for Aquaria, and another to collect and supply plants and animals for stocking them,*

commented one naturalist in 1857[16]. One could, of course, collect specimens oneself; but professional suppliers soon appeared. In the 1840s the Rev. David Landsborough had already found he could raise money to build a church and a school at Saltcoats on the Firth of Clyde by getting children to collect algae for sale to collectors.[17] Gosse became a supplier of marine specimens in 1853 when he stocked the first public aquarium at the Royal Zoological Gardens in Regents Park. This proved such an extraordinarily popular public spectacle, that other aquaria were soon set up at Crystal Palace, Surrey Gardens, the Royal Polytechnical Institution, and outside the metropolis in Edinburgh, Galway and Scarborough, and finally culminated in the enormous Brighton Aquarium of 1872 – now, alas, a seedy, sad reflection of its past glory. The entrepreneur behind most of these enterprises was W.A. Lloyd, who sold seawater at 6 pence a gallon (artificial seawater was also stocked) and who kept some 15,000 specimens in 50 tanks holding altogether 1000 gallons of water!

Like the fern craze, however, the aquarium mania abated; aquaria needed much care and attention, they could become noisome, and they were a disaster if they burst. Moreover, like fern collecting, aquaria became too common, and tastes changed. New, more artificial seaside pursuits became available. As the Gosses (both father and son) noted sadly, the anemones vanished, victims of collectors, or perhaps destroyed by the pollution of seashores by the growing resorts. Even Philip Gosse moved on to entomology and orchids[18].

Let us now return to seaside literature and by considering some of its special characteristics, find a context for Kingsley's interest.

Gosse's return to England from Alabama in the 1840s coincided with tremendous public interest in the adventures of Sir James Clark Ross's scientific voyage to the Southern Ocean. Gosse climbed on this wave of interest by writing *The Ocean* for the Society for the Promotion of Christian knowledge in 1845. It was not, however, until his illness in 1852 and his recuperation in Devon that he began to gain any first-hand knowledge of the sea-shore.

It was a characteristic of sea-side literature to be written in a high-flown style. Quotations from apposite verse is always sprinkled liberally through the text, while the reader is urged continually to exert himself and to set to the labour of collecting specimens. Both characteristics are found in *Glaucus* and in Lewes' more serious studies. Gosse was the master of the prose poem and his books *A Naturalist's Rambles on the Devonshire Coast* (1852) and *The Aquarium* (1854), which were both reviewed by Kingsley, exerted a profound influence on the style and contents of *Glaucus* and *The Water Babies*. *There are many more ways than one of studying natural history*, wrote Gosse.

> *There is Dr. Dryasdust's way, which consists of mere accuracy of definition and differenciation, statistics as harsh and dry as skins and bones in the Museum where it is studied. There is the field observer's way: the careful and conscientious accumulation and record of facts bearing on the life history of the creatures. There is the poet's way, who looks at nature through a glass peculiarly his own... with the emotions of the human mind, – surprise, wonder, terror, revulsion, admiration, love, desire, and so forth.*

This poetic anthropomorphic way – the amalgamation of *picturesque description, exact zoological statement, topographical gossip and easy reflection, combined after a fashion wholly his own*[20] – was the secret of Gosse's success with the public. It was also Kingsley's way in *Glaucus* – to create a poetry of common, but marvellous, things.

A second obvious characteristic of seaside literature is that in a pre-Darwinian world, touched only occasionally by the «absurdity» of *Vestiges of Creation*, the writers unashamedly viewed the wonders of the shore as evidence for design in nature. For example, Rymer Jones, a professional zoologist, wrote in his voluminous *Aquarian Naturalist* (1858):

> *Natural history is the appointed handmaiden of Religion, enabling us, instead of merely assenting to the existence of a Divine Providence... to feel and in some humble proportion to appreciate how closely and how carefully the well-being and happiness of all creatures has been provided for.*[21]

It is well-known that Kingsley (unlike Gosse) easily accomodated himself to the new intellectual perspectives opened up by Darwin the 1860s. One of the few significant, but rather sad, alterations made by Kingsley to the fourth edition of *Glaucus* in 1859 was a reprimand to Gosse for the obscurantism of his *Omphalos* (1858) and the withdrawal of many adjectives in praise of the fundamentalist naturalist[22]. Nevertheless, like Gosse, as the *Water Babies* and *Madame How* show, Kingsley wanted to demonstrate that there was *a quite miraculous and divine element underlying all physical nature*[23].

Another characteristic of seaside literature need not be dwelt upon. By their very nature, like tourist guide books, they were heavily derivative and symbiotic. Shirley Hibberd, for example, threatened to sue any gentleman who pirated sections f m his *Book of the Aquarium*[24]. Kingsley's *Glaucus*, his first book to be published by the science publisher, Daniel Macmillan, by its origins as a review in the *North British Review* (1854), is fairly obviously derivative, and it is possible to trace each episode as a direct quotation, paraphrase or development of an idea from a book under review[25]. On the other hand, *Glaucus* is also clearly based on first-hand knowledge of the subject. Very little new matter was added to the review for book publication – the introductory railway image which was lifted from Gosse's *Aquarium*, a section in praise of the naturalist and his occupation, and a blatant advertisement for Gosse's marine natural history classes which he proposed holding at Ilfracombe during the summer of 1855 (delightfully described through a child's eyes by Edmund Gosse at a later date)[26].

The press, lay and scientific, was unanimous in its praise for *Glaucus*. By 1856 a new edition[27] was called for, which included new sections on fly-fishing and sea-weeds, and a significant commercial plug for W.A. Lloyd, the supplier of the marvels of the sea-shore. For the Christmas, Birthday and Sunday School market, Macmillan issued a fourth illustrated edition in 1859. Finally, in 1873, Kingsley produced a fifth edition in which Darwin and Spencer were substituted for Chambers and *Vestiges*, and the interesting coda on education and natural history societies was added.[28]

The final characteristic of seaside literature was that they were illustrated books. Publishers like John van Voorst (1804-98), Lovell Reeve (1814-65) and Marshall, Simpkin Company built their reputations on the production of beautifully illustrated works of natural history[29]. Special illustrated journals, like *Land and Water, The Forest Rambler's Journal, The Field,* and Loudan's *Magazine of Natural History* catered for both scientific and amateur audiences. The chief illustrators of such books and magazines in the 1850s were Joseph Whymper (1813-1903), the father of the Alpinist, who helped Gosse, and a remarkable dynasty of naturalists named Sowerby, who collect no fewer than six entries in the *Dictionary of National Biography*. It was George Brettingham Sowerby, the younger (1812-84) who, with Kingsley's approval (and from internal evidence, collaboration), produced, for Macmillan, *A Companion to Mr. Kingsley's Glaucus* in 1858.

Here we are faced by a bibliographical problem. Briefly, what is the relationship between Sowerby's *Companion* and the illustrated editions of *Glaucus* from 1859 onwards? Some, but by no means all, of Sowerby's text was incorporated into an Appendix to the Plates for *Glaucus*, but no acknowledgement to Sowerby was made on the title-page. More startingly, Sowerby's illustrations were redrawn, renumbered and regrouped so that in an appallingly chaotic manner they failed to relate to the text and the notes at all. The explanation for this slipshod production, which Macmillan never set to rights in the Collected Edition of Kingsley's works, is a mystery.

Sowerby's coloured illustrations were clear and workmanlike, but they never reached the splendid achievement of Gosse in his *Aquarium*. Outside colour photography, his illustrations are unrivalled in beauty and accuracy. Trained as a miniaturist by his father, Philip Gosse spared no pains and no expense to reproduce his paintings of marine life by chromo-lithography. It was Gosse, too, who devised the «composite» picture or aquarium setting. which was successfully copied by Sowerby and other illustrators; i.e., instead or illustrating a single object, Gosse combined several portraits and painted-in the aquarium scenery behind. Sowerby's illustrations for *Glaucus* are of this kind, and if is not difficult to see a connection in Sir Noel Platon's well-know frontispiece for *The Water Babies*.

Once questioned on the subject of his favourite literature, Kingsley replied *Physical Science*[30]. Yet, in his own military metaphor, Kingsley was only one of *the camp-followers of science who come after the great columns of the army* (the army of marine biologist being John Ellis, George Johnston, Michael Sars, Edward Forbes and Philip Gosse), *hanging on to the outskirts, perhaps taking care of the sick and wounded, perhaps foraging and plundering a little*[31]. For as he confessed movingly to Gosse in 1856 after having been dumbfounded by the magnitude of Gosse's classification of the *Actiniae* (sea anemones), he realized that to do original work in science had become a full-time activity and it was no longer possible for someone like himself with half a dozen interests[32].

Nevertheless, there is plenty of room for Kingsley and other seashore naturalists in the Victorian scientific *community* which historians of science have in recent years enlarged so as to obtain a better perspective on the historical role of science and its contribution to economic, social and intellectual change in the eighteenth and nineteenth centuries. We must avoid seeing Kingsley and others like him as *failed scientists* because their work was openly derivative or merely popular. In any case, it must be remembered that few professional openings for naturalists existed outside the confines of Oxford and Cambridge Universities, the Royal School of Mines, and the British Museum. Provincial museums wanted only caretakers[33]. Natural history, consequently, remained *amateur* for two-thirds of the century. On the other hand, its popularity as part of general working-class, middle-class, and upper-class culture was an important ingredient for its eventual professionalization; for it created an environment, or even the economic and political conditions and motives for the institutionalization of natural history – oceanography in the 1870s, marine biological stations and fishery research stations in the 1880s, and the establishment of formal biological teaching posts in schools, universities and museums during the same decades.

This paper has attempted to place Kingsley within the scientific community by concentrating on one aspect of popular science in the 1850s. It has ignored more obvious facets of Kingsley and science, such as his attitude to evolution[34]. However, there is one general aspect of Kingsley's commitment to science which ought to be mentioned, namely his view of science as part of culture.

It will be recalled how Alton Locke sets to work under the good Dean's direction to master natural science, but privately he wonders whether the Dean would be so pleased with his progress if he knew science only confirmed Locke's democratic opinions. *The mere fact that I could understand these things... as well as anyone else, was to me a demonstration of the equality in worth, and therefore in privilege, of all classes* (chap. 15).

The fictionally expressed view was reiterated by Kingsley at the end of his life in *Town Geology* (1872) wherein he justified the study of natural history to an audience of working-class boys and girls not merely because such studies could be fun, or useful, or because they exercised fancy, imagination and polluted lungs, or that they kept one *up to date*, but for social and political reasons.

> *In becoming scientific men, in studying science and acquiring the scientific habit of mind, you will find yourselves enjoying a freedom, an equality, a brotherhood, such as you will not find elsewhere just now* (Preface)

Kingsley's point was that science was a pursuit open to talent from any class; science could be the entrée into society for groups barred from traditional élite forms of socialization such as Oxford and Cambridge or the public schools which had developed during Kingsley's lifetime.

Kingsley admired autodidacts like Hugh Miller the stonemason geologist, Charles Peach the exciseman marine biologist (both are cited approvingly in *Glaucus*). By joining the brotherhood in the freemasonry of science, Kingsley claimed, one gained the freedom to think for oneself, and the ability to work with nature by exploiting her laws. Only with such a grain of men – *good, honest, accurate, righteous, patient, self-restrained, fair, modest, scientifically-minded men* – could the freedom mill of Parliament act usefully to produce a just and Christian society. For the laws of nature were God's laws and the millenium of a perfect society would come only through scientific awareness. In this way Kingsley became a powerful spokesman for science education at a time when this was becoming an important issue among the professional scientific community.

NOTES.

(1) *Charles Kingsley. His letters and memories of his life*, 2 vols., London, 1877, vol. 1, p. 53.

(2) *Glaucus or the wonders of the shore*, 5th ed., London, 1873, pp. 140-1. The list occurs in the previous edns., and in *North British Review*, 22 (1854), 39.

(3) E.g., «The world of the sea», *Nature*, 1 (1869-70), 78.

(4) *Glaucus*, 5th ed., p. 86.

(5) For this reason John Fowles in *The French lieutenant's woman*, London, 1969, places his palaeontologist hero in Tor Bay.

(6) *North British Review*, 22 (1854), 1-56 (pp 14-15).

(7) *Madam How and Lady Why*. London, 1869, p. vii.

(8) C. Marsden, *The English at the seaside*, London 1947, p 31. The acquisitive aspects of this seaside fever date from the eighteenth century: see J.A.R. Pimlott, *The Englishman's holiday*, London, 1947, p. 135

(9) The Platonic notion that *Soul* secretes *Body* as a crab excretes its own shell,

and that body only expresses in matter the particular stage of development to which *Being* had arrived, excited Kingsley's metaphysical fancies. See O. Chadwick, 'Kingsley at Cambridge' *Historical Journal*, 18 (1975), 303-26.

(10) G.H. Lewes, *Sea-side studies at Ilfracombe, Tenby, the Scilly Isles, and Jersey*, London, 1858, p. 95.

(11) Quoted in J.W. Krutch, *Henry David Thoreau*, New York, 1948.

(12) E. Gosse, *The Naturalist of the sea-shore, The Life of Philip Henry Gosse*, London, 1890; 2nd. ed., 1896, p. 228.

(13) The word occurs in the 1st ed. of *Glaucus*, and not in the *North British Review*.

(14) D.E. Allen, *The Victorian fern craze*, London, 1969.

(15) See, for example, Shirley Hibberd, *The Book of the aquarium*, 2nd. ed., London, 1860, chap. 1. There was also much debate over nomenclature: vivary, vivarium, water-show, aqua-vivarium, aquarium, etc. See *ibid.*, p. 11.

(16) G.B. Sowerby, *Popular history of the aquarium*, London, 1857, p. 8.

(17) See entry for Landsborough in *D.N.B.*

(18) *Naturalist of the sea shore* (note 12), pp. 264-296. See also Joel W. Hedgpeth, 'De mirabili maris: thoughts on the flowering of seashore books', *Proceedings of Royal Society of Edinburgh*, 72B (1971-2), 107-14.

(19) P.H. Gosse, *Romance of natural history*, 3rd. ed., London, 1861, Preface. The French equivalent of Gosse would be J.H. Fabre.

(20) *Naturalist of the seashore* (note 12), p. 345.

(21) T.R. Jones, *The aquarium naturalist. A manual for the seaside*, London, 1858, p. 387.

(22) *Glaucus*, 4th ed., 1859, p. 12. But Kingsley did not bear a grudge and withdrew the reprimand for the 5th ed.

(23) Kingsley's comment on the aim of the *Water-babies*, quoted in *Life and Letters* (note 1), vol. 2, p. 137.

(24) Hibberd (note 15), Preface.

(25) Kingsley reviewed Rev. D. Landsborough, *A popular history of British zoophytes*, London, 1852; idem, *A popular history of British sea-weeds*, London, 1851; P.H. Gosse, *Rambles of a naturalist on the Devonshire coast*, London, 1852; idem, *Aquarium*, London, 1854; W.H. Harvey, *The Sea-side book*, London, 1849; and Ann Pratt, *Chapters on the common things of the sea-coasts*, London, 1850.

(26) *Naturalist of the seashore* (note 12), p. 257.

(27) The 2nd ed. (1855) was a straight reprint, with a few additions to the section on 'further reading'.

(28) For the editions and prices of *Glaucus* until 1887, see J. Foster, ed., *A Bibliographical catalogue of Macmillan and Co.'s publications from 1843 to 1889*, Lon-

don, 1891. In 1908 Rose Kingsley reproduced the 1st ed. (*sic*) for Dent's Everyman's Library.

(29) Commercial considerations were, naturally, paramount; consequently scientific systematists were forced to find other ways of publishing accounts of less exotic flora and fauna. In 1844, George Johnson, a marine biologist, founded the Ray Society for publishing original monographs by subscription. See R. Curle, *The Ray Society: a bibliographical history*, London, 1954.

(30) *Life and letters* (note 1), vol. 2, p. 399.

(31) *Ibid.*, p. 252.

(32) *Ibid.*, vol. 1, pp. 412-13.

(33) A.E. Günther, *A century of zoology at the British Museum*, London, 1975, p. 260.

(34) For which see Mary W. Hanawalt, *Studies in Philology*, 34 (1937), 589-611; A. Johnston, *English*, 12 (1959), 215-19; C.S. Blindermann, *Victorian Newsletter*, n° 20 (Fall 1961), 25-8; A.J. Meadows, *Theology*, 78 (1975), 15-22; Gillian Beer, *The Listener*, 17 April 1975, pp. 506-7.

VIII

Liebig's and Hofmann's Impact on British Scientific Culture

Historians would be hard put to think of any nineteenth-century chemist as well known internationally in his time as Justus von Liebig. No other chemist, not even Berzelius before him or Henry Armstrong after him, used his authority as intensely and as interestingly to disseminate his discoveries, ideas and opinions. As Grimm said, "die chemie kauderwelscht in latein und deutsch, aber in Liebigs munde wird sie sprachgewaltig."[1] That said, Liebig's principal pupil and disciple, August Wilhelm Hofmann, the heir to the Liebigian tradition of teaching, was an important figure in the popular dissemination of chemistry. He played a not insignificant role in the modernization of chemistry in the 1860s, while in his integration of academic chemistry with industrial chemistry he went beyond anything achieved by Liebig in its long term significance.

Liebig visited Great Britain many times and had over fifty British pupils. He ensured that most of his writings, academic and popular, were published in English as well as in German. Hofmann, on the other hand, spent twenty years in London, where he established an important teaching and research institution that was modelled upon his master's laboratories at Giessen.[2] One of the things an historian can do, therefore, is to compare their respective impacts on British culture and on Victorian culture generally – the one acting at a distance, so to speak, the other acting directly as a quasi-British citizen. In this paper, however, I shall restrict myself largely to Hofmann, while bearing in mind the powerful model of popularization of science and integration with culture that Liebig had provided Hofmann in the *Familiar Letters on Chemistry.* These "letters" or essays, which appeared in English in August 1843 (the principal German edition was not published until a year later) were derived from articles on chemical subjects which Liebig had begun to publish anonymously in the *Augsburger Allgemeine Zeitung.* They were translated into English by John Gardner, one of the initiators of the project to create a British College of Chemistry on the Gies-

Die Allianz von Wissenschaft und Industrie. A. W. Hofmann (1818–1892)
C. Meinel, H. Scholz (Hrsg.)

sen model and one of those responsible for encouraging Hofmann to take leave of absence from the University of Bonn in 1845 in order to direct this private institution in London. It would be interesting to know why Gardner, or possibly Liebig's German pupil, Ernst Dieffenbach, who is mentioned in Liebig's "Foreword" as having encouraged him to publish the essays collectively, chose to call them "Familiar" rather than "Popular" Letters; in all the other European language printings the plain title "Chemical Letters" is used.

Liebig added further letters from the Stuttgart newspaper, and composed new ones, for subsequent editions of his popular book which enjoyed excellent sales throughout Europe and America. The fifth and final English and German editions of 1859 contained fifty letters compared with the original fifteen in the English edition of 1843. A comparative examination of all five editions (there was a posthumous sixth German edition) and a comparative investigation of their reception and influence on the public consciousness of chemistry in Britain, France, Germany, Italy and America remains a yawning gap in Liebig studies.

As Andreas Kleinert has shown, the *Familiar Letters* had originated in 1841 at the direct invitation of the Stuttgart publisher and politician, Johann Georg von Cotta.[3] The latter, who had been tremendously excited by Liebig's *Agricultural Chemistry* in 1840, suggested that Liebig should, from time to time, discuss scientific problems, his own research, and the practical applications of science which could lead to the progress of the German States, in a form suitable for Cotta's newspaper readers. Fresh from his polemics over the state of chemistry in Prussia,[4] it was not an invitation that Liebig could resist. The resulting book text, like Alexander von Humboldt's equally successful *Cosmos* (1845), proved a superb and influential popularization of science and, indirectly, a powerful suasive factor in encouraging students to take up careers in chemistry, to commit themselves to a gospel of civic virtue and for governments to undertake financial and institutional responsibility for scientific teaching and research.[5] Who could resist Liebig's utopian prophecy?

> A more vigorous generation will come forth powerful in understanding, qualified to appreciate and to accomplish all that is truly great, and to bring forth fruits of universal usefulness. Through them the resources, the wealth and the strength of empires will be incalculably increased; and when, by the increase of knowledge, the weight which presses on human existence has been lightened, the difficulties of obtaining sustenance lessened, and man is no longer overwhelmed by the pressure of earthly cares and troubles, then, and not until then, will he be able to devote his intellect, purified and refined to the study of higher objects of investigation, and finally to the highest of all.[6]

Here was a model both for Hofmann's popular lectures and for his role as a civic scientist when, during the 1850s, he came to advise the British government on technical subjects that ranged from water supply, main drainage and the deodourization of sewage, to the pollution caused by alkali works, as well as advising on the creation of a system of examinations for technical education. Together with Liebig's British pupil, Lyon Playfair, Hofmann demonstrated to the British government that chemistry and chemists might serve not only as experts to solve particular problems thrown up by rapid urbanization and industrialization, but how chemists might serve as foils to adjudicate between what were essentially rival political positions.[7]

Although Hofmann's private life was tragic and reminiscent of the hero of Offenbach's *Tales of Hoffmann,* his public and professional life was one of triumph. A generation younger than Liebig, he lived during the period when organic chemistry (the chemistry of carbon compounds, as Hofmann may have been the first to say in 1853[8]) was transformed from being an obscure and difficult field likened to a dark forest or labyrinth without paths, light or a thread, into what Hofmann described in 1862 as a well-ordered city with streets, housing plots and market squares and (to extend Hofmann's metaphor), factories.[9] To Hofmann in the 1860s it was apparent that a revolution was occurring. In images which anticipate Th. S. Kuhn's concept of extraordinary science following the breakdown of a paradigm, Hofmann likened the state of chemistry to

> A battlefield immediately after a decisive engagement, when the vanquished squadrons, for a time dispersed, were again rallying a renewed attack. The picture of confusion which subvened was suggestive of the state of difficulty and doubt which necessarily accompanied a great revolution, and before the new rule in chemistry was generally acknowledged, there must be a period of perplexity, both to the student and to the professor.[10]

The whole purpose of Hofmann's *Introduction to Modern Chemistry (Einleitung in die moderne Chemie)*, the book which formed his leaving present to British chemists on the eve of his departure for Berlin in 1865 and which was based on lectures given at the Royal College of Chemistry since about 1850, was to introduce order and rule into chemistry again. This was to be achieved by the standardization of atomic and volumetric weights and chemical formulae and by the exploitation of the theory of types. Once this was done, Hofmann argued, it became immediately apparent (as Frankland had along argued) that elements, or atoms, possessed fixed integral powers of combination or "equivalence" – a concept soon abbreviated as "valence" or "valency".[11] For William Crookes, editor of *Chemical News* and a former pupil, the *Introduction* was "beyond comparison, the most masterly and luminous exposition of the modern philosophical doctrine of chemistry that has hitherto fallen under our notice."[12] Like the illusions made possible by the new home entertainment of lantern slides, Hofmann's lectures were like "dissolving scenes which, at a certain moment, present two landscapes, one in the act of melting away, while the other is unfolding itself to view."[13]

A new chemical regime or government demanded convincing experimental demonstration and that meant developing apparatus and equipment and teaching devices that would persuade, as well as the improvement and enlargement of laboratories' to make more students familiar with the new language of chemistry and its powers to reorder nature and society. It is one of the amusing ironies of the history of chemistry that a chemist who was by all accounts hamfisted when it came to experimental manipulation should have designed, developed and tested so much of the demonstration apparatus used in the teaching of elementary chemistry from the 1860s to the 1960s.[14]

Hofmann was also a powerful influence on the design of chemical laboratories, following here in the footsteps of his own father who had enlarged Liebig's laboratories at Giessen in 1839.[15] In 1864, while he was still directing the Royal College of Chemistry, he received an invitation to take the Chair of Chemistry at the University of Bonn and to design new chemical laboratories. This he did, though within a few months he received an "improved" offer from the University of Berlin together with a request to design a new chemical institute. Although professional architects were involved, Hofmann spent a great deal of time and effort in deciding how laboratories for elementary, advanced and post-doctoral research should be spaced and furnished. For example, although the fume cupboard had already been introduced at Giessen by Hofmann's father, Hofmann's advanced designs for the gas-jet fume cupboard remained standard objects of laboratory window furniture until comparatively recently.

Although it draws on lectures by Pettenkofer and Schödler, as well as on Humboldt's *Cosmos,* for some of its examples and arguments, probably the greatest influence of Liebig's *Chemical Letters* on Hofmann is to be found in his lecture, "On the importance of the study of chemistry".[16] This was given at South Kensington in January 1861 under the auspices of Hofmann's employers, the Department of Science and Art. (From 1853 the financially impoverished Royal College of Chemistry was subsumed within the School of Mines as part of the government's post-Great Exhibition efforts to stimulate technical education. Hofmann was, therefore, a British civil servant during the last twelve years of his career in London.)

There had been a series of lectures on education at the Royal Institution in 1854.[17] Although mathematics, physics and physiology had figured prominently in these lectures, chemistry had notably lacked an advocate. In the spirit of that series, Hofmann emphasized the merits of chemistry for mental cultivation and for the broadening of English liberal education. Of course, chemistry had always had powerful utilitarian arguments in its favour, though in the past these had usually been equally opposed by its image of dirt, smells and trade. But in the light of the 1851 Great Exhibition, the forthcoming exhibition of 1861 (actually postponed because of Prince Albert's death until 1862) and Prince Albert's strong support for science as the handmaiden of material progress, Hofmann was able to trumpet the responsibility of chemistry for the comfortable state of mid-Victorian prosperity.

Hofmann began his lecture by contrasting the fact that although nineteenth-century aesthetic values were still governed by those of ancient literature and art, the study of material nature by men of science had enormously improved the lot of human beings. Steam power had transformed transport and the nature of warfare; electricity had given man batteries, electroplating, lighthouses and the telegraph, while through Davy's isolation of sodium and potassium, chemists were poised to revolutionize the building industry through the recently-isolated metal, aluminium. All this was said in the rolling cadences that Victorians associated with Carlyle's *Signs of the Times,* as in the phrases "locomotive engines rushing with the speed of the storm over the country", or "beaming from the lofty pharos, the electric ray warns the seaman from the treacherous rock", and "as a winged messenger . . . annihilating time and space, the electric current bears our thoughts with the rapidity of their conception over land and under sea".[18] If for Liebig in the *Chemical Letters* a measure of chemical progress and of a society's state of civilization was its annual rate of production of sulphuric acid, for Hofmann it was the production of the coal (town) gas "which circulated in the arteries of our nightless towns". One of the many by-products of the manufacture of this gas included, through the genius of Hofmann's pupil, Charles B. Mansfield, the substance named benzol (benzene). The latter had undergone "at the hands of the chemist an endless number of Proteus-like transformations" which aided both the articulation of chemical theories and the creation and growth of chemical industries. It dissolved rubber, removed stains, created perfumes, but above all it could be transformed into aniline in order to produce "those beautiful mauves, pinks and crimsons in which the lovlier portion of mankind is indulging".

Other products from coal-tar distillation preserved timber, made solid pavement flagstones and fertilizers. In the latter example, after thousands of years "the nitrogen of the petrified fern-forests resuscitated in the ammoniacal liquors of the gas-works, [was made] to vegetate once more, and increase the produce of our cornfields". This point inevitably led Hofmann, in Liebig's fashion, but ignoring all controversy, to emphasize the value of chemistry to high farming. It was but a small step from agriculture to the area of Liebig's other masterpiece, the *Animal Chemistry.*

Chemists no longer searched for a philosopher's stone to prolong human life; nevertheless, modern chemistry (he claimed) was leading to a gain in the duration of human life, while every art and trade and individual human being had profited in some way from the labours of chemists. Chemistry was revolutionising medicine and leading to an improved understanding of physiology, pathology and diet. It was in this context that Hofmann introduced the word "biochemical" into English, for the action of all drugs, he claimed, was ultimately chemical.[19] Now that organic chemistry had rid itself of the bogey of vitalism – a point that he and Muspratt had argued forcibly as long ago as 1845,[20] "medicine no longer shrinks from climbing step by step the ladder of recognition, because its upper extremity, disappearing among the clouds, seems to rise for ever beyond the grasp of inquiry". The same had also been said of agriculture in the 1830s, but look how it had been transformed by Liebig's insights by

1860. The key to the transformation of medicine, like agriculture, was by the steady application of Liebig's analytical methods and through the development of a taxonomy of substances and their metamorphoses.

Hofmann concluded his lecture by arguing that chemistry was important not only for transforming man's material wealth and comfort, but that learning about it, and becoming familiar with its language, was intellectually stimulating and a valuable form of mental training. Just as the ancients had ascribed the possession of a new soul to anyone who learned to speak a foreign language, so by learning the language of chemistry, which enabled the possessor to read the book of nature, enobled common things and so elevated the mind. For example, once it is realized that respiration, burning and putrefaction are all forms of combustion in which oxygen is extracted from air and replaced by carbon dioxide which, in turn, is returned to air as oxygen after metamorphoses through the balance of vegetable and animal life, it established the fact that the universe is a world of chemical transformations. The morally-elevating thought follows that, although individuals live in order to die, there is an indestructibility to life. Chemistry thereby sows the seed of promised immortality.

Although not all of Hofmann's conceits were original, the 1861 lecture was overall a powerful argument for chemistry within the national curriculum – or at least that part of it which formed the province of his employer, the Department of Science. The probationary science teachers who heard him would have been enthralled and inspired to offer chemistry within their evening classes; more students would be encouraged to take the chemistry examinations of the Department, whose chief examiner was Hofmann, and to enter on further training at the Royal College of Chemistry or elsewhere; and his employers would have been persuaded that the teaching of chemistry deserved government patronage. The lecture reminds us that in the nineteenth century (as now) scientists did not popularize their activities altruistically and merely to translate and expound their discoveries, but that, as Kurt Bayertz has argued, much of it was "socially determined" by the ideological and professional needs to find audience support to sustain their enterprise and their individual careers.[21]

This point had been made much more explicitly in an earlier lecture he had given at the Royal College of Chemistry in order to drum up financial support from its proprietors and subscribers. In these "Remarks on the importance of cultivating experimental science in a [i.e. from a] national point of view", he cited Humboldt's famous warning in the *Cosmos* concerning the previous history of civilizations.

> Those nations who are behind in general activity, in application of mechanics and technical chemistry, in careful selection and processing of natural materials, such that regard for such enterprises does not permeate all classes, will inevitably decline in prosperity; all the more where neighbouring states in which science and the industrial arts have an active interrelationship, progress with youthful vigour.[22]

Such a statement (ironically, of course, made to the German peoples) rapidly became a handy rhetorical device throughout the remainder of the century for those who wished to argue for, and to urge greater state support for, science and its institutions of learning and research. There was, indeed, a symbolic appropriateness about the exhibition of Humboldt's bust at the Crystal Palace in 1851. And, not surprisingly, when Armstrong came to memorialize Hofmann in 1893, he referred to Hofmann's use of Humboldt's warning and not improperly deduced that a principal reason for Hofmann's eventual decision to return to Prussia in 1865 was that the importance of chemistry to national culture had not been taken seriously by the British.

Another excellent example of Hofmann's skills as a popular lecturer and of his ability to combine good chemistry with propaganda is a lecture given at the Royal Institution on 11 April 1862 on mauve and magenta.[23] In this journey from coal to colour, which passed by any foul-smelling country (i.e. difficult chemistry) quickly, Hofmann used a variety of teaching aids, including large wall-mounted diagrams, samples, models and, of course, experimental demonstrations. His use of "structural models" (Hofmann's term) in this lecture was perfected three years later when, in the same lecture theatre, he introduced coloured croquet-ball models of atoms to help his audience understand the idea of valency. Such glyptic formulae, although scorned by some sections of the chemical community, soon proved of fundamental usefulness in teaching chemistry and, by the 1870s, in understanding stereoisomerism. Atomic models were, perhaps, Hofmann's most important innovation in the popularization of chemistry.

Hofmann's 1862 lecture opened with a diagram of a gas works and a practical demonstration of the destructive distillation of coal. He then emphasized how "coal-tar is one of the most wonderful productions in the whole range of chemistry" and he whimsically confessed his deep affection for it – as well he might since it had been his principal subject of investigation since gaining his doctorate with Liebig in 1841. In a long digression, he then argued that coal was "a sort of magazine of . . . several elements". When coal was distilled these elements were shuffled into a small variety of "types of construction". Indeed, he suggested, chemists had found that all substances, animal, vegetable and mineral had been found to conform to a series of natural types or tenplates. On this basis, the existence of all chemical molecules, known and unknown, could be classified and explained as transformations or metamorphoses between these basic types. Although chemists had argued over their number and nature, Hofmann proposed the four tenplates, hydrogen, water, ammonia and methane.

Hofmann then clarified his meaning with large models of these "type moulds" consisting of wire frames bent into the shape of cubes. If it was assumed that the smallest particle of hydrogen (the molecule) consisted of two atoms, then two wire cubes represented the hydrogen type mould, two hydrogens plus oxygen a water mould, and so on. By using coloured zinc cubes which could be lowered into the wire moulds, Hofmann was able to demonstrate the hydrogen, oxygen, nitrogen and carbon types visually, and show in a visually stunning way how the substitution of a chlorine cube for a hydrogen cube in the hydrogen type, or a sulphur cube for the oxygen in the oxygen cube led to derived types: hydrogen chloride, hydrogen sulphide, and so on. Although he does not say so explicitly, the historian may speculate that Hofmann had in mind a billiards or bagatelle analogy; during distillation, as the atoms and radicals were energetically disturbed, they might "bump" a hydrogen or a nitrogen from a tenplate to form a derived type.

The audience having followed this simple exegesis was then easily moved to accept that "compound" hydrocarbon atoms might also be substituted for hydrogen atoms equivalent for equivalent. In this way he could derive the substituted ammonias, ethylamine, diethylamine and triethylamine with the ethyl group, and benzol, phenol and aniline with the "phenyl" group, C_6H_5. He could then return to the interpretation of distillation as variations on typical themes, while elsewhere emphasizing that the two-volume weight system recommended by Cannizzaro was "essentially necessary to represent with the requisite clearness, the vast and daily multiplying class of substitution changes".

We need not follow the essence of the lecture in any more detail since it will be plain enough how Hofmann, in a further dazzling sequence of aids and practical demonstrations, was able to reach the aniline dyes. And although he freely confessed that (in 1862) the exact chemical nature of mauve and other dyes was not known, he convincingly showed what a powerful conceptual tool the type theory could be in the days before the theory of structure.[24]

What, then, was the social/cultural message, the propaganda element, in this vividly entertaining lecture which Hofmann modestly concluded with his apologies for the imperfections of his guided tour: he hoped the scenery would not be forgotten even if the guide was. The moral arose from the fact that mauve and magenta were really "Royal Institution colours" insofar as benzol had been discovered by Faraday in 1825. Now if one had asked Faraday in 1825 (and, of couse, Faraday was in the audience in 1862) why he had found it, he would probably have said, "pure delight in the elaboration of the truth". The same would have been said by Faraday's successors and Hofmann's predecessors who had investigated the properties of benzene.

> Whenever in future one of you chemical friends, full of enthusiasm, exhibits and explains to you his newly-discovered compounds, you will not cool his noble ardour by asking him that most terrible of all questions, "What is its use? Will your compound bleach or dye? Will it shave? May it be used as a substitute for leather?" Let him quietly go on with his work. The dye, the leather will make their appearance in due time. Let him, I repeat it, perform his task. Let him indulge in the pursuit of truth, – of truth pure and simple, – of truth not for the sake of Mauve, not for the sake of Magenta – let him pursue truth for the sake of truth![25]

Like Liebig, who in his famous remark to Faraday about British society's respect only for practical matters "while the purely scientific works which possess far greater merit are almost unknown",[26] Hofmann paid lip-service to the idea that pure research was the handmaiden of applied science. Interestingly, in the same year that Hofmann drew this moral, he was persuaded otherwise by his experiences as a Juror for the 1862 Exhibition and, as Travis has demonstrated, by the stimulus of the dyestuffs made by his former pupil and research assistant, Edward C. Nicholson.[27] From then on Hofmann became an advocate of a symbiosis between pure science and trade – a view that he did much to encourage on his return to Germany.

It will have been clear from some of the phrases and passages quoted in this essay that Hofmann loved language and that the secret of his ability as a populariser was his ability to hone metaphors. Because Hofmann loved metaphor, of which he was a master in English and German, this relationship between pure and applied chemistry and between academic chemistry and society deserved its own special image and analogy, that of the tree – one of mankind's fundamental religious and imaginative symbols.[28] He seems to have first used this image in 1865 at the end of the *Introduction to Modern Chemistry* where, on looking back at the subjects he has covered from the vantage point gained, he hoped readers would be tempted to follow up each topic further, just as "the climber, whose appointed aim is the top of the tree, is tempted by the fruit-laden branches he passes in his ascent."[29] In the same year, at his farewell dinner before leaving London for Berlin, the tree came to symbolize the Royal College of Chemistry he was leaving when he warned that "no external props will sustain the [College] tree unless the vital juices circulate freely through roots and stems and branches."[30]

This tree soon became a type, a tenplate capable of transformations and substitutions and significant social messages. In his delightful after-dinner speech in praise of Kekulé's hexagonal formulae for benzene in 1890, Hofmann brilliantly elaborated the metaphor into the benzene tree whose harvest was both intellectual and material.

> Mit der Konzeption des Benzolringes erschien die Anzahl der organischen Verbindungen mit einem Schlage, man kann sagen, bis ins Unendliche gesteigert. In dem Benzolkern war ein Boden gegeben, aus welchem wir mit Erstaunen die bislang bekannte organische Chemie von neuem, nicht ein- oder zweimal, nein drei-, vier-, fünf- und sechsmal, gleich eben so vielen Bäumen, emporwachsen sahen. Welche Fülle von Arbeit war plötzlich nötig geworden, und wie schnell haben fleißige Hände sich gefunden, sie zu verrichten! Zunächst verfolgt das Auge wohl noch die sechs Stämme, aus denen sich der gewaltige Benzolbaum entfaltet! Aber schon haben sich die Äste der benachbarten Stämme verschlungen, und es hat sich ein Blätterdach gebildet, welches umfangreicher und umfangreicher wird in dem Masse, als der Riese in den Lüften emporstrebt; die Krone des Baumes aber ragt bis in die Wolken und bleibt dem Blicke noch verborgen. Und wie drängt sich an diesem herrlichen Benzolbaum Blüte an Blüte! Allerorten blickt aus dem Blättermeere die schlanke Hydroxylknospe hervor; kaum seltener die Gabelblüte, die wir

> die Amidgruppe, am häufigsten die schön geformte Kreuzblume, welche wir die Methylgruppe nennen. Und zwischen diesem Blütenschmucke, welcher Reichtum an Früchten! Einige in wunderbarer Farbenpracht erglänzend, andere fast betäubenden Wohlgeruch ausströmend! Auch fehlt es begreiflich nicht an fleissigen Arbeitern, welche emsig bestrebt sind, die herrliche Ernte einzuheimsen. Kühne Kletterer sind schon bis zum dritten und vierten Astansatze emporgeklommen, einige gewahren wir bereits in schwindelnder Höhe an der Arbeit. Am dicksten aber sitzen die in dem unteren Gezweige des Benzolbaumes. Einige haben schon genug gesammelt und sind im Begriff, herabzusteigen, andere können sich von der reichen Ernte noch nicht trennen, wieder andere liegen sich bereits ob des Erntegutes mit ihren Nachbarn in den Haaren.[31]

The dyestuffs chemist, Heinrich Caro, who like Hofmann spent a number of years in England working in the Manchester dye industry, compiled a history of dyestuffs in 1892 at the time of Hofmann's death. In this he hoped no storms would blow the leaves from the tree of German chemical industry; but whatever individuals thought of the tree, he urged them not to forget the loyal gardener who had watched over the sappling from the time it was a mere sprout for almost fifty years. "Remember always with love and respect our August Wilhelm Hofmann."[32]

Finally, in Britain, Hofmann's tree became an image of neglect. According to some, notably Henry Armstrong, one of Hofmann's four English obituarists, the tree of England's chemical industry was dying. Hofmann had done his best to raise up an English tree, but it had been found to have cankerous roots.

> No one could be brought to understand that it not only required frequent careful pruning, but also the addition continually of new grafts; it was left too much in the charge of mere uninstructed labourers, and competent forresters who could [have appreciated] its requirements and forward its growth were considered unnecessary.[33]

How ironic that one of Hofmann's best metaphors for popularising the support of chemistry and the integration of pure and applied science should have become a diagnostic tool for the identification of its supposed industrial decline!

1 Jacob Grimm, ["Vorrede,"] in: Jakob and Wilhelm Grimm, *Deutsches Wörterbuch* (Leipzig 1854), col. xxxi.

2 G.K. Roberts, "The establishment of the Royal College of Chemistry: An investigation of the social context of early Victorian chemistry," *Historical Studies in the Physical Sciences* 7 (1976), 437–485.

3 Andreas Kleinert, *Justus von Liebig "Hochwohlgeborner Freyherr": Die Briefe an Georg von Cotta* (Mannheim/Heidelberg/Wien 1979).

4 J. Liebig, "Der Zustand der Chemie in Preussen", *Ann. Chem. Pharm.* 34 (1840), 97–136. Cf. R.S. Turner, "Liebig versus Prussian chemistry: Reflections on early institute-building in Germany," *Historical Studies in the Physical Sciences* 13 (1982), 129–162.

5 See Robert H. Kargon, *Science in Victorian Manchester* (Manchester 1977), pp. 101–108.
6 J. Liebig, *Familiar Letters on Chemistry,* 3rd ed. (London 1851), p. 20.
7 See Christopher Hamlin, *A Science of Impurity: Water Analysis in Nineteenth Century Britain* (Berkeley 1990).
8 "Were I called upon to express myself still more explicitly" concerning the failure "to establish a definite boundary-line between inorganic and organic chemistry ... I would say, I will give you 'the chemistry of carbon and its compounds'". Hofmann, "Lectures on organic chemistry," *Medical Times* (1853 February 5), p. 134.
9 From a speech given by Hofmann in Berlin in 1865 and cited by H. E. Armstrong in his "Notes on Hofmann's scientific work [1893], in: *Memorial Lectures Delivered Before the Chemical Society, 1893–1900* (London 1901), pp. 637–732, at p. 703.
10 Report of Hofmann's Chemical Society lecture on "Lecture illustrations," *Chemical News* (1865 February 10), p. 66–67. Also Hofmann's *Introduction to Modern Chemistry* (London 1865), preface; idem, *Einleitung in die moderne Chemie* (Braunschweig 1866; 6th ed. Braunschweig 1877). Cf. Thomas S. Kuhn, *The Structure of Scientific Revolutions* (Chicago 1962).
11 C. A. Russell, *The History of Valency* (Leicester 1971).
12 *Chemical News* (1865 October 27), p. 202.
13 ibid., p. 203.
14 A. W. Hofmann, "On lecture illustrations," *Chemical News* (1865 July 28 and August 4), pp. 42–45, 55–57. Lists of "Hofmann's Apparatus" appeared in dealers' catalogues well into the twentieth century. Cf. [W. B. Jensen,] "Reinventing the Hofmann sodium spoon," *Bulletin for the History of Chemistry* 7 (1990), 38–39; cf. ibid. 8 (1990), 30–31; idem, "To demonstrate the truths of 'Chymistry': An historical and pictorial celebration of the art of lecture demonstrations in honor of Dr. Hubert Alyea," ibid. 10 (1991), 3–15.
15 J. P. Hofmann, *Acht Tafeln zur Beschreibung des chemischen Laboratoriums zu Giessen* (Heidelberg 1842); August Wilhelm Hofmann, *The Chemical Laboratories in Course of Erection in the Universities of Bonn and Berlin* (London 1866). The Bonn material was reprinted in *Chemical News* (1866 October 26), seriatum.
16 *Chemical News* (1861 July 6 and 20), pp. 7–11, 34–36. This formed the seventh lecture in *Lectures Addressed to Teachers on Preparation for Obtaining Science Certificates, and the Method of Teaching a Science Class* (London 1861). The lecture drew considerably an Hofmann's previous "On the importance of cultivating experimental science in a national point of view" given at the RCC and printed in *Reports of the Royal College of Chemistry, and Researches conducted in the Laboratories,* vol. 1 (London 1849).
17 *Lectures on Education Delivered in the Royal Institution* (London 1855).
18 *Chemical News* (1862 July 6), p. 8; the following quotations from the same source.
19 "Medicine some years ago found itself in a predicament very similar to that of agriculture at the same period; its resources appeared to be in a state of exhaustion, the rich capital of facts accumulated in the department of organic morphology by the industry of the anatomist, and by the acumen of the physiologist, could not yield its full fruits until an equivalent of knowledge had been drawn from the study of bio-chemical phenomena." Ibid., p. 35.
20 A. W. Hofmann and J. S. Muspratt, "On toluidine, a new organic base," *J. Chem. Soc.* 2 (1843–45), 367–383; *Ann. Chem. Pharm.* 54 (1845), 1–29. The anti-vitalistic remarks were highlighted in the *Pharmaceutical Journal* (1845), 455–456.
21 See especially Kurt Bayertz, "Spreading the spirit of science: Social determinants of the popularization of science in nineteenth-century Germany," in: T. Shinn and R. Whitley (eds.), *Expository Science: Forms and Functions of Popularization* (Dordrecht 1985), pp. 209–227.

22 I have used Mrs Sabine's translation of *Cosmos* (London 1846), vol. 1, p. 38. Hofmann's version, which does not differ materially, is quoted in Armstrong (1901), pp. 638–639.
23 *Chemical News* (1862 August 16), pp. 90–95; the following quotations from this same source.
24 On this point, see the important essay by Tony Travis, "Science's powerful companion: A. W. Hofmann's investigation of aniline red and its derivatives," *The British Journal for the History of Science* 25 (1992), 27–44. Note also W. Hornix's essay in this volume.
25 *Chemical News* (1862 August 16), p. 95.
26 Liebig to Faraday (1844 December 19), in: L. P. Williams (ed.), *The Selected Correspondence of Michael Faraday* (Cambridge 1971), vol. 1, pp. 429–430.
27 Travis (1992).
28 See Roger Cook, *The Tree of Life: Symbol of the Centre* (London 1974).
29 Hofmann (1865), p. 204.
30 *Chemical News* (1865 May 5), 210–216 at p. 213.
31 G. Schultz, "A. W. Hofmanns Tischrede bei der Kekuléfeier," *Zeitschrift für Farben- und Textil-Chemie* 1 (1902), 231–233. For an English translation, see William H. Brock, O. Theodor Benfey and Susanne Stark, "Hofmann's benzene tree at the Kekulé festivities," *J. Chem. Educ.* 68 (1991), 887–888.
32 H. Caro, "Ueber die Entwickelung der Theerfarben-Industrie," *Ber. dtsch. Chem. Ges.* 25 (1892), 955–1105, at p. 1105. To B. Rayman, Hofmann had been the tallest tree in the forest of chemical workers, albeit his height had not been appreciated until death cut him down. See S. Štrbáňová's essay in this volume.
33 Armstrong (1901), p. 730.

IX

Humboldt and the British: A Note on the Character of British Science

Summary

Through his Romanticism, aesthetics, 'religiosity', the escapism which he offered urban readers, and the appeal that his search for unification, order, association, and simplicity had during a period of growing cultural fragmentation, Humboldt's translated writings asserted their magic on Regency and early Victorian lay and scientific minds.

When I was but thirteen or so
I went into a golden land,
Chimborazo, Cotopaxi,
Took me by the hand.

My father died, my brother too,
They passed like fleeting dreams.
I stood where Popocatepetl
In the sunlight gleams.

I dimly heard the master's voice
And boys far-off at play,
Chimborazo, Cotopaxi
Had stolen my heart away.

Romance, W. J. Turner[1]

As the twentieth-century poet and critic Walter Turner realized in his schoolboy's romantic reverie, there is a magic, poetic quality in the three words Chimborazo, Cotopaxi and Popocatepetl—melodious names of Andean and Mexican mountains and volcanoes which were first made well known to Europeans through the publication of Alexander von Humboldt's South American travel books and reinforced by the English translations and editions of the *Ansichten der Natur* and *Kosmos* in the 1840s. The poem suggests two reasons for the appeal and fascination which Humboldt's writings asserted on British readers, namely their romantic, poetic quality and what we should nowadays recognize as their cinematographic or televisual quality. Humboldt's writings, much as Darwin's *Journal of Researches* later, conveyed urban Britons on a magic carpet to places of extraordinary colour, beauty and strangeness.

As is well known, during the Romantic Period, 1780–1830, English artists and poets and composers looked to the diversities, complexities and subtleties of Nature for

[1] Walter James Turner (1889–1946), 'Romance', quoted by kind permission of Sidgwick & Jackson Ltd. For Turner, a Georgian poet and music critic, see *Dictionary of National Biography* [hereafter *DNB*].

religious analogies and moral truths, valuing the power of their personal, individual *imaginations* as the chief tool for unveiling these truths. Nature, meaning the whole created universe (or Cosmos as Humboldt was to recall it) was, according to the Romantic ideology, alive with spiritual truths. It is not surprising therefore, that Humboldt's portrayal of another of Nature's scenarios, thousands of miles from Tintern Abbey or the English Lakes, should have stirred the imagination. And although Humboldt personally was always at pains to 'number, weigh and measure'—the Newtonian perspective which many Romantics violently rejected, his ability to compromise and combine fact-seeking with generalization and word painting of a high literary quality (in which he was well served by his translators) made him very attractive to the exponents and disciples of the English Romantic tradition. Humboldt sought a synthesis of the physical and moral, the mathematical Newton with the morphological Goethe.[2] Nature considered *rationally*, as Humboldt emphatically puts it in the *Kosmos*, turns out to be a moral, poetic messenger, for it is 'a unity in diversity of phenomena; a harmony, blending together all created things, however dissimilar in form and attributes'.[3] Such a programme of synthesis and unification also fitted in well with the notions of chemists, such as William Prout, who sought unity and simplicity in the large number of different Lavoisieran elements and Daltonian atoms, or with the explorations of physicists like Michael Faraday and William Grove who hoped to find a 'correlation' between the diverse physical powers or forces of matter.

Ut pictura poesis. Horace's epigram that 'as is painting, so is poetry' reminds us, too, of the tradition which the English Romantics had absorbed from Claude Lorraine: the appreciation of landscape by reference to painting. This cult of the 'picturesque', with its associated Grand Tour or journey to the English Lakes (or even to the furnaces and industry of the iron country of Shropshire), together with the development of landscape gardening, quite naturally took to its heart Humboldt's fine word painting of Nature (as well as his own informative digressions in the *Ansichten* and the *Kosmos* on the importance of landscape painting and landscape and greenhouse gardening for the appreciation and discovery of Nature). As Janet Browne has described in her book, *The Secular Ark*,[4] Humboldt's Hippocratic 'Airs, Waters, Places'-type of environmentalism interpreted the art of nations as ultimately a function of vegetation, temperature, soil, mountains, and rivers.

For the *Quarterly Review* of 1818, praising the publication of further volumes of the *Relation historique du voyage* in the previous year, Humboldt's skill lay in his 'ability to seize a subject and exhibit it in the most striking point of view; and by a happy faculty of grouping, or contrasting, the meanest and the most familiar objects, to give to them an interest to which separately considered they could have no pretension'.[5] The reviewer went on explicitly to praise Humboldt's technique of '*keeping*'—an eighteenth-century technical word in painting which referred to an artist's ability to maintain a proper or harmonious balance between representations or near and distant objects. Humboldt gave light and shade to his prose pictures. 'He is what artists would call a mannerist', the

[2] Adolf Meyer-Abich, 'Alexander von Humboldt' in *Alexander von Humboldt 1769–1859* (Bonn/Bad Godesberg: Inter Nationes, 1969), p. 92. Meyer-Abich points to parallels between Goethe and Humboldt throughout his essay.

[3] *Cosmos*, translated by E. C. Otté (London: H. Bohn, 1849), I, pp. 2–3. Compare Mrs Sabine's more opaque rendering of the same passage: 'The aspect of external nature, as it presents itself in its generality to thoughtful contemplation, is that of unity in diversity, and of connection, resemblance and order, among created things most dissimilar in their form;—one fair harmonious whole'. *Cosmos*, translated by Mrs Elizabeth Juliana Sabine (London: Longmans, 1846), I, p. 5.

[4] Janet Browne, *The Secular Ark* (New Haven and London, 1983).

[5] John Barrow, *Quarterly Review*, 18 (1818), 135–58 (p. 136).

reviewer concluded, meaning that he had a distinctive style. Similarly, John Herschel, in his long *Edinburgh Review* of the *Kosmos* in 1848, saw its author, in striving to be extensive without suppressing detail, as being in the tradition of the *picturesque*. No matter how beautiful or striking an object Humboldt was describing, Herschel noted, his aim was always to associate it with other things—its true significance lay entirely in its associations and relationships.[6]

It is no accident that Charles Darwin was to use this associationist technique in the remarkably beautiful final paragraph of *The Origin of Species* in which we 'contemplate an entangled bank clothed with many plants of many kinds, with birds singing on the bushes, with various insects flitting about, and with worms crawling through the damp earth, and to reflect that these elaborately constructed forms, so different from each other, and dependent on each other in so complex a manner, have all been produced by laws acting around us'.[7] Nor is it an accident of style that Darwin, in the Beagle Diary and the published *Researches*, like Wordsworth or Coleridge or Humboldt, fixes a 'spot of time'—an image of Nature—on his and the reader's mind. Consider, for example, the passage for which Humboldt himself praised Darwin in a letter he wrote to him on 18 September 1839:

> Who when examining the cabinet of the entomologist the gay exotic butterflies, and singular cicadas, will associate with these objects, the ceaseless harsh music of the latter, and the lazy flight of the former,—the sure accompaniments of the still glowing noon day of the Tropics.[8]

Science, or Nature, was valuable to Humboldt, as it was for Wordsworth in *The Excursion*, a poem which the young Darwin enormously admired, precisely because it stimulated the *imagination* to search for higher truths. And it fell to Darwin, following Humboldt's method, to discover in the undergrowth of the entangled bank of beauty, cruelty, death, decay and transfiguration—The 'Rhodian Genius' of Humboldt's *Ansichten*[9]—the evolutionary secret which, by a great leap of the imagination, could lead man to perceive life in a nobler way.

Darwin, who had been much 'stirred' by reading Humboldt's *Personal Narrative* when he was an undergraduate at Cambridge, agreed with Joseph Hooker that it was right to 'call Humboldt the greatest scientific traveller who ever lived ... the parent of a grand progeny of scientific travellers'.[10] In his study of eighteenth-century travel literature,[11] the literary historian Charles Batten has argued that the traditional ideal

[6] *Edinburgh Review*, 87 (1848), 170–229. For the attribution to Herschel, and for the other attributions of anonymous reviews cited in this essay, see *The Wellesley Index to Victorian Periodicals, 1824–1900*, edited by W. E. Houghton (Toronto: University of Toronto Press, 1966), I.

[7] C. Darwin, *On the Origin of Species*, final paragraph of the first edition (London: John Murray, 1859).

[8] C. Darwin, *Journal of Researches* (London: John Murray, 1839; reprint New York, 1952), p. 394. The passage, which is slightly revised in the second edition (1845; reprint New York, 1956), may be traced back to the even more remarkable prose of the Beagle diaries. For a stimulating discussion of this passage, see Stanley E. Hyman, *The Tangled Bank: Darwin, Marx, Fraser and Freud as Imaginative Writers* (New York, 1962). For Humboldt's appreciation of the passage, see Paul H. Barratt and Alain F. Coreos, 'A Letter from Alexander Humboldt to Charles Darwin', *Journal History of Medicine*, 27 (1972), 159–72.

[9] 'Vital Force, or the Rhodian Genius', in *Views of Nature*, translated by E. C. Otté and H. G. Bohn (London: Henry Bohn, 1850), pp. 380–85. For the influence of Wordsworth see Hyman (footnote 8) and E. Manier, *The Young Darwin and his Cultural Circle* (Dordrecht: Reidel, 1978), pp. 89–96.

[10] C. Darwin to J. D. Hooker, 6 August 1881 in *The Life and Letters of Charles Darwin*, 3 vols (London: J. Murray, 1887), III, p. 227.

[11] Charles L. Batten, *Pleasurable Instruction: Form and Convention in Eighteenth-Century Travel Literature* (Berkeley: University of California Press, 1978). Compare John Tallmadge, 'From Chronicle to Quest: The Shaping of Darwin's *The Voyage of the Beagle*', *Victorian Studies*, 23 (1980), 325–45.

account of a journey had to balance the empirical and the aesthetic. If too much factual detail was given, the author was thought dry and pedantic; if too much stress was laid on personal adventures, the author was thought egotistical. However, by the end of the eighteenth century, facts were becoming relegated more and more from travel books to guide books, for which publishers like John Murray and Karl Baedeker were to gain fame. This excision, together with the impetus of Romanticism and the cult of the picturesque, encouraged the transformation of the travel book into a record of the traveller's personal responses to 'exotic landscapes, peoples and experiences'. It is probably not accidental, therefore, that Helen Maria Williams, the first translator of Humboldt's *Relation historique du voyage* in 1814 gave it the English title *A Personal Narrative*, a title used again by the second translator Thomasina Ross in 1852.[12]

So far I have deliberately chosen to emphasize the literary appeal which Humboldt had for British readers, both because it provides an important insight into the talents of this multi-faceted man and because it provides an example of the important rapprochement which is at last taking place in English literary studies between the examination of literary forms and conventions and the study of scientific prose.[13] If Humboldt's success as a scientific communicator lay, as the *Blackwood's Magazine* reviewer of *Kosmos* explained in 1845, in the combination of keen observational powers, 'an ardent poetical temperament', and 'eloquence and intellectual acumen',[14] then it seems to me that Humboldt is well worth studying as much as a literary, as a scientific, figure.

Humboldt had many different points of contact with British people and English culture;[15] not only did he visit Britain six times, number over a hundred English correspondents, and gain the coveted Copley medal of the Royal Society in 1852, but he was the inspiration for the 'magnetic crusade' of the 1830s, the general enthusiasm for 'Humboldtian science' in Britain and for the successful sectionalization and specialization of the British Association for the Advancement of Science (BAAS). Although he personally disliked Queen Victoria's husband, Albert, both the organization of magnetic observatories and the sectionalization of the BAAS were examples of the division of labour which Prince Albert repeatedly urged the British people to accept as the key to industrial and cultural progress.[16]

[12] A. Humboldt, *Personal Narrative of Travels to the Equinoctial Regions...*, translated by H. M. Williams, 7 vols in 9 (London: Longmans, 1814–1829; reprint New York: AMS Press, 1966); ibid., translated by T. Ross, Bohn's Scientific Library (London: Bohn, 1847–1852; reprint New York: B. Blom, 1971). Helen Maria Williams (1762–1827) was a prolific novelist, poet and political writer (*DNB*); of Thomasina Ross, who translated several German, French and Spanish works during the period 1820–1850, nothing is known.

[13] Hyman (footnote 8); Tallmadge (footnote 11); *Victorian Science and Victorian Values. Literary Perspectives*, edited by J. Paradis and T. Postlewait (New York: New York Academy of Sciences, 1981); Tess Cosslett, *The 'Scientific Movement' and English Literature* (Brighton and New York: 1982); Gilian Beer, *Darwin's Plots* (Cambridge, 1983); S. Shuttleworth, *George Eliot and Nineteenth-Century Science* (Cambridge, 1984). Note also E. C. Hentchel's essay, 'Alexander von Humboldt's Synthesis of Literature and Science' in *Humboldt* (footnote 2).

[14] *Blackwood's Magazine*, 58 (1845), 541–59. The reviewer was Archibald Alison (1792–1867), historian (*DNB*).

[15] See Jean Théodoridès, 'Humboldt and England', *British Journal for History of Science*, 3 (1966), 39–55.

[16] Victor W. von Hagen, 'The Fate of Humboldt's Library', *Isis*, 41 (1950), 164–7, referring to *Briefe an Varnhagen von Ense aus den Jahren 1827 bis 1858* (Leipzig, 1860), which appeared in English the same year. Albert himself said of Humboldt's death: 'People of this kind do not grow upon every bush and they are the grace and glory of a country and a century'. (Quoted by Theodore Martin, *Life of the Prince Consort*, 5 vols (London, 1875–1880), IV, p. 288.) For Albert's commitment to the division of labour and to 'Humboldtian Science', see his Presidential Address to the British Association for the Advancement of Science, *BAAS Report* (Aberdeen, 1859), and *The Principal Speeches and Addresses of His Royal Highness The Prince Consort* (London, 1862). The term 'Humboldtian Science' is due to Susan Faye Cannon, *Science in Culture. The Early Victorian Period* (New York, 1978), pp. 73–110.

Many illustrations of Humboldt's generosity and magnanimity towards British men of science could be given; for example, when the Manchester engineer William Fairbairn visited Berlin to advise on the design of a bridge over the Rhine at Köln, it was, to Fairbairn's amazement, the 80-year-old Humboldt who turned out to greet him, and not Fairbairn who went to seek him out as he had intended.[17] 'There are few heroes who lose so little by being approached as Humboldt', said Lyell[18]—adding for good measure that the same could not be said of Baron Cuvier. Perhaps the best example of the way in which Humboldt's name, and his programme, sank into English consciousness comes from the optimistic remark of the Royal Commission on the state, discipline, studies, and revenues of the University and Colleges of Cambridge in 1852. Concluding that science must be taught to the poll-men—that essentially dangerous class of undergraduates 'in the social system of the place' who did not take honours degrees, the Commissioners concluded that they should be taught how to handle 'meteorological and magnetic instruments, and cameras', so that 'they may become Humboldts'.[19]

Nevertheless, it would be a mistake to think that Humboldt was received uncritically by the British. One of the earliest types of criticism is found in the four notices of the *Personal Narrative* which appeared in the *Quarterly Review* between 1816 and 1821. In, perhaps, deliberate contrast to John Allen's reviews in the *Edinburgh Review* (which found Humboldt wanting only in meteorological skills), John Barrow refused to be 'ravished' by the 'whistling' of Humboldt's name as 'the first of all travellers... a rare union of all that Plato, Thales and Pythagoras taught among the ancients...'.[20] No none of that! The title was wrong, for it was not a *personal* account, but full of quotations from other writers, nor was it a *narrative* since 'the most trite and trivial matters' were 'meretriciously tricked out in the garb of philosophy and mathematics'. Not only was Humboldt at fault for failing to write in the (actually already outmoded) tradition of travel writing, he was also at fault for continually generalizing and breaking up the story of his experiences by long-winded dissertations. In other words, by mapping things together—by artificially putting things together and not necessarily describing them in the way they occurred, as Acosta and Forster had previously done, Humboldt was taking the *personal* out of the story. Of course, this was precisely the technique which Darwin so successfully used in writing up his 'Beagle' notes, but it was evidently not a technique universally admired in 1816.

Humboldt's prolix tendencies were another problem for all reviewers, including David Brewster in the *North British Review* in 1849.[21] As the *Quarterly* put it, one had to say of Humboldt in physics 'what was said of [Isaac] Barrow in divinity, that he never quits a subject until he has exhausted it'.[22]

Another related criticism was Humboldt's decision to bring his writings out in instalments. 'To spend twenty years upon acquisitions made during five cannot be a

[17] *The Life of Sir William Fairbairn, Bart*, edited by William Pole (London, Longmans, 1877; reprint Newton Abbott, 1970), pp. 217–37.

[18] Quoted by Théodoridès (footnote 15), p. 45.

[19] Quoted by J. G. Crowther, *Statesman of Science* (London: Cresset Press, 1965), p. 195.

[20] *Quarterly Review*, 14 (1816), 379. Barrow's authorship is cited by H. Shine, *The Quarterly Review under Gifford* (Chapel Hill: University of North Carolina Press, 1949). Compare John Allen's review of the *Tableau Physique* (1807) in *Edinburgh Review*, 16 (1810), 223–53 in which he stated that 'no name stands higher than that of Humboldt, among the lovers of geographical and physical science'.

[21] David Brewster, review of Mrs Sabine's translation of the *Ansichten* (*Aspects of Nature*) in *North British Review*, 12 (1849), 225–64. Sabine's translation has been reprinted as *Aspects of Nature* (New York: AMS Press, 1970). For other reviews of Humboldt's writings by Brewster, see *North British Review*, 4 (1845), 202–54; 5 (1846), 454–503; *Meliora*, 3 (1860), 273–95.

[22] *Quarterly Review*, 21 (1819), 320.

desirable arrangement', thought the physicist and Alpine traveller, James D. Forbes; 'it encourages a prolix and embarrassed style'. 'Even his handwriting', Forbes went on, 'bears testimony to the drudgery of continued labours for the press, and the minute conglomeration of half-formed characters betrays the secret of writing a volume with the least possible amount of muscular exertion.'[23] In fact, Humboldt's British contemporaries, still lacking what David Allen aptly called 'the corset of professionalism', could be just as slow; John Herschel, for example, took some twenty years to publish his astronomical observations of the southern heavens. Nevertheless, although there were special reasons for the delayed production of Humboldt's South American researches, one remains puzzled, even today, why *Kosmos* was produced so slowly (and, of course, incomplete). Faced by Humboldt's seriatim *Kosmos*, Herschel himself must have felt like some modern reviewer of Joseph Needham's *Science and Civilisation in China*, for it was difficult from seeing only the first volume what the complete plan and purpose was. However, he felt intuitively that *Kosmos* was a cry for both freedom of investigation and for man to conform to the laws of nature.[24]

The most singularly vituperative review of *Kosmos* was published by James Forbes in the *Quarterly Review*. Whereas David Brewster and John Herschel had seen Humboldt's *Ansichten* and the *Kosmos* as superior Bridgewater Treatises,[25] and interpreted him as writing within the British tradition of evidence for the providential design of nature, Forbes was angered by Humboldt's failure to make explicit the design of a superintending Providence. 'We call it the highest pedantry of intellect', Forbes wrote, 'to put to silence suggestions which arise spontaneously in every mind, whether cultivated or not, when engaged in such contemplations'.[26] For Forbes, God was too conspicuously absent!

It was for this reason, no doubt, that he spent two-and-a-half pages rebutting Humboldt's little joke about British magnetic observers in Tasmania failing to observe the progress of a world-wide magnetic storm because it occurred on a Sunday when it was sinful (sündhaft) to read a scale. Mrs Sabine, whose husband was directly involved in the organization of international magnetic observatories, craftily left that sentence out (Forbes was reviewing the German text) and was then attacked by Elise Otté and Henry Bohn in their rival translation for censoring Humboldt.[27]

It may be thought paradoxical that, on the one hand, some reviewers could praise Humboldt for fitting into the British tradition of natural theology and drool over such exquisitely beautiful passages as 'In the jungles of the Amazon, and on the heights of the Andes I became aware that One Life is poured out into stones, plants and animals and in the throbbing breast of mankind, as if inspired by one breath,'[28] and, on the other hand, Forbes could suspect him of playing into the hands of sinners like the

[23] J. D. Forbes, review of the first German volume of *Kosmos*, *Quarterly Review*, 77 (1845), 154–91.

[24] *Edinburgh Review*, 87 (1848), 170–229 (p. 174).

[25] W. H. Brock, 'The Selection of the Authors of the *Bridgewater Treatises*', *Notes and Records of the Royal Society*, 21 (1966), 162–79. Both Herschel and Brewster were considered, but not chosen, as authors.

[26] *Quarterly Review*, 77 (1845), 164. Forbes had met Humboldt in Paris in 1835. Théodoridès (footnote 15), p. 47.

[27] Henry Bohn's criticisms are found in a Preface to *Cosmos*, Bohn edition (1849), and in an eight-page leaflet dated 28 February 1849 usually bound in with the Bohn edition of *Views of Nature* (London, 1850). For magnetism, see J. Cawood, 'The Magnetic Crusade: Science and Politics in Early Victorian Britain', *Isis*, 70 (1979), 493–518.

[28] This striking quotation is, in fact, from a letter to Karoline von Wolzogen and would not have been known to English readers until 1873 when it appeared in Jane and Caroline Lassell's translation of *Life of Alexander von Humboldt*, edited by Karl Bruhns, 2 vols (London: Longmans Green, 1873), I, p. 417. However, equally graphic and emotional passages may easily be found in the introductory sections of the *Kosmos* and *Ansichten*.

phrenologist George Combe or Robert Chambers (the then anonymous author of *Vestige of Creation*) who were appealing to the need to obey the secular laws of nature rather than for mankind to seek for moral analogies within nature. However, this only goes to show the universality of Humboldt's appeal. Like Bacon before him, or his chemical protégé, Justus Liebig in the 1840s, Humboldt could be all things to all men and a useful 'name' with which to associate often quite diverse rhetorics and concerns.

A good example of this arises from a criticism which seems to have been peculiar to Forbes, and which involved an intense nationalism which the cosmopolitan Humboldt himself had always taken pains to avoid. According to Forbes, Humboldt's 'physics have been written for the longitude of Paris, and the geology for that of Berlin'. It was not enough, he asserted, that Humboldt should cite English books

> as mere authorities of a fact, as Dr Buckland's Bridgewater Treatise is not infrequently. We miss the recognition of the place which our geologists are entitled to hold in the history of science, which was never so conspicuous as within the recollection of those now alive.[29]

In view of Hanno Beck's demonstration of the centrality of physical geography to Humboldt's interpretation of nature, his citation of von Buch and other German physical and chemical geologists in preference to British geologists, with their essentially palaeontological and stratigraphical viewpoint, is readily understandable.[30] However, if Forbes's comment is put within the long-standing context of the debate over whether British science was declining relatively to French and German science—a debate which had formed a backcloth to the foundation of the British Association for the Advancement of Science in 1831[31]—his nationalist perspective is equally comprehensible. For Forbes, by the 1840s a non-declinist British science was as good as, if not better than, German science, and Humboldt was being ungracious in failing to recognize the glories of British heroic geology or Edinburgh/Cambridge mathematical physics.

On the other hand, British men of science who did continue to feel that, despite the advent of an Association for the Advancement of Science, the state of British science left something to be desired, could seize enthusiastically upon Humboldt's warnings concerning the rise and fall of civilization:

> Those nations who are behind in general activity, in application of mechanics and technical chemistry, in careful selection and processing of natural materials, such that regard for such enterprise does not permeate all classes, will inevitably decline in prosperity; all the more where neighbouring states in which science and the industrial arts have an active interrelationship, progress with youthful vigour.[32]

Such a statement (ironically, of course, made to his own countrymen) might well have been made by, say, Lyon Playfair, following the Great Exhibition,[33] and was to

[29] *Quarterly Review*, 77 (1845), 191.

[30] Hanno Beck, *Alexander von Humboldt* (Wiesbaden: F. Steiner Verlag, 1959–61).

[31] Jack Morrell and Arnold Thackray, *Gentleman of Science* (Oxford, 1981), especially pp. 47–52.

[32] *Cosmos*, Sabine translation (1846), I, p. 38; Otté translation (1849), I, p. 34; Herschel's own translation, *Edinburgh Review*, 87 (1848), p. 181 is freer, but in the same spirit.

[33] 'The progress of science and of industry in countries which have reached a certain state of civilisation ought actually to be synonymous expressions... As surely as darkness follows the setting of the sun, so will England recede as a manufacturing nation, unless her industrial populations become much more conversant with science than they are now', L. Playfair in *Lectures on the Results of the Exhibition* (London, 1852), quoted by D. S. L. Cardwell, *The Organisation of Science in England*, second edition (London, 1972), p. 81.

prove a handy rhetorical device throughout the remainder of the century for those who wished to argue for and to urge greater state support for science and its institutions of learning and research. There was, indeed, a symbolic appropriateness about the exhibition of Humboldt's bust at the Crystal Palace in 1851.[34]

While I have drawn attention to the fact that British reviewers were not uncritical of Humboldt's writings, there was always universal admiration for his character; even Forbes expressed his high regard for Humboldt's being and attainments while wishing 'he had not devoted himself to [raising] a literary pyramid whose mass, like that of Egypt, [was] itself a passport to immortality'.[35] At a time when the liberal culture of Englishmen was poised to fragment into a dozen sub-cultures and disciplines, each with its own educational institutions, professional organizations and literature, Humboldt was welcomed as a nineteenth-century Pliny; or, as Goethe had suggested, he was seen as a fountain gusting forth streams of different knowledge.[36] Through his Romanticism, aesthetics, 'religiosity', the escapism which he offered urban readers, and the appeal that his search for unification, order and simplicity had during a period of growing cultural fragmentation, Humboldt's translated writings asserted their magic on the Regency and early Victorian mind.

As a boy of eight, Humboldt is reputed to have been asked his name by Frederick the Great who, on learning that it was Alexander, asked him whether he wanted to conquer the world like his Greek namesake. 'Yes Sire,' he is supposed to have replied, 'but to conquer with my head.'[37] Several generations of Englishmen, including Darwin and Wallace, were to be inspired by that head, which guided them to and from the golden lands of Chimborazo, Cotopaxi and Popocatepetl and stole their hearts with a story of unity and stability amidst 'the vacillating ever-recurring alternation of physical metamorphoses'.[38]

[34] The bust, by the German sculptor Christian Rauch, is now housed in the offices of the Alexander von Humboldt-Stiftung, Bonn.

[35] *Quarterly Review*, 77 (1845), 159.

[36] Er gleicht einem Brunnen mit vielen Röhren'. Goethe, as quoted by E. R. Brann, *The Political Ideas of Alexander von Humboldt* (Madison: University of Wisconsin Press, 1954). Note also R. M. Young, 'Natural Theology, Victorian Periodicals and the Fragmentation of the Common Context', *Darwin to Einstein. Historical Studies on Science and Belief*, edited by C. Chant and J. Fauval (London: Open University, 1980), pp. 69–107; reprinted in R. M. Young, *Darwin's Metaphor* (Cambridge, 1985).

[37] The anecdote is given by Brann, (footnote 36), p. 5.

[38] *Cosmos*, Otté translation, I, p. xiv.

X

SCIENCE EDUCATION

1. HISTORIOGRAPHY

Like the history of science, history of education is part of the wider story of the history of society. In the words of Asa Briggs, it is 'social history broadly interpreted with the politics, the economics and . . . the religion put in'.[1] Such a definition explains why historians of education are interested in the history of science, and vice versa. Science and technology have played major roles in shaping modern society, and although the historian may choose to emphasise either the intellectual or the social processes which stimulated or hindered the development of western society, the growth of modern science and medicine and their technical exploitation cannot be fully understood without considering how their curricula are learned and acquired by one generation after another. The study of scientific and technical institutions, or of the educational systems and facilities of different countries, or of the financial support given to scientific teaching and research, helps the historian of science to understand the dynamics of scientific development. Accordingly, historians of science share some territory with historians of education. Like the latter they can approach territorial problems through local studies, comparative national history, quantitative (cliometric) analysis, administrative history, working-class social history ('history from below') and intellectual history.

To avoid an exhaustive but dreary catalogue approach, or the production of a series of vignettes, historians of science education continually need to raise fundamental questions about society's ideas of, and attitudes towards, education at different periods. Why teach science? How is science teaching linked with innovation in mathematics, technology or medicine? To what extent does a society's interest in science teaching (or lack of it) depend on religious, economic or political factors? What is the status of the scientist, or science teacher, and what careers are open to them? Was science taught formally, or by example, or heuristically, and at what cost? Should the curriculum be specialised or general?

What particular sciences are the most popular for instruction purposes and why? How does one country's system affect another's? To what extent is there agreement over methods of instruction and examination? Which are the influential textbooks? What do they indicate concerning contemporary attitudes? Can the influence of past science teaching be seen in the contemporary organisation and development of the subject? By composing its history around such questions the historian of science education will ensure that the specialism becomes an integral part of history like the history of science itself.

A potent example of the way in which a contemporary British debate is both informed and misinformed by knowledge of the history of education is 'declinism'. Ever since Jacob Bronowski's and C. P. Snow's analyses of Britain's relative decline as a world scientific power in terms of two opposed artistic and scientific cultures, it has been fashionable to use 'education' as a unique explanation for national decline. Barnett, Roderick and Stephens, and, most notably, Wiener, have all argued that the failure of Britain's governing and industrial classes to develop a more practical, applied and technical system of education in the 1880s is the direct cause of Britain's present position in the world. Hobsbawm's explanation that Britain's present-day position is an inevitable consequence of colonial imperialism and of being the first European country to industrialise, is rejected as unsatisfactory.[2]

Such 'declinism' is historically and historiographically too simplistic. It has yet to be proved convincingly that there is a direct correlation between 'technical' education and improved economic performance, fashionable though this equation may be. It also has to be remembered that the fundamental role of any educational system – up to the secondary level at least – is conservative: the transmission of a culture's heritage. Indeed, before 1914 all European nations had developed remarkably similar educational systems for their elites. The declinist thesis also lays too much stress upon 'aristocratic', 'pre-industrial' and 'countrymen's' cultural values while failing to consider the possibility that scientists, engineers and industrialists may not themselves have deliberately distanced their activities from manual and craft images of science and industry in order successfully to enhance their status as professional men and women. Finally, there has been a failure by economists and historians of education to recognise the perennial rhetorical nature of declinist arguments.

It is mainly because of class attitudes, *laissez-faire* and narrow views of liberal education as 'superior' to technical training, which had worked satisfactorily in the first phase of industrialisation, that educational and economic historians have tended to view scientific and technical education in Great Britain as somehow 'held back'. Support for such a view is, of course, readily found in nineteenth-century scientific rhetoric of 'decline', beginning with Charles Babbage's peevish *Reflections on the Decline of Science in England* (1830). Babbage held up France as the model for Britain to follow while ignoring Scotland and the richness of English provincial science. Later men of science, particularly those like Lyon Playfair and Edward Frankland, who had studied in Germany with Liebig or Bunsen in the 1840s, praised the German states for the way they had created (following Prussia's example at the beginning of the nineteenth century) a highly organised, technically-trained society through a state system of education. In fact, as historians of French and German science education have pointed out, the 'decline' argument was relative and rhetorical: both French and German scientists often complained that their respective scientific and technical educational systems were dangerously inferior to a 'foreign' system and even commented enviously on the mixed state and voluntary/philanthropic systems which generally prevailed in Great Britain until after the First World War.[3] No doubt there have been significant differences in the responses to, and the impact of, the argument in different decades and in different countries, and the history of education would be greatly enriched by a detailed comparative study of the argument's continuity and historiography.

Whatever the contemporary justification for the rhetoric of 'decline', it (along with social Darwinism) was undoubtedly effective in making European nations competitive during the second phase of the Industrial Revolution when electricity began to complement steam power and gas as an energy source and the chemical industry became dependent upon engineering. A detailed understanding emerged of synthetic pathways and thermodynamics rather than the empirical, analytical, 'boilerman' approach of most chemical technology before 1860. The German infra-structure of pharmaceutical education had a tradition of practical workbench experience in privately-run Chemical Institutes long before Justus von Liebig exploited it for the production of chemists who might apply their analytical knowledge to the relief of man's estate by improving agriculture, fighting disease, maintaining health and exploiting chemical laws for commercial gain. This gave the German states some competitive advantage over other nations. However, given the large numbers of British chemists whom Liebig taught, his tremendous popularity in Britain and the way the Giessen method of training was copied and exploited by A. W. Hofmann in the privately-financed Royal College of Chemistry (founded in 1845), it is doubtful whether Britain really 'lagged' behind in chemical education.[4] Nor, given the fact that most German *Gymnasium*-educated men who entered university had received a purely classical education, was the British schools' anti-scientific bias all that unusual. Where Germany differed was in her manufacturers' willingness to absorb graduates and to spend time transforming them into industrialists.

Historians of science and education have done a good deal of work on French, German and American science and education, but apart from some partial studies of Japanese and Russian science teaching, little has been done on other countries. No one has yet attempted a global synthesis of the history of science education, or even a comparative analysis of European development, and apart from a major study of the emergence of the female scientific community in America, surprisingly little has been published on the history of the educational development of women scientists in Europe.[5]

2. SCIENCE EDUCATION IN GERMANY

After its defeat by Napoleon in 1806, Prussia, followed by the other German states, mobilised itself for European spiritual and intellectual leadership. It did this by formally developing government-controlled segmented systems of primary, secondary and tertiary education which would ensure that the state had an elite and an efficient class of officials and learned professions. To this end, the *Gymnasium*, or grammar school, rather than the *Volkschule*, or primary school, became the focus of attention. For it was the ancient-languages-based curriculum of these *Gymnasien* that formed the basis of the school-leaving examination (*Abitur*) which allowed middle-class pupils to matriculate to universities or obtain direct entry into various state examinations for the civil service. Other types of secondary school, which failed the severe accreditation test for the *Abitur*, became known as *Realschulen* (modern schools). Although the latter developed science teaching, their pupils were unable to matriculate at universities until legal battles had been fought in the 1860s and tended instead to take their higher education in the *Technische Hochschulen* (technical schools or polytechnics), whose examination was considered inferior to that of universities. As in Great Britain, therefore, until 1900, when all secondary school pupils were admitted to the *Abitur*, most German scientists of the top rank had had severely classical educations and first experienced science at university.

The whole purpose of education in Germany's many universities was to achieve success in state examinations which allowed entry into government service, including the professions of pharmacy, medicine and academic teaching. However, with the expansion of faculties of philosophy, which became imbued with the ideology of *Wissenschaft*, or the search for knowledge, an increasing number of students throughout the nineteenth century, particularly foreign students, stayed on at university to take a doctoral degree. This, like the higher degree of *Venia Legendi* or *Habilitation*, called for independent research and the composition of a thesis. It was in this atmosphere that teaching and research laboratories emerged, beginning with Liebig's internationally renowned chemical laboratory at the University of Giessen in 1824. The development of experimental research was greatly aided by the competition for prestigious professorial staff that occurred between the various German state governments prior to unification in 1870.

The first subject to be put in this laboratory mould was chemistry, principally because pharmacy and the growth of agricultural and industrial chemistry, after

1848, created new career opportunities for trained chemists. A second avenue of expansion was provided by physiology, which became regarded as a necessary part of medical knowledge from the 1840s onwards. At first teaching was carried out by professors of anatomy; however, between 1855 and 1874 some 26 German scientists were appointed to new chairs in physiology. By the mid-1870s, however, this period of expansion was over; consequently physiological research began to lose momentum and the brightest *Privatdozenten* (qualified university teachers without chairs who were allowed to give extra-curricular lectures) began to move into other specialisms, such as pathology, biochemistry and psychology.

Germany remained a continuing seat of productivity in laboratory-based sciences up until the First World War. By then, Germany's *Hochschulen* had become accredited 'technical universities', able since 1899 to award the doctorate to students for research in applied science. The Berlin *Hochschule* at Charlottenburg, which was established in its final form only in 1875, played a decisive role in the rationalisation of the several scientific and technical institutions of South Kensington in London into the Imperial College of Science and Technology in 1907. In using the 'decline' argument, British scientists and commentators repeatedly looked to Germany; however, Germany's output of science graduates, impressive and important though it was, must always be kept in proportion. Science graduates were always heavily outnumbered by those graduating in law, medicine, the humanities and (before 1911) in theology.

3. SCIENCE EDUCATION IN FRANCE

The revolution of 1789 destroyed the old universities and church-based schools and replaced them with an integrated, centralised education system which was, and remains, unique in Europe. Until the 1850s, the curriculum of French state schools (*lycées*) was severely classical and, like the German *Gymnasien*, it culminated in an intensely difficult examination, the *baccalaureate.* In 1852, under the influence of industrial demands for a more technically competent work-force, bifurcation was permitted whereby some students could choose to take a science-based *baccalaureate.* The same pressure led to the regional development of faculties of applied science in former university towns. Although these developments were extremely important for France's industrial development, Paris, and the specialised *grandes écoles* like the Ecole Polytechnique, which had replaced the universities as training academies for government servants, continued to be powerful magnets drawing off provincial scientific talent, as the careers of J.–B. Dumas, Claude Bernard and Louis Pasteur illustrate. Like Britain, French education throughout the nineteenth century was also complicated by intense religious debates over the content and purpose of education, debates which scientific modernists and secularists had decidedly won by 1914.

4. SCIENCE EDUCATION IN THE UNITED STATES

Unlike in Europe, a state examination system controlling entry into the professions and government service was never developed in the United States; consequently, higher education never came to confer status upon individuals in the way it always demarcated elite groups in Europe. Private fee-paying local academies, based upon English Anglican and nonconformist models, were the principal sources of education in eighteenth-century America; like their English counterparts, they provided an all-round curriculum which included mathematics and science for a commercial vocation.

Although academies continued to provide a secondary education in the nineteenth century (some of them developing into degree-conferring institutions), they were generally superseded by free public high schools which were maintained by boards of local ratepayers. This system, which originated in Massachusetts in the 1820s, acted as a stimulus to the establishment of the school board system of secular schools established in Britain in 1870 and their successors, the local authority schools, after 1902. As in English schools, American school science before the 1870s was predominantly 'catechistic' and didactic, and only the wealthier academies and school boards possessed demonstration apparatus. The introduction of laboratory instruction into schools, colleges and universities was promoted by a number of factors. These were: the great influence of Herbert Spencer's essays on education (which portrayed science as *the* crucial subject in the educational curriculum of an industrialised society); the effects of the Morrill Land Act of 1862 (whereby Congress donated federal land for the establishment of vocational training colleges of agricultural, mechanical and industrial arts); the publicity given to the curricula of German *Realschulen* and of Swedish and Russian industrial schools which exhibited at the Philadelphia World Fair in 1874 and the growing numbers of American scientists who, by 1880, had received a German university education. After 1890, as in Great Britain, few American public schools were built without chemistry and physics laboratories. The so-called 'Harvard' lists of pre-matriculation experiments in physics and chemistry which were recommended for use in American schools in 1886 closely resembled those issued by Edward Frankland and Frederick Guthrie for Department of Science and Art classes in Britain some years earlier.

5. SCIENCE EDUCATION IN ENGLAND AND WALES

For centuries, education in England and Wales was the responsibility of the Church of England, and it was not until the end of the Victorian period that the clerical monopoly of schools and universities ended and secular education was accepted as a legitimate alternative to 'godliness and good learning'.[6] It is for this reason that so much of English scientific activity took place within an essentially religious context. Until the early nineteenth century, university students were of what we should now regard as school age, and before the seventeenth century there is little point in distinguishing between school and university. Scientific instruction, as far as it went, consisted in the main of Aristotelianism imbedded in the curriculum of 'the seven liberal arts'.

Britain's pre-eminence in the first phase of industrialisation was founded upon a combination of capital, good natural mineral resources, the development of transport systems for both internal and external export trades and a native inventiveness which produced innovations which gave Britain an overwhelming lead over European nations. This industrial success had little to do with the gentry's traditional educational institutions, though its effectiveness undoubtedly owed much to many informal educational agencies such as the networks of eighteenth-century itinerant lecturers who spread intelligence, and to the 'modern' and commercially-minded curriculum of the noncomformist academies and the many proprietory schools which were established to provide education and training for the children of dissenters who were unwelcome at Oxford and Cambridge or at the Anglican endowed schools.

Both Oxford and Cambridge came in for a good deal of criticism in the early part of the nineteenth century for their religious exclusiveness, their privileges and the narrowness of their curricula. It was not until the 1870s, however, prompted both by internal reformers such as William Whewell and Charles Daubeny and by government interference, that college-teaching Fellows at these universities ceased to be clergymen before anything else and became, instead, dons in the modern sense, concerned with advancing as well as preserving knowledge. It is noticeable that in Europe and the United States, and to some extent in Scotland, the absence of a clerical monopoly led to an earlier emergence of a secular academic profession.

Nevertheless, mid-century reforms at Oxford and Cambridge did have important ramifications for science: in 1850 Oxford created Honours examinations schools in mathematics and science which ended the monopoly of classics, while in 1851, Cambridge complemented its already important mathematics degree course with one in the natural sciences. These new science degrees, despite their origins in 'liberal education' ideology, laid a solid foundation for the more laboratory-based and research-oriented science degrees of the 1880s.

Meanwhile, religious tests and the narrowness of the Oxbridge syllabus had led to the foundation of the secular University of London in 1826. Its eclectic degree included an astonishing array of sciences; moreover, its charter of 1836 recreated the University as an examining body which, in the next 50 years, allowed the affiliation of dozens of London, provincial and overseas institutions. One of the first to affiliate, and to offer London degrees, was Owens College, Manchester, which, once its science classes had begun to attract the interest of its industrial community, became, in turn, the model for other civic university colleges. The federal structure of the University of London also became a model for universities in Ireland, Canada, India, Australia and New Zealand, all of which followed its 1859 precedent of establishing separate bachelor degrees in arts and sciences.

The British national system of education was established barely a century ago, in 1870. Before then, well over a million working-class children between the ages of six and ten years received no education whatsoever, and it is estimated that a further half-million aged between ten and thirteen also never attended school. The 1870 Education Act was a belated recognition by the state that the voluntary elementary schools established by the Anglican National Society for Promoting the Education of the Poor in the Principles of the Established Church (founded in 1811) and the mainly nonconformist British and Foreign School Society (founded in 1808), whose sectarianism bedevilled nineteenth-century elementary education, were quite unable to cope. Nevertheless, it must be borne in mind that elementary education was not made compulsory until 1876 and then only up to the age of ten. Providing a child aged between ten and thirteen had passed the fourth standard (*i.e.* that expected of a ten year-old) in the '3R's, he or she was allowed to enter the labour market. Although the minimum age for school exemption was raised to twelve years in 1899 (with crucial exemptions in agricultural districts) it is clear that the chances of an English or Welsh working-class child adopting a scientific career in the nineteenth century were extremely low. This is not to deny the fact that a few British scientists emerged from the voluntary schools before 1870 and, more positively, from the non-sectarian Board schools after that date. Indeed, the principles of self-help and informal education ran deep in Victorian society. For the determined child and ill-educated adult, educational chances were offered by apprenticeship, Mechanics' Institutes and mutual improvement societies. In addition, the ideologies offered by phrenology, socialism and even spiritualism, the cheap books and periodicals made possible by the steam press and the repeal of 'taxes on knowledge' after 1855, and, above all, the teaching and examination system offered by the government's Department of Science and Art after 1853, undoubtedly offered routes into scientific careers. This is shown by the famous case of Oliver Lodge, and also by John Perry and Richard Wormell, who made valuable contributions to the cause of science teaching. Nevertheless, the point remains that, for most of the nineteenth century, there was no royal highway to a scientific career for working-class children, but only a few ladders. Indeed, it should be remembered that the vast majority of British children left school at the age of fourteen until 1947 and until then (and even since) have had to rely on 'night school' and 'day release' systems for any further educational advancement.

Great Britain was scarcely exceptional in this respect. Although both France and Germany had established elaborate state systems of education at the beginning of the nineteenth century (Italy not until 1878), in practice, recruitment into secondary and higher education and hence into significant scientific careers, was restricted, as in Britain, to the middle- and upper-classes. For these classes, entirely different systems of education operated, ranging from home education to the *lycées* of France, the *Gymnasien* of the German states and the endowed, grammar and proprietory schools of Britain, which commonly educated children up to the ages of sixteen or nineteen.

Not until 1902 did the British government provide, through local authority taxation, a system of secondary education for all, thus giving a university route for ordinary working-class children via a 'free place' system based upon scholarly prowess. The independent academically-orientated private schools became the models for these local authority schools by providing a ready-made tradition of standards of scholarship, internal organisation and an ideal of corporate ('team') spirit, as well as aspirations for university entrance and for the learned professions. Unfortunately, although this model made science teaching obligatory, the new state schools, by adopting it, discouraged the technical and vocational training which had begun to emerge in the higher classes of state elementary schools before 1900. In the absence of a sufficient number of state secondary schools, 'free places' were bought by local authorities in endowed grammar schools – a system which led later in the twentieth century to the controversial '11-plus' examination system. Fees were finally abolished by the 1944 Education Act which mandated local authorities to provide and maintain elementary and secondary education for all children aged between five and fifteen. (Compare the situation in the United States, where the last state to abolish fees was New Jersey in 1871.) Authorities were then encouraged to divide secondary pupils by attainment and vocational aptitude between grammar schools, which were academically orientated and well-equipped with science laboratories; technical schools for those less academically gifted whose craft skills were high; and secondary modern schools for the residuum. Another option of large, multilateral or comprehensive schools which catered for varieties of academic ability, was taken up by very few local authorities before the 1960s, but is now the nearly-universal state system in England and Wales, as in the United States.

6. SCIENCE EDUCATION IN SCOTLAND

For the historical reason that the Church of England was never established in Scotland, the education system there developed differently, and this may well have affected the production of scientific manpower from this part of the United Kingdom. Many historians of science have been interested in the study of styles of science in different countries, or in particular local studies of science and have analysed the ways in which scientific culture is related to the educational facilities and structures of the national and local community. Attention has focused particularly on Scottish science, its birth during the Enlightenment, and its maintenance, expansion and decline within specific institutions, especially the universities. The Scots, despite their union with the English kingdom in 1707, retained a very considerable educational independence. During the eighteenth century their five universities (and particularly those at Edinburgh and Glasgow) challenged and overtook Oxford and Cambridge in science and medicine. Because they allowed matriculation without imposing a religious test, the Scottish universities proved particularly attractive to English dissenters, most of whom had been educated in the nonconformist academies until the foundation of the University of London (by Scots-educated reformers) in 1826. The magnetism of the University of Edinburgh, in particular, was systematically strengthened by the Town Council which successfully established an influential medical school. This, together with the ancillary sciences like chemistry which it encouraged, was significant for the production of men of science from the 1720s onwards. These developments owed much to Edinburgh's decision, in 1708, to abolish 'regenting' (the system whereby arts students were tutored in all subjects by one academic), thus allowing the appointment of specialist professors. This decision, taken initially to raise the intellectual attainment of Scottish clergy, gave students the opportunity to specialise and, more significantly, stimulated dedicated teaching from professors whose incomes depended upon the sizes of classes they were able to attract. Like English university education, whose purpose was to 'enlarge' the mind through the study of classics and logic at Oxford, or mathematics and classics at Cambridge, Scottish universities also emphasised the mental, rather than the vocational, value of study. However, they enacted it through what has been described as 'democratic individualism', for these universities were open to privileged and under-privileged students alike; they taught a wide range of subjects at elementary level and knitted the miscellany together with doses of philosophy. (See art. 53) Consequently, scientists like William Thomson (later Lord Kelvin) and James Clerk Maxwell, who attended Scottish universities, as well as the University of Cambridge, went through a good general education in Scotland, including natural philosophy, which could be capped by specialised mathematical studies in England.

7. TECHNICAL EDUCATION

As we have seen, class attitudes have prevailed in British education; each form, or system, of education was felt to be appropriate to a particular class. Grammar schools had been endowed for the middle classes, some of whom might also enjoy a university education which was otherwise reserved for the wealthy. Technical education was a lower form of education appropriate only for artisans, not for gentlemen. Engineering was a craft; its pioneers were self-taught, and to their practical knowledge they had coupled a 'self-help' determination. Industry was essentially a collection of craft-based firms. This empirical approach had worked and from it had grown the strong tradition that engineering was something to be learned on the job.

The state itself was disinclined to intervene in educational matters and most educational philosophy (e.g. that of Herbert Spencer) began with the assumption that state intervention – as in France after the Revolution – led to state control. This *laissez-faire* attitude was complemented by that of most British manufacturers (though exceptions, like Bernhard Samuelson, can be cited) who had little use for, and no understanding of, the value of research or of scientific education. For them, a scientifically-trained workman seemed an unwise investment since the artisan would therefore understand the principle of the process and would possibly move to a competitor or even raise capital to begin a rival business.

The situation on the continent was different; there the image of the engineer as a highly-educated professional man had originated in the mining academies and military schools of France, Germany and Hungary in the eighteenth century and had been transposed to the Ecole Polytechnique in 1796. This, in turn, had set the pattern for later technological institutions (*Technische Hochschulen* or 'Polytechnics') founded in Germany and Switzerland.

The doctoral degree, which is now the *sine qua non* of scientific professionalism, has only been awarded for post-graduate research in British universities since 1918. Before then the higher, examined degree of D.Sc. was available to graduates of a particular university but rarely taken. Post-graduates from other universities (as at Cambridge University's Cavendish laboratory after 1895) had to make do with a demeaning 'research' B.A. or B.Sc. Therefore, British students in search of research experience tended to attend German universities, as did Americans, where their efforts were rewarded by the Ph.D. It should be noted, however, that before the 1870s the German doctorate awarded to foreigners was often of an honorary nature and frequently worthless as a measure of competence. Nevertheless, research degrees on the German model were strongly advocated by members of the British and American scientific communities who gave evidence to educational inquiries and who obtained positions of influence after the 1870s. In the event, it was the economic and political necessity of colonial cohesion made necessary by the First World War, the Foreign Office's concern to wean American students away from German contamination, and the availability of post-graduate research grants by the Department of Scientific and Industrial Research which prompted universities to introduce the doctoral degree.

Although the British government had patronised the sciences in many ways before 1914 – notably through the administration of an annual grant to several universities and colleges from 1889 – the creation of the DSIR was a landmark in the relationship between science and government in Britain and a model for the Commonwealth. Until its demise in 1964, the DSIR encouraged and financed research in universities and colleges, awarded stipends to post-graduates and established and maintained a host of useful research institutions. Until the second half of the nineteenth century it had never been considered a duty of European universities to produce trained men for industry. However, the expanding industries at the end of the nineteenth century – industries based on metallurgy, chemicals and convenience foods – were precisely those in which science graduates were needed, if only for quality control of raw materials and products, let alone for research and development. The fact that the social and economic processes, which compensated for the diminishing imperial advantage which Great Britain had enjoyed, were the very sorts of changes which 'declinists' and the emerging professional scientists vociferously demanded, has made the connection between economic success and investment in science education appear more casual than it really was.

The 'civic universities', technical colleges and polytechnics of Manchester, Sheffield, Leeds and Birmingham, were largely founded by industrialists who were alarmed and disturbed by the position of Britain's industries in the face of German and American competition. (There are comparisons to be made here with the regional Science Faculties of France and the German and Swiss *Hochschulen*.) That these science colleges were founded in provincial cities, rather than as extensions of Oxbridge, was due to the physical growth of Victorian cities, their concomitant civic pride and their coexistence with industry. They also represented a reaction against the Oxbridge commitment to non-vocational 'liberal' education, though in so far as the provincial colleges were forced to take London external degrees, even the most vocational of them, like Mason's College, Birmingham, admitted the Arts into their curricula. As it was, the coming of competitive examinations for the professions after 1860 meant that by 1900 even Oxford and Cambridge had come to accept financial aid from industry, and were cultivating industrial careers for their graduates through their newly-established Appointments Boards.

That business and industry were able and willing to recruit more graduates than ever before during the Edwardian era was due to the increasing scale of industrial enterprise, the shortage of managerial staff and the gradual introduction of research methods and laboratories in industries that were under competition from large-scale German and American firms. Even so, the number of graduates involved remained small compared with Germany, the majority being forced to choose schoolteaching as a career. However, the coming of the Ph.D. and the consequent gain in research experience, coupled with the appearance of many new science-based technologies thrown up or catalysed by the First World War (such as radio, aircraft and fibres, all of which needed technical expertise to maintain and develop them), led to the firm establishment of research and development laboratories in a number of industries. These became increasingly willing to employ Ph.D. or B.Sc. students from the universities.

In the 1920s and 1930s, consequently, industries were increasingly well disposed and philanthropic towards the universities and (in the 1950s) towards private schools. Thus, industrial money poured into Cambridge where (despite some difficulties over which piper, academic or industrial, should call the tune) research in physics, chemistry and biochemistry remained basic and unapplied in excellent laboratory facilities endowed by industry.

Inevitably, there were some voices raised against the increasingly closer symbiosis of university and industry. For example, despite his American experience, Abraham Flexner, in his *Universities, American, English and German* (1931), deplored the way specific technologies, like brewing and glass-making, had come to dominate civic universities such as Birmingham and Sheffield. Although both these departments made fundamental contributions to biochemistry and the physics of materials, for Flexner they were neither 'liberal nor university quality'. Socialist critics, like J. D. Bernal and Julian Huxley were similarly critical; albeit for the different reasons that industry's support was, in their eyes, leading to the neglect of the more socially-relevant biological sciences. But whether existing departments were admired or criticised, one point, already fairly clear, was put beyond doubt by the developments of the Second World War. Universities could specialise in their teaching of science, as they did in research, and still produce graduates who would be as acceptable to industry as they were to the civil service and the academic world.

NOTES

1. A. Briggs, 'The study of the history of education', *History of education*, 1 (1972), p. 5.
2. J. Bronowski, *Science and human values* (London, 1961); C. P. Snow, *The two cultures and the scientific revolution* (Cambridge, 1959), reissued as *The two cultures: a second look* (Cambridge, 1964, 1978); C. Barnett *The collapse of Britain's Power* (London, 1972) and 'Technology, education and industrial and economic success', *Journal of the Royal Society of Arts* (No. 527, 1979), 117–30; G. Roderick and M. Stephens, *Education and industry in the 19th century: the English disease* (London and New York, 1978); Idem. (eds.) *Where did we go wrong? Industrial performance, education and the economy in Victorian Britain* (Lewes and Philadelphia, 1981); M. J. Weiner, *English culture and the decline of the industrial spirit 1850–1980* (Cambridge, 1981). (Weiner's original title, 'English culture and the *containment* of industrialism', was sensationalised by his American publisher; see *New society*, 17 (1983), 274–5.)
3. H. W. Paul, 'The issue of decline in 19th century French science', *French historical studies*, 7 (1972), 416–40; P. Lundgreen, 'Education for the science-based industrial state? The case for nineteenth-century Germany', *History of education*, 13 (1984), 59–67.
4. See, however, R. F. Bud and G. K. Roberts, *Science versus practice. Chemistry in Victorian Britain* (Manchester, 1984). For Liebig's method, see J. B. Morrell, 'The chemist breeders: the research schools of Liebig and Thomas Thomson', *Ambix*, 19 (1972), 1–46.
5. M. W. Rossiter, *Women scientists in American: struggles and strategies to 1940* (Baltimore and

London, 1982); R. MacLeod and R. Moseley, 'Fathers and daughters: reflections on women, science and Victorian Cambridge', *History of education*, 8 (1979), 321–33.

6. C. M. Heward, 'Industry, cleanliness and godliness: sources for and problems in the history of scientific and technical education and the working classes, 1850–1910', *Studies in science education*, 7 (1980), 87–128.
7. G. E. Davie, *The democratic intellect: Scotland and her universities in the nineteenth century* (2nd edition, Edinburgh. 1964).

FURTHER READING

W. H. Brock, (ed.), *H. E. Armstrong and the teaching of science, 1880–1930* (Cambridge, 1973).

——'From Liebig to Nuffield. A bibliography of the history of science education, 1839–1974', *Studies in science education*, 2 (1975), 67–99.

D. S. L. Cardwell, *The organisation of science in England* (London, 1957; revised ed. London, 1972).

C. M. Heward, 'Industry, cleanliness and godliness: sources for and problems in the history of scientific and technical education and the working classes, 1850–1910', *Studies in science education*, 7 (1980), 87–128.

E. W. Jenkins, *From Armstrong to Nuffield. Studies in twentieth-century science education in England and Wales* (London, 1979).

E. W. Jenkins, 'Some sources for the history of science education in the twentieth century, with particular reference to secondary schools', *Studies in science education*, 7 (1980), 27–86.

D. Layton, *Interpreters of science. A history of the Association for Science Education* (London, 1984).

C. E. McClelland, *State, society and university in Germany, 1700–1914* (Cambridge, 1980).

H. W. Paul, *From knowledge to power. The rise of the science empire in France, 1860–1939* (Cambridge, 1986).

XI

'OBSERVE, EXPERIMENT AND CONCLUDE'. FINSBURY COLLEGE'S NEW COURSE OF EXPERIMENTAL PHILOSOPHY IN 1879-1880

In September 1925 the American *Journal of Chemical Education* commissioned an essay for its second volume from the 77-year old English elder statesman of organic chemistry and chemical education, Henry Edward Armstrong. Extravagantly entitled "The Epistle of Henry the Chemist to the Uesanians", Armstrong delivered a characteristically hyperbolic, paradoxical, aphoristic and infuriating sermon.

For his epigraph, or text, Armstrong gently parodied selections from the first and fifth chapters of St. Paul's Epistle to the Thessalonians.[1]

> Henry, unto the Church of Chemists on the great continent. Grace be unto you and the peace of understanding. For the gospel is come unto you not in word only and also in practice, but in much assurance, so that ye are ensamples to all.

In other words, since chemistry had reached adulthood in America, it was consequently in a position to influence other countries.

> For this cause, when we could no longer forbear we send to know your faith, lest by some means the tempter hath tempted you.

Who was St. Henry's chemical or scientific *tempter*? Armstrong's essay had, in fact, any number of devilish targets: education, the ethics of belief, the rottenness of the press, the poor literary values and jargon of most contemporary scientific writing, the over-emphasis on science as *useful* knowledge and the under-emphasis on science as culture, and Armstrong's particular *bêtes-noires*, the ionic theory and the enthronement of physical chemistry over organic chemistry. But there is a serious *leit-motif* in the essay: dogmatism in science is the negation of science.

The religious parody of the epigraph is, of course, partly characteristic of Armstrong's literary style, which was strongly influenced by reading Carlyle and Ruskin; but it was also suggested in this case by the fact that he was writing during the Tennessee Scopes evolution or "monkey" trial; at a time when the British journal *Nature*, in a special summer supplement, was pleading that the "intellectual freedom" meant (following St. Paul's mandate to the Thessalonians) "Prove all things and hold fast that which is good". Alas, he claimed, this was patently *not* what contemporary school courses, or chemistry departments in either the Old, or the New Worlds, did. Instead, like William Bryan, the anti-Darwinian lawyer, most science teachers were "mere ranters of unproved statements" who were failing totally to inculcate the scientific method in their students. It was precisely because of this failure to diffuse the spirit of science generally among human kind ("Science for All") that ignorant stump-orators like Bryan would "furiously rage together" against the scientific establishment.[3] Unlike the Bible, which could be read as literature, most twentieth-century science books were unreadable and filled with technical trivialities like the ionic hypothesis, which had never been proved. "In forcing the ionic doctrine upon students, chemists are as guilty of intolerance as Bryanites—they are seeking to impose an unproved faith."[4]

However, it is not Armstrong's rejection of the ionic theory which I want to examine, interesting though that is. Nor do I want to explore whether Armstrong was really justified in being so apparently disillusioned with the state of science education in Britain and America, as he made out in this Epistle and in numerous other essays during his long retirement from 1912 until his death in 1937. I use the Epistle only as a reminder of Armstrong's total commitment to self-education through laboratory research summarised in the injunction of my title—'observe, experiment and conclude'; a methodology which, from 1884, he and others called *heuristic.*[5]

On returning to England in 1870, from postgraduate training with Kolbe at Leipzig, Armstrong had been appointed to teach chemistry to medical students at St. Bartholomew's Hospital. He coached students here for London University First M.B. Examinations for twelve years, "with no little gain in experience" of students and the influence of examinations. At the end of 1870 he was also appointed to succeed Frankland's great rival in water analysis, Alfred Wanklyn, at the London Institution. His duties were fairly nominal—the delivery of evening lectures to artisans on "analytical chemistry and the methods of original investigation".

The students at the London Institution were very different in character from the medical students of St. Bartholomew's. In addition, he was not tied by examination syllabuses, but was free to devise methods of teaching that would develop the fundamentals of chemistry in a relevant way for the practical trades followed by members of his classes. Gradually he found ways of interesting his audiences by encouraging them to tackle problems experimentally in the confined and dusty Institution laboratory.

It was during Armstrong's period with the London Institution that Huxley published several essays on education, and began to criticise the wealthy City Guilds for their neglect of their historical role as technical educators. In 1878 a Committee of Livery Companies reported on technical education and recommended a practical plan for the creation of a central London institution which would serve as a model trade school. Eleven companies together financed this scheme which became officially known as the City and Guilds Institute. Anxious to make an immediate start, at the suggestion of Trueman Wood and Richard Wormell, the Institute created two lectureships in applied chemistry and physics. In October 1879 Armstrong was appointed to the former on the basis of his experience at the London Institution (whose proprietors were all City men), and he began teaching in temporary premises loaned by the Cowper Street Middle Class School.

Immediately he planned a new kind of chemistry course, beginning *not* like DSA syllabuses with the preparation and properties of metal and salts and what he angrily rejected as *test-tube* chemistry but with elementary principles and the observation of familiar phenomena such as combustion, proceeding to pure and applied inorganic and organic chemistry, and concluding with a thorough outline of industrial chemistry; the whole scheme was to be illustrated by experiments. After a long struggle he and his colleague, William Ayrton, persuaded the Council of the Institute to agree to have a purpose-built composite building (opened in February 1883 as the Finsbury Technical College[6]) in which all the first-year subjects (mathematics, engineering drawing and elementary workshop practice, chemistry and physics) were common to all students; specialization was deferred until the second year. There was an entrance examination, but no externally controlled examinations.

Since it was the Finsbury College experience of non-specialization and teaching by a practically-oriented curriculum designed to make students *think* which formed the basis of all Armstrong's future zeal and propaganda for heurism, it will be worth examining the Finsbury course in more detail. Here, I shall only be concerned with the two-year day (not evening) course begun in March 1880 with 100 pupils from the Cowper Street Middle Class School, a day school which transferred, in 1883, to the 56-place teaching laboratory on the 2nd floor of the new Technical College; moreover, I shall only be concerned with the *chemical* portions of the course, but it should be borne in mind that students also did substantial practical work in the physics and mechanics laboratories of W.E. Ayrton and John Perry, as well as in Perry's engineering drawing classes.

The course was developed jointly by Armstrong and a young Welsh Senior Demonstrator, John Castell-Evans, who was appointed in 1880. In 1892 the latter conveniently published the course substantially as *A New Course of Experimental Chemistry, including the Principles of Qualitative and Quantitative Analysis, being a systematic series of experiments and problems for the laboratory and classroom,* together with a separate *Key* for instructors.

As Castell-Evans, and Armstrong in his several reports to the British Association Committee on Chemistry Teaching both made plain, the guiding principle of the Finsbury course was:

> not to furnish the student with the knowledge of chemistry, but to help him to acquire that knowledge in the most thorough and scientific manner; to have as little teaching as possible but good honest learning on the part of the student, and systematic guiding and helping to learn on the part of the instructor; and it is therefore not a collection of facts, but of methods and means of discovering facts and of drawing conclusions from them. [Evans, p.2]

The way to learn chemistry and to train chemical students was not to list and accumulate facts, but to proceed "systematically from the known to the unknown", by asking questions. Experiments were merely "questions put to Nature"; but because Nature is very open-ended, the scientific tyro had to learn to close the system by asking definite questions. Further questioning might be needed in order to define the information gleaned still more precisely. Clearly, for Evans, chemical instruction and learning involved both tacit and explicit knowledge.

Thus, if we begin with the observation of the phenomenon of combustion and inquire into its cause and effects, the air will have to be rigidly confined, and so confined that we can see what happens, and a fuel chosen which can be ignited after the system is closed or confined. Hence the choice of phosphorus and a stoppered glass flask. Furthermore, however, in order to insure a non-ambiguous answer to the student's inquiry (i.e. no hidden variables), both flask and phosphorus have to be as clean and dry as possible.

The student received printed instructions to ignite the phosphorus and to describe what happened ("a great change...with the evolution of light and heat") and also what happened when the flask was opened with its mouth under water ("the air has diminished in volume" EITHER "by mere condensation or by the actual loss of some part"). The student then learned to quantify what had happened by fairly roughly calibrating the volume of the flask into 5 equal parts and repeating the experiment—finding, of course, that roughly a fifth of the air had disappeared and that the 4/5ths remaining would not burn more phosphorus.

Only at this point was the student told to call the 4/5ths portion "nitrogen", for it was a general principle of the Finsbury Scheme:

> not to employ any chemical *name* or *term* until [a student] has discovered by himself the *thing* or *process* represented by it. (Evans, p.3).

In subsequent experiments the student sought to discover whether phosphorus's reaction with air was unique and led to discover, by systematic testing, that other substances, such as iron and copper would also reduce air by 1/5th of its volume. By the eighth experiment the student was ready to isolate this 1/5th portion, to call it "oxygen", and to move on from air to water "and discover a new fuel" called "hydrogen". At this point the student was introduced to carbon, the distinction between compounds and elements, and the laws of chemical combination. But only by experiment 91 was he ready for equivalent weights, the notion of the atom, the determination of molecular and atomic weights and chemical formulae.

Once that stage was reached the course became more conventionally systematic, for the new pedagogic aid of the *natural* classification of the elements or periodic law[7] could be used to treat elements in families or groups; at the same time the practical work went on to the *artificial* classification of elements by the experimentally-determined groupings and separations of qualitative analysis. However, in the spirit of "prove all things", this was not mere *test-tubing* since the Group Tables had to be deduced from chemical reactions and not absorbed or copied from such well-known aids as Fresenius's, *Introduction to Qualitative Chemical Analysis.*

All this practical work was accompanied by parallel lectures on what might be called the "physics of chemistry"—weights and measures, calorimetry, gas laws, calculations, etc., as well as regular tutorial problem classes.

Here, then, was an innovatory course of chemical instruction emphasising the practical and the quantitative; here was the course upon which Armstrong drew for his remarks to the International Conference on Education "On the Teaching of Natural Science as Part of the Ordinary School Course" in 1884 which launched the heuristic movement that so happily triggered off the school laboratory movement.[8]

Armstrong left Finsbury College in 1884, but the method and curriculum for First Year teaching, which he and Castell-Evans had established, was continued by his successor Raphael Meldola. But the College itself fell on increasingly hard times after the 1st World War in competition with the new municipal technical colleges for whom it had been the model. Indeed, within a year of Armstrong's pessimistic Epistle to the Uesanians in 1925, Finsbury Technical College closed its doors. It is now almost totally forgotten as the Institution which pioneered a whole host of innovations in science and mathematics teaching,[9] not the least of which was the heuristic method.

REFERENCES

1 *Journal of Chemical Education*, 2 (1925), 731-6.

2 *Nature*, July 1925.

3 Armstrong's point echoes that of Michael Faraday when he justified the more widespread teaching of science by citing the lack of judgement of people duped by table-turning. See my "Prologue to Heurism" in History of Education Society, *The Changing Curriculum*, London, 1971, pp.71-85.

4 *Journal of Chemical Education*, 2 (1925), 736.

5 See W.H. Brock (Ed.), *H.E. Armstrong and the Teaching of Science 1880-1930*, Cambridge, 1973.

6 W.H. Brock, "An Experiment in Technical Education", *New Scientist*, vol. 84, 22 November 1979, pp.622-3.

7 The periodic law, which was first formulated by Mendeleev in 1869, was adopted enthusiastically by Armstrong in his articles on "Chemistry" for the *Encyclopaedia Britannica*, 9th ed., 1876, vol. 5, 467-544.

8 *International Conference on Education Proceedings*, London, 1884, vol. 2., pp.69-82; partly reproduced in Brock, *op. cit.* (note 6).

9 W.H. Brock and M.H. Price, "Squared Paper in the 19th Century. Instrument of Science and Engineering, and Symbol of Reform in Mathematical Education", *Educational Studies in Mathematics*, 11 (1980), 365-81.

XII

Geometry and the Universities: Euclid and his Modern Rivals 1860-1901

> We did not find him out,
> Which I bitter grieve,
> For I've not the least doubt
> That he'd placed up his sleeve
> Mr Todhunter's excellent Euclid,
> The same with intent to deceive.[1]

One of the hallmarks of the organization of science during the nineteenth century was the appearance and formation of scientific pressure groups—lobbies of men with roughly similar ideas on education, all well acquainted with one another, whose voices and opinions were listened to and studied not only by their fellow scientists but also by a much wider public. During the 1860s, as scientists reached their intellectual ascendancy in Great Britain, they began to argue that a classical education in the older endowed schools, public schools and ancient universities, did not, and could never, provide a complete training for the Mind. These external attacks on the concept of a Liberal Education, as it was understood in mid-Victorian Britain, were increasingly coupled with sympathetic internal attempts to inject science into schools and universities. Although the final work on this battle has not yet been written, the story of the strategy and campaign and its qualified success is fairly familiar. Less familiar, however, are the attempts of mathematicians and mathematics teachers between 1870 and 1901 to reform the study of geometry in British schools and, perforce, to reform the curriculum of mathematical studies as a whole. How and why was the organization known as the Association for the Improvement of Geometric Teaching (hereinafter AIGT) set up? How did it set about achieving its aims? What were its successes and failures in its relationship with university examination bodies?

These are some of the questions that will be examined in this essay. It will be shown that the newly-written geometrical textbooks of the 1870s (the so-called 'rivals of Euclid'),[2] and the tactics of the AIGT had little effect on the universities, who, as with science teaching, continued to call the tune of mathematics syllabuses into the twentieth century. It will appear that three conditions were necessary before the reformers of Euclid had effect on the universities. Firstly, the focus of Euclid's rivals and the tactics of the AIGT had to be adjusted and their field of view widened to embrace the reform of the whole of elementary mathematics teaching, not just geometry. Secondly, the rather élitist and formal arguments and interests of the AIGT had to be modified and reinforced by another lobby of teachers who were less concerned with formal training of the mind than with a vocationally-based mathematics helpful to engineers. Thirdly, the appearance of other logical geometries—synthetic, projective and non-Euclidean—helped to undermine the authority of Euclid as an educationally unique mind-trainer. Finally, as is always the case, the success or failure of a reform movement is very much dependent on the characters of its leading personalities.

Although the exact implications of the words might differ considerably from one educational authority to the next, there was general agreement in the nineteenth century that an education only qualified for the title 'liberal' if it developed aspects of the mind and the character. In this respect the classical curriculum was thought to be nearly perfect, and before the end of the nineteenth century little mathematics was taught in the endowed schools of Great Britain. To be sure, the University of Cambridge had begun to emphasize mathematical teaching in the eighteenth century, and this century saw the establishment of a plethora of utilitarian mathematical schools and academies.[3] But the gradual extension of mathematics teaching into the education of 'gentlemen' in the public schools in the first half of the nineteenth century was undoubtedly a most significant change in the concept of 'liberal education'.[4]

Here, the major influence was William Whewell's advocacy of mathematics from the vantage point, not of a school, but of the University of Cambridge. In 1837, as a result of his reflections on the history and philosophy of the 'inductive' sciences, he distinguished between permanent subjects (those with a static curriculum content) and progressive subjects (those where the content changed with time). The former group, which consisted of classics and mathematics (meaning, as we shall see, Euclidean geometry)—provided a suitable basis for a liberal education; whereas the latter (which included the sciences) did not.[5]

The extent of the introduction of mathematics into the curricula of public schools must not be over-emphasized. The evidence gathered by the Clarendon Commission set up to investigate the state of such schools showed that mathematics had not begun to be compulsory at Eton until 1837, and that even in 1864, only three hours a week was devoted to mathematics at Eton, Harrow, Rugby and Shrewsbury.[6]

Similarly, the Taunton Commissioners on Endowed Schools, reporting during the same decade, found little evidence of mathematics teaching in such schools beyond arithmetic and the first four books of Euclid. Mathematics at Eton (which began in 1836 when Stephen Hawtrey, a cousin of the then headmaster, was appointed to teach the subject) was taken on a purely voluntary basis until the 1850s, and Hawtrey had to provide both the classrooms and assistant teachers out of his own pocket. Clearly, although defenders of the existing public schools' curriculum in the 1860s might speak of classics *and* mathematics where earlier they would have referred to classics alone, mathematics retained a markedly inferior position. This can be seen both in the number of hours per week allotted to it—usually three hours in school and about the same time in preparation during the 1860s (thus perforce restricting the scope of the syllabus), and in the small weight allotted to mathematics in assessing a boy's ability. It was a common grouse of mathematics teachers (as it was to be later of the science teachers who formed the Association of Public School Science Masters in 1901) that the form, or set, to which a boy belonged was determined almost entirely by how he performed in classics.[7] As a result, the mathematical ability within a group might vary considerably. It is hardly surprising that when Bartholomew, the Sedlerian Professor of Natural Philosophy at Oxford, was asked his opinion of the mathematical attainments of public school entrants to Oxford, he replied: 'I do observe a very marked difference between young men coming to this university from the great public schools and from other schools or from private tutors, as to their mathematical attainments. The young men from public schools are far worse prepared.'[8]

However, the imposition of matriculation requirements by universities in the 1860s, together with the creation of competitive entrance examintions for the Army and Indian Civil Service in which mathematics and classics counted for the majority of marks, ensured that mathematics in the guise of Euclidean geometry had found a firm place in the curriculum by the 1870s.[9] This is not to deny that the historian of education can find examples of teaching which included some geometry other than straight-Euclid (trigonometry, or solid, practical, and co-ordinate geometry). However, such activities were not part of liberal education as ordinarily conceived; nor was such instruction demanded by the universities for matriculation.

The limited time-table advantages apart, why did Euclidean geometry form the most significant element in British mathematical teaching? On the one hand, it had the advantage of being a classical text—one that, in principle, could be studied in Greek. It has been said that the importance of a great book is determined 'by the number of previous publications it makes superfluous to read'.[10] Euclid had assembled the geometrical ideas and theorems of his predecessors and arranged them in a logical order derived from simple axioms. His work and achievement as represented by the *Elements* (if not the man himself) were contemporaneous with all that was thought best and 'liberal' in Greek culture—the humanizing world of Plato and Aristotle.[11] On the other hand, far outweighing the humanistic propensities of Euclid was its *formal* strength—the fact that it could be used in inculcate reasoning power and logic. As Whewell argued in *Of a Liberal Education* (1845) it was precisely this aspect that made geometry a *necessary* part of liberal education:

> There is no study by which the Reason can be so exactly and rigorously exercised. In learning Geometry . . . the student is rendered familiar with the most perfect examples of strict inference; he is compelled habitually to fix his attention on those conditions on which the cogency of the demonstration depends . . . He is accustomed to a chain of deduction in which each link hangs from the preceding, yet without any insecurity in the whole; to an ascent, beginning from the solid ground, in which each step, as soon as it is made, is a foundation for the further ascent, no less solid than the first self-evident truths. Hence he learns continuity of attention, coherency of thought, and confidence in the power of human Reason to arrive at the Truth.[12]

A book of such power was not to be dislodged easily; indeed, could not be dislodged in the absence of anything better. Yet, might not a modern mathematics book have been preferred over an ancient Greek one? Whewell had no truck with this possibility:

> Perhaps it may sometimes appear, both to teachers and students, that it is a waste of time and perverseness of judgement to adhere to the ancient kind of mathematics, when we have, in the modern analysis [*sc.* calculus], an instrument of greater power and range for the solutions of problems; giving us the old results by more compedious methods an instrument, too, in itself admirable for its beauty and generality. But to this we reply 'We require our present Mathematical studies [*sc.* of Euclid, and at University, of Newton's *Principia*] not as an instrument (for the solution of today's mathematical problems) but as an exercise of the intellectual powers; that is, not for their results, but for the intellectual habits which they generate that such studies are pursued.[13]

Euclid, then, was an instrument for training the reasoning faculty. It taught a gentleman logic, how to think, rather than geometrical techniques *per se*. Did this ideal work in practice? From the large number and tremendously varied editions of Euclid that were produced during the nineteenth century we may judge that teachers experienced some difficulties. Of course, many of these editions are explained by the development of school textbook publishing in the nineteenth century; but some, without examining the root cause of the problem, were undoubtedly attempts to make Euclid more lucid and palatable for pupils.[14]

As early as 1849 Augustus De Morgan wrote critically of Euclid's style,[15] though he was quick to defend the continued use of Euclid in schools in the 1860s when the criticisms began to mount. In 1866, two years after the Clarendon Commissioners had published their final reports, and while the French were wondering whether their scientific culture was declining, two Frenchmen, Jacques Demogeot and Henry Montucci were sent to study Anglo-Saxon secondary education. In their Report (1868) they indicted the British for divorcing education from instruction, and noted disapprovingly that the only distinctive thing about British mathematical teaching was that it called on the memory of pupils and not their intelligence. We can see this in the vivid picture of geometry teaching at a training college of the 1860s painted by Richard Wormell:

> The tutor sat in the chair and was usually silent while the student went through the proposition, but if there was the slightest deviation of phrase or almost if there was an effort to vary the proposition by putting an emphasis on a particular word it was checked at once, and the student bade to begin again.[16]

Euclid's logic was no doubt iron-like, Demogeot and Montucci supposed, but they felt certain that the same results could be, and were being currently obtained in France in more elegant and simple fashion by the use of Legendre's geometry, rather than Euclid's. It was far better to allow teachers to develop their own method of presenting geometrical theorems to students than to instil them by rote following the Euclidean order without understanding.[17]

Although Todhunter dismissed Demogeot and Montucci's report as worthless, mainly on the ground that both were literary men who knew no mathematics,[18] several English teachers had already reached the same conclusion as the Frenchmen and announced their disquiet to the Taunton Schools Inquiry Commissioners. For example, although the Commissioners found that Colenso's *Arithmetic* and Euclid's *Geometry* were more or less the only textbooks used in English endowed grammar schools,[19] they were told by George Griffith, the Deputy Professor of Natural Philosophy at Oxford and Secretary of the British Association, that Euclid occupied too great a proportion of school mathematics lessons, and that such studies did not lead to an understanding of geometry.[20] Griffith urged the adoption of a more practical geometry; nevertheless, he did not wish to see Euclid abandoned altogether Yet this was the hope of the Headmaster of University College School, Thomas Hewitt Key (1799–1875), who wanted to teach 'plain geometry, not by Euclid. I think the book is a thorough mistake. It is . . . a most illogical book.'[21] Now Key, who was also Professor of Comparative Grammar at University College, had not taught mathematics since the period 1825–27 when he had been Professor of Pure Mathematics at the University of Virginia. From where, then, came his dislike of Euclid? The probable answer is from Thomas Archer Hirst (1830–92), who, before his elevation to the chair of Mathematical Physics at University College in 1865, had been Key's chief mathematical master at the College School.[22] Between 1860 and 1865 Hirst had widened the school's mathematics syllabus to include plane and spherical trigonometry, conic, sections, analytical geometry, the theory of equations, and determinants.[23]

Hirst had begun his career as a surveyor in Halifax, before learning geometry in Germany from teachers like Stegmann and Steiner, and from Chasles in France. He was, therefore, equipped not merely with a more practical outlook on geometry, but he also possessed a practised familiarity with Continental methods of higher geometry—especially projective geometry and the theories of correlation and transformation. Moreover, as a teacher, despite his forbidding physical appearance,[24] he had the patience of Socrates and an impressive heuristic technique—as the following quotation from his *Journal* will indicate. The story concerns a stuttering, shy boy named Pilkington of about twelve or thirteen.

> I broke the silence at length, and putting my arm around his neck quietly asked him, 'Can you tell me what an angle is, my boy?' 'Ye-e-es, sir, an angle's er-er-er.' He could not get any further, and turned his eyes up to me beseechingly, as much as to say, 'Don't be cross, sir, I'm sure I know it, but I cannot tell you.' 'Well,' I replied to his silent answer, 'Go and pull two long grass stalks and come and show me.' This he did, and with the grass stalks he continued to answer my questions about the geometrical definitions . . .[25]

This episode had taken place at the extraordinary Quaker School, Queenwood College, Hampshire. One of the other mathematics teachers there was Richard Wright (d. 1892?) who, in 1868 (when the Taunton Commissioners wondered 'whether Euclid be the proper textbook for beginner and whether boys should not commence with something easier and less abstract'[26]) published an original *Elements of Plane Geometry*. This book was, in fact, closely based on Hirst's school lessons at Queenwood and University College School, together with material drawn from various French textbooks. In his Preface to Wright's book, Hirst noted challengingly:

> The superiority [of Wright's use of intuitive proofs,[27] rearrangement of Euclid's theorems and use of superposition] for educational purposes will be contested by no one who has observed either the permanently pernicious effects of the discouragement produced by initial vagueness, tediousness, and difficulty, or the permanently beneficial influence of the encouragement arising from early successes, and from the fulfilment of the pupil's natural expectation that every intellectual effort will be followed by a conscious acquisition of knowledge. To secure and sustain the pupil's interest from the first is also a point of unquestionable importance; and although personal qualification in the teacher are here indispensable, this end is undoubtedly promoted by a manual wherein difficulties are judiciously tempered to the pupil's capacities, and the subtleties of the subject to his powers of appreciating them.[28]

Again in this statement we catch an echo of the heuristic, or Pestalozzian approach used at Queenwood College. Euclid had never been designed for school use, or any form of elementary teaching; but the reformers' approach was to be more child-centred and age-graded—the geometry had to fit the child and not vice-versa.

Key's statements to the Taunton Commissioners fell upon at least one sympathetic ear—that of Frederick Temple, the Headmaster of Rugby on whose staff was James Maurice Wilson (1836–1931), a former wrangler who had little respect for Euclid:

> It did not bring out originality of thought, and it was plainly disliked by the boys and the amount of knowledge obtained was very small.[29]

Supported by Temple, and Demogeot and Montucci's visits to Rugby, Wilson began to examine Continental and American geometrical textbooks, and to draw up a revolutionary syllabus which Macmillan published as *Elementary Geometry* in 1868. In his Preface, Wilson significantly acknowledged help from Hirst (whom he had met through their joint activities for the British Association—Hirst as its General Secretary, and Wilson through his membership of the important Committee on Scientific Instruction [1867]). He used the Preface, however, primarily to launch a devastating attack on the use of Euclid in schools. His indictment, which was repeated in an address to the London Mathematical Society in 1868,[31] was fiercely attacked by De Morgan in the *Athenaeum*,[32] and continued to evoke astringent defensive reactions from the Cambridge mathematician Isaac Todhunter in 1870.[33] Although Wilson's views were not unique, it seems that he good naturedly bore the brunt of the Establishment's criticisms in the 1870's.

Wilson had four grounds for dissatisfaction. Firstly, Euclid's whole approach was both geometrically and logically *artificial*. Euclid's aim was not to do creative geometry but merely to show 'on how few axioms and postulates the whole could be made to depend'.[34] Euclid obscured the important logical distinction between the artificially impossible and the necessarily impossible. (For example, the trisection of an angle was only impossible within the terms of Euclid's constructions.) The force of such a logical criticism was probably helped in the last decades of the nineteenth century by a growing awareness among school teachers of higher, and non-Euclidean, geometries. Secondly, the invariable syllogistic character of the demonstrations made the book obscure and unsuitable for beginners. Intuitive proofs were not permitted and the artificial restrictions stultified creativity, and rendered the subject of geometry stiff, obscure and barren.[35] Thirdly, the prolix length of many Euclidean demonstrations rendered geometry an exercise of memory instead of intelligence, and (echoing Demogeot and Montucci's criticism) through its tediousness made English education inferior to that on the Continent.[36] Finally, Euclid was unsuggestive, insofar as all his theorems appeared to possess equal weight, and those which might be strategically significant or fruitfully used in other geometrical contexts, were not emphasized.[37]

On the other hand, Wilson recognized that there was a case for using Euclid which, however, he felt could be destroyed. One argument was a fear, later expressed most loquaciously by Dodgson[38] that if teachers abandoned Euclid, chaos would result from the advantage of uniformity being lost. Scores of rival textbooks would appear and any fair examination of a pupil's knowledge would become difficult, thus undermining the conforming social function of examinations. But, Wilson claimed, in practice the rival textbooks would not differ widely from one another, or ever as widely as, say, chemistry books did without prejudicing the uniformity of chemistry examinations. Furthermore, French experience suggested that there were advantages in abandoning Euclid insofar as the examiner could then truly test understanding rather than memory.[39]

As we have seen, the more serious argument in favour of Euclid, as Wilson recognized, was formal; Euclid did not teach geometry but orderly thinking. It was an educational masterpiece because it concentrated on manipulating *things* deductively (and not merely symbols of things, as in algebra).[40] Such mind-training would prove useful in later life and in science. To this powerful argument Wilson replied that geometry could still be presented in simpler ways without any loss of rigour. To be fair to De Morgan, Todhunter and Dodgson, they all accepted this possibility; however, they agreed unanimously that none of Euclid's rivals was both similar and as equally rigorous as Euclid. It was also Wilson's experience that there did not seem in practice to be a transfer of learning (a major premise in the Victorian argument for science teaching), for boys who were adept at memorizing Euclid seemed unable to apply it to original problems. To this, Todhunter could only imply that Wilson must be a bad teacher[41]—a possibility that his evidence to the Clarendon and Devonshire Commissions does not make credible.

Wilson's *Geometry* was a success—by 1878 it had reached a fourth edition.[42] Moreover, its first appearance in 1868, together with Wright's text, had two organizational consequences. In September 1871, the British Association (which by then had become alive to its educational responsibilities) established a Committee to consider the 'possibility of improving the methods of instruction in Elementary geometry' and to examine Euclid's rivals and recommend which deviations from Euclid were or were not desirable.[43] There is no doubt that the inspiration for the establishment of the Committee came from the formal establishment of the AIGT eight months previously.

In March 1870, Robert Tucker, Secretary of the London Mathematical Society[44] and Hirst's successor at University College School, complained to *Nature*[45] that the working of the syllabus for London University Matriculation insisted on the use of the first four books of Euclid. This, he said, made it difficult to use Wright's textbook, which he clearly preferred. In May 1870, his plea for reform of the rubric was supported from King Edward's School, Birmingham by Rawdon Levett, who suggested the formation of an 'Anti-Euclid Association' which would exert pressure on examining bodies to frame questions independently of particular textbooks. Tucker was enthusiastic and believed that the members of such an association could write a co-operative anti-Euclid textbook.[46] Finally, in June 1870, Tucker, Levett and Richard Wormell (of Cowper Street, Middle Class School and the author of another Euclidean rival[47]) urged interested readers of *Nature* to send their names to Levett, who *ipso facto* thereby became the Association's first secretary. He, together with the second master of King Edward's School, the Reverend E. F. M. McCarthy, Tucker and Wilson then circulated the mathematics teachers of Britain calling them to a meeting at University College, London, on 17 January 1871, with Hirst in the Chair.[48]

At this meeting Hirst suggested that they should constitute themselves an Association for the *Improvement,* not Reform, of Geometrical Teaching, on the grounds that 'reform' implied only short-term aims, whereas 'improvements' would always be needed. It was Hirst, too, as the Association's first President, who urged that they should not rush to follow Wilson's suggestion that the Association should compile a new geometrical syllabus, despite some support from the membership.[49] Good teaching was more important than any good textbook; reforms, to be effective would have to be gradual; the Association should not be precipitate. As we shall see, Hirst's tactics were too idealistic, and the AIGT's 'leisurely' approach to reform probably lost it prestige and diminished its energies.

By January 1873 there were a hundred members of the AIGT, most of whom (as a poorly answered questionnaire suggests[50]) were probably more interested in improving formal or scientific geometry rather than in the introduction of vocational, practical geometry for application in, say, surveying or engineering. It would not be unfair, then, to describe the first stage of the AIGT as élitist, both because it was concerned to outdo Euclid, or bring his formal definitions up to date, and because its members were principally drawn from the ancient universities and older endowed schools. Despite Hirst's background, it had little to do with the more practically-oriented geometry examined by the Department of Science and Art.[51] A new syllabus, covering the ground of Euclid Books I–III, together with an introductory section concerning definitions and the principles of logical deduction, was compiled by Hirst, Wilson, R. B. Hayward of Harrow School and J. F. Moulton, and presented to the Association in January 1873.[52] The extension of the syllabus to Euclid Books IV and V took a further two years. Finally, in 1875, the complete syllabus was felt to be ready for printing by Macmillan and to send to the British Association Committee for their approval, the plan being to use the accolade of the British Association's approval as an advertisement in the battle against the examination authorities.[53] Unfortunately, although the intentions of the AIGT had received warm praise from the mathematician H. J. S. Smith at the British Association in 1873, the BA Committee (despite a considerable overlapping of membership[54]) did not report on this syllabus until 1876, by which time nearly 1000 copies of it had been sold. No collective approach to the examination authorities was therefore possible until the Spring of 1877. It thus took six years for the AIGT to begin to challenge the rubrics that had led to their constitution as a pressure group! To be fair, there had been difficulties. 'On the one hand', Lovett and MacCarthy reported in 1874, 'it was necessary that the work should be self-consistent, and should be so thoroughly done that it would bear the inspection of even the hostile critics without; on the other hand, . . . the opinions of the members of the Association on some important principles, as well as matter of detail, was far from unanimous.'[55] However, granted that there were important and sincerely debated scholarly issues at stake over such matters as the treatment of proportion, the historian is nevertheless left with an impression that the organization of the AIGT was not well adapted for rapid committee work. One is also forcibly reminded of the significance of the typewriter, for each section of the syllabus, or a member's suggested amendment, had to be cumbersomely set up in type for circulation and criticism.

In the Spring of 1877 the AIGT Council circulated its syllabus, the BA 'testimonial', together with its own memorandum that the syllabus replace Euclid in examination rubrics, to examination bodies at the Universities of Oxford, Cambridge, London, Durham, Aberdeen, St Andrews, Glasgow, Edinburgh, Dublin, Queen's, the Civil Service Commissioners, the India Office, the Education Department, the Department of Science and Art, the College of Preceptors, the Incorporated Law Society, the Pharmaceutical Society, the Royal College of Surgeons, and the Royal Veterinary College. The response from this impressive list of Euclid examiners was dismal; only Cambridge, London and Durham Universities, together with the Government Departments, seem to have bothered to reply.[56] Cambridge simply shunted the memorandum from Board to Board, though at the end of 1877 the Cambridge Local Examiners conceded that non-Euclidean proofs could be used (one suspects they were meeting them more frequently), *providing* they were based on Euclidean axioms and followed Euclid's logical order of proof.[57] This was an impracticable concession, for the new syllabus had deliberately altered the Euclidean order in many places. London, more agreeably pointed out that since 1876 it had allowed free choice of textbook,[58] but Durham reported significantly that it could do nothing until it saw a textbook based upon the syllabus, while the Government Departments equally significantly stated that they would follow, not lead, the universities in the matter. At Oxford, mathematicians (including Dodgson) memorialized the University in favour of the retention of Euclid against the uncertainty of his rivals—the theme of Dodgson's *Euclid and his Modern Rivals,* which ridiculed the AIGT syllabus.[59] In retrospect, the whole episode was a miserable and costly mistake. The AIGT should have concentrated its energies from the beginning on a textbook and on the examiners of Oxford and Cambridge in the sure knowledge that if they changed their ruling, other examiners would follow.

Hirst, who was appointed Director of Studies at the Royal Naval College at Greenwich in 1873 (where he introduced instruction in higher geometry) resigned the Presidency of the AIGT in 1878. The second phase of the AIGT saw a change of tactics under the Presidency of R. B. Hayward,[60] a mathematics teacher at Harrow School. In 1880, Levett commented:

> I believe . . . that as the Examining Boards have practically refused to recognize our work on the ground that we have not issued a Textbook, it is necessary to take such a step as that proposed [to prepare proofs of the propositions in the *Syllabus* of 1875], there being great danger that those who already use the Syllabus for teaching purposes may find it impossible to continue to do so in face of the tacit disapproval of examiners.[61]

In fact, Wilson had already revised his *Geometry* and based a new edition on the AIGT syllabus,[62] but it was not adopted officially by the AIGT, presumably because they, or Wilson, felt it would be unethical to do so. Indeed, the AIGT was altogether too complacent and liberal over Euclid's rivals, neither recommending nor condemning them. Instead a new committee was established in 1881 to prepare proofs to add to the bare bones of the syllabus. Three years' work was needed.[63] But publication of the syllabus (which had sold 2033 copies by 1881), and now a textbook, cost money; yet in 1881 the membership was still only 113,[64] and the balance at the bank only £3–19–6d.[65] Therefore, under Hayward's leadership the Association began a recruiting drive and to widen its interest in order to accommodate the different concerns of Membership.

1871	1881	1882	1883	1884	1885	1886	1887
61	113	108	128	138	136	153	178

Partly to overcome the universities' indifference, partly to attract more members, the AIGT reconstituted itself in 1882, retaining its former title for historical reasons, but now using the term *Geometrical* in the French sense of *Mathematical.* Thus, although it was not to call itself the 'Mathematical Association' until 1897, it was this in everything but name fifteen years earlier. The revitalized Association began to publish *The Mathematical Gazette* in April 1894.

The first part of the AIGT's new geometrical textbook, which was drawn up jointly by Hayward, John Fletcher Moulton and Levett, was published in October 1883, but the second part was not ready until 1886, making a gap of eleven years between the appearance of the Association's syllabus and its complete textbook. In the light of their experience a decade previously the Association decided to concentrate their forces of persuasion on the Universities of Oxford and Cambridge alone. Indeed, the favourable adjustment of the Cambridge Statutes in 1877 and 1882 had persuaded the Association to establish a local pressure group within Cambridge.[66] This 'Cambridge Committee', which was established in 1885, consisted of about a dozen AIGT members who were resident in Cambridge, three of whom were on the Board of Mathematical Studies.

Once again, however, the immediate results were disappointing. In the Spring of 1887, the AIGT sent its textbook to the Oxford and Cambridge Schools Examination Board, and to both the Oxford and Cambridge Syndicate for Local Examination. In all three cases the book was accompanied by a long memorial signed by 168 mathematicians, 86 of whom were not members of the AIGT.[67] The Oxford and Cambridge Joint Board, and the Oxford Syndicate did agree that proofs other than Euclid could be used providing the Euclidean order was not violated. But this was less a concession than a move to consistency with the Cambridge Local Examinations.

To be fair, rather more was conceded by the Cambridge Board of Mathematical Studies, namely, that 'rigid adherence to Euclid's text is prejudicial to the interests of education'. With this admission, which seems to have been opposed seriously only by Arthur Cayley (who was then Sadlerian Professor of Mathematics[68]), came the implication that much depended on the spirit with which the regulations were interpreted by examiners.[69] Nevertheless, despite these admissions from the two ancient universities, formal relaxation of the Euclidean order of proof was not to be given in the nineteenth century.

It seems, therefore, that the achievement of the primary aim of the AIGT—its ultimate purpose during the first two decades of its existence—was a signal failure. By 1890 only the Universities of London and Edinburgh permitted the use of any geometrical textbook and *any* theorem order in their matriculation requirements; and only Edinburgh was apparently persuaded to change by the AIGT.[70] Their mistake in not compiling a textbook at the start of their existence was now compounded by the failure to insist that the AIGT text, and that alone, could effectively replace Euclid. An adamantine stand on this principle might have overcome the universities' fears of examination chaos if Euclid were replaced. As it was, the AIGT was too fair-minded in its free-market attitude towards competing textbooks.

What, then, did the AIGT achieve, beyond permitting students to prove a theorem in any manner providing it was deductive and based on previous theorems which, in Euclid's *Geometry,* also preceded the one to be demonstrated? Two general answers can be suggested. In the first place, their protracted campaign of persuasion does seem to have produced a shift of focus among mathematicians and teachers away from the view of liberal educationalists that the teaching of geometry had formal, logic-inducing qualities. By the 1890s we find instead the view developing that geometry was of interest in its own right as one of many branches of mathematics, while the work of Hirst and Cayley showed British teachers that Euclidean geometry was itself only one aspect of modern geometry. This altered viewpoint is well illustrated by Hirst's statement in his Presidential Address to the AIGT in 1878:[71]

> Elementary Geometry is no longer regarded as a long-since perfected branch of knowledge; it is no longer classed with the seven orders of architecture . . . that cannot be touched without being spoiled. On the contrary . . . the elements of Geometry, so far as principles and methods of exposition are concerned, constitute not a dead but a living science, susceptible still of being improved, and still capable of furnishing new matter of thought to both Teacher and Student.[72]

Secondly, this same shift of focus enabled the AIGT to see that other branches of elementary mathematics required improvement in their methods of instruction and presentation. This realization became explicit in 1897 when the AIGT became the Mathematical Association. If the AIGT's first phase was formal, instrumentalist, and even élitist, by 1880 its failures encouraged it to join forces with those mathematics teachers whose aims were more vocational—namely teachers who received payments from the Department of Science and Art, and especially those teachers who were to fulfil their roles within the burgeoning technical college movement in the 1880s and 1890s. Such teachers formed what has been called 'the Perry Movement'—named after John Perry (1850–1920), the mathematician who taught at Finsbury Technical College (where the AIGT syllabus was followed), and who eventually became Professor of Engineering at the Central Institution of the City and Guilds of London. Many of these teachers were caught up in the exciting development of electrical engineering—a field full of vocational possibilities for young men and women,[73] but one in which a moderate competence in more than elementary arithmetic or geometry was required.

At the British Association meeting in Glasgow in 1901, in a joint symposium between its new Educational Section L, and the Mathematics–Physics Section B, Perry delivered an entertaining and pungent attack on contemporary syllabuses, and demanded a far more vocationally-oriented teaching. As for the tradition of geometrical teaching, he asked:

> Where would be the harm in letting a boy assume the truth of many propositions in the first four books of Euclid, letting him accept their truth partly by faith, partly by trial? Giving him the whole of the fifth book of Euclid [on proportion] by simple algebra? Letting him, in fact, begin his severer studies where he is now in the habit of leaving off.[74]

It can be argued that Perry's plea had been an original motive of Hirst, Wright and Wilson; but in the institutionalization of their motive, in concern for a logically-constructed geometry syllabus that would satisfy the examination machinery, the AIGT somehow lost sight of it.

As a result of the Perry debate at Glasgow the participants agreed unanimously that Euclid should be abandoned once and for all; that teachers should make greater use of intuitive and practical methods of proof in geometry, on the grounds of simplicity and because such methods *did* promote deductive reasoning; and that the Mathematical Association should form a liaison committee with the British Association that would again approach the University examination authorities.[75]

The Perry debate attracted considerable publicity, and encouraged Bertrand Russell, who was more than familiar with the existence of non-Euclidean geometry, to make a stylish demolition of Euclid. 'His definitions,' Russell wrote, 'do not always define, his axioms are not always indemonstrable, his demonstrations require many axioms of which he is quite unconscious' while his value as a logic-trainer was 'grossly exaggerated'.[76]

Faced by a physical assault of paper, petitions and representations from the Mathematical Association, the British Association, Russell, and a notably strong-worded letter from a group of mathematics teachers,[77] the universities capitulated. The end was, in fact, swift and simple. In 1903, during A. R. Forsyth's presidency of the Mathematical Association and while he served on the Cambridge Local Examination's Syndicate, the University of Cambridge agreed to accept instead of Euclidean proofs 'any proof of the proposition, which appears to the Examiners to form part of a systematic treatment of the subject.' Examining Boards at Oxford and elsewhere immediately followed the Cambridge lead.[78]

In the long run the twentieth century has witnessed a decreasing stress on geometry in school education, and a greater emphasis on the more symbolic and analytical languages of co-ordinate geometry and trigonometry. Concomitant with this change, and undoubtedly causally linked with it, is the fact that during this century the language of scientific research has not been geometry, but analysis. Thus although as late as 1915 Einstein cast his work on general relativity in a geometrical framework, one would be hard put to think of another *significant* twentieth-century paper which uses a geometrical focus.

In tracing the way in which Euclid's hallowed book was slowly replaced by rival geometrical texts, some fifty years of education and mathematical history have been covered. It is clear that the AIGT, which was founded in 1871, was only a modest pressure group for much of this period. It suffered from weak organization, deployed bad tactics against examination boards, failed to write a textbook as a first priority and failed to insist on its exclusive use instead of Euclid. Only when it began to widen its scope to include the whole of possible mathematical teaching and to embrace the more proletarian and utilitarian 'Perry Movement' teachers, were its initial aims and ambitions achieved. By about 1903 the way was open for the publication of fresh systematic geometrical texts, like the hugely successful ones by Godfrey and Siddons, and Durell.[79]

In many ways the story outlined in this essay complements that of the introduction of science into schools. In the 1850s there had been a conflict of aims between those who wanted a 'science of common things' taught for social and utilitarian reasons, and those who wanted a formal, mind-training, specialized kind of science teaching. Contrast the possibility of a vocational, technical, and useful teaching of geometry and mathematics with the more formal approach adopted by the AIGT. Of course, the scientists had no battle to wage against a particular textbook (though they conflicted over didactic and heuristic modes of instruction). In the case of science teaching the formal approach seemed to the young-guard of scientists like Tyndall, Huxley and Carpenter the best way to professionalize and institutionalize their subjects. Generally speaking, they were successful, though by 1900 the impact of science teaching on public schools was still minimal and demanded close attention. Such a formal approach to their subject's objectives seems to have helped mathematics teachers little in their dialogue with university examiners when they wished to transform the contents of their curriculum without losing the instrumental assumption. Success did come, however, when Perry re-emphasized the importance of a 'mathematics of common things' at the end of the century.[80] Science and mathematics teachers could then begin slowly to join forces in a common enterprise. It is surely significant that the Association of Public School Science Masters was founded at the height of the Perry Movement, that it should have modelled itself on the Mathematical Association, that it should have rejected instrumentalism and the belief in learning transfer, and that its fundamental platform in 1916 had become 'science for all'.[81]

Victorian Studies Centre,
University of Leicester.

References

1 A. C. Hilton, 'The Heathen Pass-ee', 1872 (in imitation of Bret Harte) in Michael Roberts (ed.), *The Faber Book of Comic Verse* (London, 1942), 240–2. I am grateful to Professor A. J. Meadows for this reference.

2 To avoid misunderstanding, by 'rivals of Euclid' (C. L. Dodgson's phrase) are to be understood those textbooks which tried to make Euclidean geometry clearer to tyros by the improvement of Euclid's definitions, the rearrangement of his theorems, and the use of new constructions and proofs not in Euclid. 'Non-Euclidean geometries', on the other hand, refers to the new geometries of Lobatchevsky and Bolyai which used a different definition of parallelism from Euclid.

3 N. A. Hans, *New Trends in Education in the Eighteenth Century* (London, 1951).

4 Significant because it paved the way for the consideration of science and modern languages.

5 W. Whewell, *On the Principles of English University Education* (London, 1837), 5, 42. The Appendix reprints an essay of 1835, 'Thoughts on the Study of Mathematics as a Part of Liberal Education'. Note also Baden Powell (Savilian Professor of Geometry at Oxford), *The present state and future prospects of mathematical and physical studies in the University of Oxford* (Oxford, 1832).

6 *Public Schools Commission* [*Clarendon*] *Report*, Parliamentary Papers 1864 [3288] xx. My citations are to the Irish University reprint, *British Parliamentary Papers* (*Education General*), vol. 10, 456–86. Cited hereafter as *PSC Report*.

7 Ibid., vol. 11, Q.6332 (Reverend S. T. Hawtrey). Cf. *James M. Wilson; an autobiography, 1836–1931* (London, 1932), 57–9. The deficiencies in mathematics teaching were also to be found in newer foundations, like Marlborough College, which copied older foundations.

8 *PSC Report*, vol. 10, 23.

9 The University of London matriculation examination (1838) demanded a knowledge of Euclid, as did the later established Oxford and Cambridge Local Examinations (1858).

10 David Hilbert, as quoted by O. Neugebauer, *The Exact Sciences in Antiquity*, 2nd ed. (New York, 1962), 145.

11 For modern estimates of Euclid see T. L. Heath, *The Thirteen Books of Euclid's Elements*, 2 vols. (London, 1926), Dover (ed.), 3 vols. (New York, 1956); B. L. van der Waerdon, *Science Awakening* (Groningen, 1954); J. Itard, *Les livres arithmétiques d'Euclid* (Paris, 1961); and Ivor Bulmer-Thomas and John Murdoch, 'Euclid' in *Dictionary of Scientific Biography* (New York, 1971), vol. 4, 414–59.

12 W. Whewell, *Of a Liberal Education in General and with a particular reference to the leading studies of the University of Cambridge* (London, 1845), 30.

13 Ibid., 65.

14 See, for example, Oliver Byrne's extraordinary *The First Six Books of Euclid with coloured diagrams* (W. Pickering, London, 1847) which used colours instead of letters (copy Maths. Association Library, Univ. of Leicester). The *British Museum Catalogue of Printed Books*, sub. Euclid, lists 238 editions of Euclid between 1801 and 1900; but this ignores the large number of editions by Simson and other editors, as well as the large number of other Geometries or Mathematical texts containing sections on geometry. Thus, P. J. Wallis, *A check list of British Euclids up to 1850* (typescript, Dept. of Education, University of Newcastle, 1967) found 287 Euclids between 1801 and 1887.

15 *Companion to the British Almanack . . . for 1849* (London, 1848). Of course, criticism of Euclid did not begin in the 1860's. See D. K. Wilson, *The history of Mathematical teaching in Scotland to the end of the eighteenth century* (London, 1935), 53–6.

16 *Association for the Improvement of Geometrical Teaching. Nineteenth Report* (Bedford, 1893), 15. These Reports will be cited as *AIGT Report*.

17 J. Demogeot and H. Montucci, *De L'Enseignment Secondaire en Angleterre et en Ecosse. Rapport addressé à son exc. M. Le Ministre de l'Instruction Publique* (Paris, 1868), 581ff. See R. D. Anderson, 'French views of the English Public Schools: some nineteenth-century episodes', *History of Education*, 2 (1973), 159–72. The mathematician Legendre (1752–1833)

produced his *Eléments de géometrie* in 1794 (note the Engish trans. by the young Thomas Carlyle in 1824). It was revised continually by others during the nineteenth century and reached a 21st ed. in 1876.

18 I. Todhunter, *The Conflict of Studies* (London, 1873), 137–40. Demogeot taught literature at the Sorbonne, but Montucci, who was a modern languages teacher in a Parisian lycée, had studied mathematics. For Demogeot see *Dictionnaire de Biographie Française*, x, 991–2. Todhunter (1820–84), a Fellow of St John's College, Cambridge, produced one of the most successful of all the editions of Euclid used in the nineteenth century, *The Elements of Euclid* (London, 1862). Cf. Hilton's parody above.

19 J. W. Colenso, *Arithmetic designed for the use of schools* (London, 1843), much reprinted; sales slumped after Colenso's notorious trials in the 1860s.

20 *Schools Inquiry [Taunton] Commission*, PP 1867–8 [3966], XXVIII, Pt. III. I have used the Irish University reprint *Brit. Parl. Papers (Education General)*. See vol. 19, Q.1664. Cited hereafter as *SIC Report*.

21 Ibid., Q.2992. Key became a founder-member of the AIGT. Cf. the evidence of the Astronomer Royal, G. B. Airy, *PSC Report*, vol. 12, 402, and further contrast the pro-Euclid evidence of the Scots engineer, W. J. M. Rankine, *SIC Report*, vol. 19, Q.2325 and Thomas Selby of Taunton, ibid., vol. 20, Q.12444.

22 In 1867 Hirst succeeded De Morgan as Professor of Pure and Applied Mathematics at University College. At the time of the Taunton inquiry in 1868 he had become Professor of Pure Mathematics.

23 *SIC Report*, vol. 19, Q.2992.

24 See H. Hale Bellot, *University College London 1826–1926* (London, 1929), 321–2.

25 Thomas Archer Hirst, *Journal*, 7 January 1854, ff. 1101–2 (Royal Institution Archives). See *The Journals of Thomas Archer Hirst, FRS (1830–92)*, ed. W. H. Brock and R. M. MacLeod, in press. The episode is mentioned by John Tyndall (who had access to Hirst's *Journal*) in his lecture to the Royal Institution in 1854. See *Lectures on Education Delivered in the Royal Institution* (London, 1855), 181–2.

26 *SIC Report*, vol. 17, 30.

27 Intuitive proofs were *any* sound methods of reasoning—'if a direct proof can be found of any theorem, which naturally arises out of the data of the theorem, it is to be preferred to a circuitous proof which depends on other [previously-proved] theorems. (J. M. Wilson, *Elementary Geometry*, 3rd ed. (London, 1873), vi).

28 Richard P. Wright, *The Elements of Plane Geometry* (London, 1868), vii. For Hirst's co-authorship, see p. vi. Hirst himself published an anti-Euclid syllabus of *Lectures on Plane Geometry* (London, 1870), which he had used in lecturing to the Ladies' Educational Association in 1869.

29 Wilson, *Autobiography* (note 7), 66, and his 'Early history of the Mathematical Association', *Mathematical Gazette*, 10 (1921–2), 239–44.

30 J. M. Wilson, *Elementary Geometry* (Cambridge, 1868), 2nd ed. (Cambridge, 1869), 3rd ed. (London, 1873). Both Macmillan, and Longman (who published Wright's text), were in the vanguard of textbook publishing in the 1860s.

31 *Educational Times*, Sept. 1868. Cf. Reverend Joshua Jones (headmaster of King William's College, Isle of Man), *On the Unsuitableness of Euclid as a Textbook of Geometry* (Liverpool, 1870).

32 *Athenaeum*, 18 July 1868.

33 Todhunter, *Conflict* (note 18), Essay 5, 136–92; Charles L. Dodgson, *Euclid and his Modern Rivals* (London, 1879; reprint Dover Publications, 1974), 220–6 for a reproduction of De Morgan's review. Dodgson's Carrollean book went through an enlarged 2nd ed. in 1885. Note Dodgson's *First Two Books of Euclid* (Oxford, 1882).

34 Wilson, *Geometry* (note 30), 3rd ed., v. Methodologically, his arguments for and against Euclid remind one of those used by him for and against science teaching. See F. W. Farrar, ed., *Essays on a Liberal Education* (London, 1867).

35 Wright and Hirst also agreed (note 28, vii); but Dodgson, *Rivals* (note 33), 188, thought the syllogism had merits.

36 This comment led Todhunter, who slightly missed the point, into an interesting defence of British mathematics, which he thought as good as any in Europe (*Conflict*, note 18, 146). For Hirst's views on Continental geometry, see *AIGT Report*, I (1871), 9–11 and ibid., II (1872), 9–13.

37 Dodgson thought this a trivial objection since Euclid had survived 2000 years.

38 *Rivals* (note 33), 11–12, *et passim*.

39 Wilson, *Geometry* (note 30), 3rd ed., ix; Todhunter, *Conflict* (note 18), 180, did not agree.

40 This was one reason why geometry was so strongly entrenched in the Scottish universities. See G. E. Davie, *The Democratic Intellect* (Edinburgh, 1961), chap. 7, and R. Olsen, 'Scottish Philosophy and Mathematics 1750–1830', *J. History Ideas*, 32 (1971), 29–44.
41 *Conflict* (note 18), 139–42. Perversely, Todhunter challenged Euclid's rivals to produce wranglers—which, of course, they were forbidden to do (173).
42 Wilson gave sixpence to every Rugby boy who spotted an error. J. M. Wilson, 'Early history . . .' (note 29), 241.
43 *Brit. Ass. Reports*, XL (1871), lxxii. The Committee's first report appeared in ibid., XLII (1873), 459–60; note also ibid., 45 (1876), 8–13.
44 On Tucker (1832–1905), see J. A. Venn, *Alumni Cantabrigienses* (Cambridge, 1954), Pt. II, vi, 240. For the LMS, which was founded by De Morgan, Hirst and others in 1866, see E. F. Collingwood, 'A Century of the London Mathematical Society', *Journal LMS*, XLI (1966), 577–94.
45 *Nature* 1 (1869–70), 534. The Headmaster's Conference, meeting at Sherborne in December 1870, expressed disquiet at the continued use of Euclid. See *AIGT Report*, I (1871), 13.
46 *Nature* II (1870), 141. For a sketch of Levett, see A. W. Siddons, 'Progress', *Mathematical Gazette*, 20 (1936), 7–26 (13–15).
47 Richard Wormell, *An Elementary Course of Plane Geometry* (London, 1868), 2nd ed. 1870.
48 *Nature* III (1870–1), 169. Hirst was obviously elected Chairman because of his association with Tucker and because, apart from Cayley (a devoted Euclidean) he was the most distinguished geometer in England.
49 *AIGT Report*, II (1872) 23 (Wilson), 24 (Hirst, 'no really valuable textbook would ever be produced by a committee or an association').
50 *AIGT Report*, III (1973), 11. Only 13 Members replied! Only one of these was in favour of teaching 'technical' geometry.
51 The only AIGT members to stand out for technical geometry and, equally significantly, for the admixture of arithmetical and algebraic matter to textbooks of geometry, were the Reverend W. H. Laverty, a Fellow of Queen's College, Oxford, and J. F. Iselin, an Inspector of science schools for the Department of Science and Art. See *AIGT Report*, II (1872), 21–2, 28–9.
52 *AIGT Report*, II (1872), 17 and ibid., III (1873), 12.
53 Note the assumption that the British Association possessed educational authority. Approval of the Headmaster's Conference was also sought.
54 See *Brit. Ass. Reports*, XLV (1876), 9. Of the 15 BA members, 7 were from the AIGT. Not unexpectedly the BA concluded that the AIGT syllabus was 'drawn up with such care, and with regard to the essential conditions of the problem, as to render it highly desirable that it should be considered in detail by authorised representatives of the universities and the other great examining bodies of the United Kingdom'.
55 *AIGT Report*, V (1875), 11, quoting the circular of 21 November 1874, a copy of which is bound in J. S. Mackay's copy of early AIGT Reports in the Mathematical Association Library, Leicester (MA 510.4/135).
56 *AIGT Report*, VI (1878), 18–19.
57 I.e., the proof of a theorem appearing in Euclid Book II could be done in any way providing it did not involve using theorems that appeared after that theorem in Book II or later Books. The Cambridge Previous Examinations followed this rule after 1882.
58 Hirst was Assistant Registrar of the University of London between 1870 and 1872, but I have found no evidence to suggest that he influenced the Senate's decision of 1875–6.
59 For Hirst's anonymous review of Dodgson, see *Nature* XX (1879), 240–1. He thought most of Dodgson's points were 'mere verbal quibbles' over apparent inconsistencies which arose from the different standpoints of the rival authors.
60 On Hayward (1829–1903), see *DNB*. He and Hirst had corresponded on geometrical topics.
61 Circular to members from R. Levett dated 31 December 1880, bound in Mackay (note 55). There were no meetings of the AIGT in 1879 or 1880.
62 J. M. Wilson, *Elementary Geometry. Books I–V. Containing the Subjects of Euclid's first six books: following the syllabus of geometry prepared by the Geometrical Association* [sic], 4th ed. (Cambridge, 1878).
63 See *AIGT Report*, VII (1881), 28 and ibid., X (1884), 19.
64 Ibid., VII (1881), 5–10, list of members.
65 Ibid., 35.
66 On 12 December 1882 the Syndicate of the University of Cambridge recommended 'some deviations from the old practice of merely asking questions directly from the propositions of Euclid as such, and that there should be questions on the subject matter of Euclid, as

well as the propositions themselves', *AIGT Report*, IX (1883), 16–17. This admission referred to examinations 'previous' to the tripos.

67 *AIGT Report*, XIII (1887), 21–2.

68 Cayley did not sign the Cambridge Board of Mathematics' recommendation. See *AIGT Report*, XIV (1888), 28.

69 For the Cambridge attitude, see ibid., 23. For reminiscences concerning the AIGT's negotiations with examiners, see Siddons (note 46), 19–21.

70 Oxford was petitioned again in 1892—to J. J. Sylvester's embarrassment (R. C. Archibald, 'Unpublished letters of J. J. Sylvester', *Osiris*, I (1936), 152–3); Oxford declined. See *AIGT Report*, XIX (1893), 7–8.

71 *AIGT Report*, VI (1878), 12.

72 For a summary of Perry's position, see *Teaching Mathematics in Secondary Schools*, Min. Educ. Pamphlet 36 (HMSO, London, 1958), 7–9.

73 For opportunities for women, see Evelyn Sharp, *Hertha Ayrton 1854–1923* (London, 1926), 146–7. Hertha Ayrton was the wife of the electrical engineer W. E. Ayrton, a close friend of Perry.

74 J. Perry, ed., *British Association Meeting at Glasgow, 1901. Discussion on the Teaching of Mathematics* (London, 1901); *Mathematical Gazette*, II (1901–1904), 81–3.

75 See *Mathematical Gazette*, II (1901–1904), 105–6, where E. M. Langley also spelt out where the AIGT had gone wrong. The Perry campaigners also wanted the power of examiners and school inspectors reduced.

76 B. Russell, 'The Teaching of Euclid', *Mathematical Gazette*, II (1901–1904), 165–7.

77 This letter to the BA Committee and the Mathematical Association from a ginger group of 23 mathematics school teachers demanded reforms of *every* aspect of school mathematics. It was first published in *Nature*, LXV (1901–2), 258–9 and rep. *Mathematical Gazette*, II (1901–1904), 143–6. For Forsyth's connection with the campaign, see *Teaching Mathematics* (note 72), 9.

78 For the capitulation, which included the Civil Service Commission, see *Mathematical Gazette*, II (1901–1904), 197–201, 349–51; ibid., III (1904–1906), 146–57; and Siddons (note 46), 19–21. On Forsyth, see *DNB*.

79 Charles Godfrey and A. W. Siddons, *Elementary geometry* (Cambridge, 1903); for Godfrey's contributions to school geometry, see A. C. Howson, 'Milestone or Millstone', *Mathematical Gazette*, LVII (1973), 258–266. Clement Vavasor Durell, *A Course of Plane Geometry for Advanced Students* (London, 1909–10); *A Concise Geometry* (London, 1920). For later developments in mathematics teaching see the resumé in *The Teaching of Mathematics in Secondary Schools*, Assistant Masters Association, 2nd ed. (Cambridge, 1973), 9–38; and A. G. Howson, 'Milestone'.

80 My reflections are the result of reading D. Layton, *Science for the People* (London, 1973). For one link between the Perry movement and the PSSMA see W. D. Eggar, 'On the teaching of science and mathematics', *Conference of Public School Science Masters*, 19 January 1901 (priv. prin., London, 1902).

81 'Science for All', *School Science Rev.*, II (1920), 197–212. See W. H. Brock, ed., *H. E. Armstrong and the Teaching of Science 1880–1930* (Cambridge, 1973), 53–5.

XIII

SQUARED PAPER IN THE NINETEENTH CENTURY: INSTRUMENT OF SCIENCE AND ENGINEERING, AND SYMBOL OF REFORM IN MATHEMATICAL EDUCATION

Abstract. The paper traces the gradual adoption of squared paper from its exclusive use as a research tool in the early 19th century to its universal use for a variety of purposes in mathematical education by the end of the first decade of this century. Three underlying causal factors are explored – the growth of new educational philosophies; the development of science teaching and the associated need for mathematics correlation; and the growing demands of engineering and technical education. Whilst the focus on squared paper is a narrow one, it is argued that its adoption in education generally was symptomatic of a much wider transformation of mathematical curricula in response to various demands which, significantly, arose *outside* the academic mathematical community.

Squared paper, a marvellous instruction which ought to be in the hands of every one who works in mathematics from the kindergarten to the university. (Laisant, 1904, p. 23.)

The graphical presentation of experimental data has a number of advantages for the physical sciences, as Dr. Laura Tilling (1975) indicates in her useful paper on early experimental graphs. She elaborates upon this 'graphical method' as follows:

Its greatest strength lies in the clarity and succinctness with which it displays the information contained in tabulated results: for the experimenter a graph provides a rough and immediate check on the accuracy and suitability of the methods he is using, and for the reader of a scientific report it may convey in a few seconds information that could only be gleaned from a table of measurements by hours of close study. (p. 193.)

A graph's power to summarize data can also be utilized in economic, social and other quantitative fields but, as Tilling's paper makes clear, the graph – something that is nowadays part of the visual furniture of every newspaper and *Punch* cartoonist – was something only rarely used by eighteenth- and early nineteenth-century experimentalists. Two examples may be cited. A paper published by the physician and chemist, William Prout (Prout, 1813, p. 332) in 1813 on human carbon dioxide output over a 24 hour period uses squared paper, and the French chemist and physicist, Gay Lussac (1819), published a graph of solubilities in 1819. However, as Tilling notes (Tilling, 1975, p. 210), John Herschel (1833) explicitly recommends the usefulness of squared paper in his 1833 paper on revolving stars as if few scientists had adopted the technique.

Reprinted by permission of Kluwer Academic Publishers.

Educational Studies in Mathematics **11** (1980) 365–381.

> . . . let a sheet of paper be procured, covered with two sets of equidistant lines, crossing each other at right angles, and having every tenth line of each set darker than the rest. By these, the whole surface of the paper will be divided into large squares by the dark lines, and each of these sub-divided into 100 smaller ones by the faint ones. Such charts may be obtained, neatly engraved; and are so very useful for a great variety of purposes, that every person engaged in astronomical computations, or indeed, in physico-mathematical inquiries of any description, will find his account in keeping a stock of them always at hand. (p. 178.)

It is clear that until the 1820s and 1830s the graphical method and, in particular, squared paper were *rarely* used. Moreover, until the 1870s, as we will show, such methods remained in relatively uncommon use in Great Britain and were not employed by students as part of their elementary scientific and mathematical education.[1] One obvious inhibiting factor was the cost of squared paper. Before 1876 the price of a single sheet of paper engraved in the manner described by Herschel was as much as (old) 8d. Yet by 1887 it could be obtained for a little over ¼d a sheet or 8d. a quire (a quire being a printer's term for 24 sheets of paper); by 1899, 7½d a quire; by 1903, 7d a quire; and by 1913, 6d a quire (exactly ¼d a sheet) (Ayrton, 1887, Perry, 1899). By 1903 a number of firms were marketing various forms of squared paper in both single sheets, for as little as 2¾d a quire with ¼ inch rulings, and variously mixed with normally ruled papers in the form of exercise books. It might also be added that polar graph paper, published in books of 40 sheets at 1s a book, first appeared in England in 1902 (Salmon, 1903). Needless to say, this dramatic fall in price and increasing availability coincides with the adoption of various forms of squared paper for a variety of purposes in both science and mathematics within technical, elementary and secondary education. The publication of textbooks like Richard Wormell's *Plotting or Graphic Mathematics* (1888) provides an early illustration of the trend in mathematics, though Charles Godfrey (1912), in his report on a survey of teaching methods in English secondary schools, prepared for the International Commission on the Teaching of Mathematics (ICTM), claimed:

> The use of graphical methods in elementary algebra teachings is universal and entirely a 20th-century development. Other aspects of the same movement are the adoption of descriptive geometry by the mathematicians, the use of handy 4-figure tables, and of graphical methods in statics, and, though, in these cases, the victory is less complete than that of the 'graph', it is remarkable and equally modern.

Godfrey also found that the use of squared paper in connection with the topic of area was a standard part of the curriculum for lower and middle forms by this time. In 1912 also, the Board of Education's (1912) *Suggestions for the Teaching of Arithmetic* noted that squared paper was to be found in any well-equipped elementary school, reflecting the tendency towards 'concrete methods'. However, the Board went on to issue the following warning concerning its uses:

Any risk of injury to eyesight by the excessive use of squared paper should be avoided. Its use for practical geometry, etc., will only be occasional and little danger to health is likely if no paper with rulings less than one-tenth of an inch apart is used. The use of paper ruled in squares for the working of arithmetical examples has no real educational advantages and might well be discontinued. (p. 8).

What caused the change from a fairly unusual and always exclusively research use of squared paper to it becoming a versatile, familiar and elementary tool of the mathematical tyro? We should like to suggest three underlying factors:

(1) The break-up of the narrow 'instrumentary' tradition in elementary education, which Selleck (1968) has described, and the parallel erosion of the dominant classical tradition in secondary education, within which mathematics was regarded as an 'art', which disciplined the mind, rather than a 'science'.

(2) The rise and development of science teaching in schools, and the associated problem of correlating mathematics and science teaching (and, indeed, finding mathematicians who could teach mathematics for scientists).

(3) The development of engineering in its various forms and the associated movement for technical education which brought about the resurgence of a broader conception of mathematics, 'practical mathematics', to replace the dominant, narrow and inward-looking 'academic mathematics'.

Now there may well be other factors that others would emphasize, such as the decline of British industrial supremacy and attempts to alleviate it, or the availability in the stationer's printing shop of cheaper dividing engines for producing squared paper from copper engravings or by machine rule, but these seem to us the most important. Moreover, for once the historian finds a simplifying thread linking these factors, namely the careers of just three men and their pioneering work in just two geographically distant institutions. These are Henry Armstrong, William Ayrton and John Perry, and the Royal Engineering College, Tokyo and Finsbury Technical College in London (Brock, 1979).

1. NEW EDUCATIONAL PHILOSOPHIES

The various features of the 'new education' in the period 1870–1914 have been disentangled and ably discussed by Selleck (1968). The new thinking influenced teaching methods from the kindergarten through to the secondary stage and, for our purposes, some of the important arguments and influences are those which emanated from the 'practical educationists'. Armstrong, Perry and many other reformers pressed for more practical approaches in education, which implied a more active role for the learner, whose education should be based upon 'concrete' experiences and related to the circumstances of life. The training of hand and eye through learning by observation and 'doing' also came to be valued as an end in itself. Drawing on squared paper was one valuable activity for such purposes, even at the kindergarten stage, where it might go along with other exercises such as paper folding, tearing and laying, bead laying, string measurement and division, clay modelling and colour sorting (Selleck, 1968, p. 111). By 1902, the educational reformer J.J. Findlay (Findlay, 1902), was able to single out 'self-activity' as one of the principles governing the reform movement in mathematical education generally. The principle could be pursued:

by 'using' squared paper, by measuring, by plotting curves, etc. (p. 185.)

In 1908, Benchara Branford (1908), a London County Council Inspector, observed:

Attention is now beginning to be paid in rapidly increasing measure to mathematics on its experimental and graphical side, and is exemplified by the use of drawing-boards, improved mathematical instruments, squared paper and spherical blackboards.

Branford went on to mention one further and more dramatic sign of a changed outlook, namely the provision of 'mathematical laboratories':

. . . well stocked with clay, cardboard, wire, wooden, metal, and other models and material, and apparatus for the investigation of form, mensuration and movement . . . .

The gradual adoption of 'practical geometry' within a liberal secondary education and the overthrow of the strictly Euclidean treatment of geometry have been considered by Brock (1975). The gradual transformation of other branches of mathematics was more subtle and cannot be adequately discussed here. It is, however, significant that the International Commission on the Teaching of Mathematics chose as a major focus for international comparisons, at Cambridge in 1912, 'methods of intuition and experiment in secondary schools'. Such methods embraced the graphical study of statistics, functions, vectors and statics, as well as the use of slide rules and mathematical tables. It is clear that a more practical approach had been adopted universally by this time and as American David Eugene Smith (Smith, 1913) noted:

Of the value of squared millimetre paper there is no question anywhere. (p. 614.)

One important argument for the use of squared paper, and indeed for practical tendencies generally, has yet to be considered.

In the nineteenth century the common justification for practices within both the instrumentary and classical traditions in mathematical education was in terms of the disciplinary value for both the mind and the character. The mind was modelled as a set of discrete faculties (like mental muscles) which could be exercised, and thereby developed, by working upon suitable subject matter. Thus the formal literary study of Euclid could be justified as a training for general faculties such as those of reasoning and memory, taking transfer of training as axiomatic. Algebra as an elaborate symbolic game divorced from both arithmetic and reality, and apparently leading nowhere, could be similarly justified. No consideration was given to the learner's perceptions and interests. Like physical training, the educational process might be very uncomfortable but it was considered to be good faculty training for the learner. However, this extreme doctrine of mental and moral discipline was gradually undermined towards the latter part of the nineteenth century. For younger pupils, alternative rationales emphasizing the idea of natural growth from within, rather than the process of forming the mind from without, were derived from the writings of Pestalozzi, Froebel, and subsequently Montessori in particular. More generally, the writings of Herbart provided further inspiration for the reformers, and shifted the focus to questions of motive and interest in education and to the growth of ideas in the individual (Selleck, 1968, pp. 179–272). In particular, mathematical educators such as Godfrey, Branford and Percy Nunn reflected the influence of Herbart and all developed philosophies before the First World War, which emphasized the role of motives in the historical development of mathematics and in the learner, and stressed the value of the outward-looking aspects of mathematics in particular (Branford, 1908; Price, 1976). In this shift away from a crude and simplistic rationale of mental discipline squared paper provided one means for implementing the doctrine of interest in mathematics. In 1902, Godfrey (1902) made a number of detailed suggestions for preparatory schools on the uses of squared paper, and a square ruled blackboard, for work in fractions, decimals, areas and algebra. He boldly claimed:

> GRAPHS have found their way into elementary work, and are now recognised as quite the most valuable instrument in our possession for awakening interest. (p. 289.)

2. MATHEMATICS AND SCIENCE TEACHING

The development of 'practical' or 'experimental' mathematics as an approach to teaching the subject involving the use of apparatus of various kinds, already mentioned, and perhaps a special room – a mathematical laboratory such as that devised by Godfrey *et al.* (1905) and Bell (1912) at Winchester – has obvious parallels with the somewhat earlier emergence of practical science teaching. John Perry (1850–1920) taught physics at Clifton College between 1870 and 1873 and claims to have established the first workshop within a public school in England, where boys could make apparatus to repeat experiments for themselves (Perry, 1900, pp. 73, 94). At the same school, from 1875, A. M. Worthington pioneered a course of practical work in elementary mensuration and hydrostatics which subsequently became generally adopted as the first stage in practical physics – 'elementary physical measurements' in public schools, and science classes and schools under the Department of Science and Arts (Fawdry, 1912; Jenkins, 1979). The growth of both physics and chemistry teaching, as well as demanding the use of various materials and apparatus, was accompanied by the rise of a new approach to teaching, 'heurism', or the method of discovery, which was so strongly advocated by H. E. Armstrong (1848–1937). The search for functional relationships between quantities, and hence the use of squared paper, was an important component of the heuristic method in the laboratory (Salmon, 1903; Ostwald, 1894). However, apart from practical plane and solid geometry and practical mensuration, which were supported by examinations such as those of the Department of Science and Art (DSA) and the City and Guilds of London Institute, the elementary mathematics curriculum in nineteenth century institutes was predominantly 'academic' in character, limited in scope, and out of step with the increasing practical uses of mathematics in science (and engineering, as we will show). The situation was well captured by Professor G. H. Bryan in his response to Perry's agitation and the British Association debate of 1901, and Bryan (1902) deserves quoting in some detail at this point:

> Formerly our schools and colleges were given over mainly to the study of classical and literary subjects, and mathematics was looked upon as a portion of an arts course. The rigid deductive system was undoubtedly admirably suited to the object then held in view. With the development of experimental science new teachers have been appointed all over the country for physics, chemistry and biology, *but next to nothing has been done to meet the greatly increased demand for mathematical teaching thus produced.* The same teachers who provided efficiently for the teaching of mathematics on the classical side have now thrust upon them an influx of new pupils having quite different requirements. (p. 90) (Bryan's stress.)

As well as the new demands of science, there also arose the more tangible demands of the Civil Service Commission's examinations, particularly for entrance to the army, which stimulated the growth of modern sides and army classes in British public schools, and virtually forced a more practical mathematics curriculum upon certain classes, including laboratory work (Montgomery,

1965; Boyt, 1906). By 1902, Findlay (1902) was able to distinguish the issue of 'correlation' as one of the factors behind the general reform of mathematical education:

This whole movement [correlation] is the outcome of the desire to rank mathematics in its place as the handmaid of science. Not, be it observed, as a tool for engineers, but in close correlation with the needs of the science syllabus right through the school. And, in another aspect, it emphasizes the necessity of bringing the various parts of elementary mathematics in correlation with each other. (p. 185.)

The issue of correlation was pursued further by mathematics and science teachers over the first decade of the twentieth century and the debate involved the Mathematical Association and the Association of Public School Science Masters (APSSM), but we need not consider the detail here. What is more important for our argument is the earlier shift from a state of *no* correlation to one which demanded at least *some* correlation, and a new outlook from teachers of mathematics in particular. Finding teachers who were able to adapt to the new demands was, however, problematic, given the dominance of the Cambridge tradition in mathematics. Thus, as Perry (1902) scathingly remarked in 1902:

We want mathematical masters from Cambridge, but we must do without them, and indeed severely reject them, as candidates for posts, if they will not give boys such teaching as befits the twentieth century. (p. 203.)

It could be argued that Perry subsequently pushed his case too far by crudely opposing the 'mere mathematician' and the practical man with wider scientific interests. In 1908, he remarked (Perry, 1908):

. . . the marriage of mathematics and science seems to me like that of December and May – the marriage of a man of seventy with old bachelor habits to a bright young virgin of seventeen. (p. 461.)

However, it is significant that pioneers in mathematical education like Wormell, Findlay, Nunn and W. D. Eggar of Eton College (a founder member of the APSSM in 1900), were trained in both mathematics and science, and could teach both subjects heuristically (Findlay, 1902; Nunn, 1903; Wormell, 1900). Findlay and Nunn both devised correlated curricula for their schools, before they entered the field of teacher training, and Eggar (1903) in the preface to his *Practical Exercises in Geometry*, an early innovatory textbook for general educational use, emphasized:

This book is an attempt to adapt the experimental method to the teaching of Geometry in schools. The main object of this method, sometimes called 'heuristic', is to make the student think for himself, to give him something to do with his hands for which the brain must be called in as a fellow-worker. The plan has been tried with success in the laboratory, and it seems to be equally well-suited to the Mathematical classroom.

The book demanded the use of squared paper for work on areas and included a separate chapter on graphs. Graphs subsequently became subsumed under algebra, and indeed the rise of a new subject called 'graphic(al) algebra' was a source of some confusion for schools generally, The board of Education (1909) attempted to clear the air with *Circulars* on the subject in 1909 and 1914. The Board acknowledged that one of the roots of graphical work lay in science teaching:

> Even where graphs are neglected in the mathematical classroom they inevitably make their appearance in laboratory work. Generations of boys had indeed been familiarised with graphs in connection with physical observations before ever graphic work was recognised as an essential element in the teaching of algebra.

However, there were other roots of graphical work, and indeed of the reform of mathematical education in general. As a perceptive member of the College of Preceptors (1903) commented in the *Educational Times* in 1903:

> . . . the change was inevitable . . . When teachers began to teach science – as well as other subjects – scientifically and with more due regard to the pupils' share in education, when the demands of the laboratory and the workshop, of technical schools and colleges, made themselves felt, reform was bound to come in the *teaching* of mathematics, as of logic and languages and other subjects. It is part of a general movement. (p. 465.)

The demands of engineering and technical education remain to be considered.

3. ENGINEERING, TECHNICAL EDUCATION, AND INTRODUCTION OF SQUARED PAPER

In the older engineering colleges only civil engineering was taught and the mathematical demands were strictly limited. However, the requirements increased somewhat with the development of shipbuilding and mechanical engineering, and increased greatly with the advent of high-speed machinery and the development of electrical engineering (Perry, 1912). What is particularly interesting is that the teaching of graphical methods, within a reorganized and broader conception of mathematics, 'practical mathematics', first arose in the later nineteenth century in response to engineering's various demands and that the subsequent general pattern of adoption was a common one internationally. The American D. E. Smith (1913) was able to claim at Cambridge in 1912 that:

> Graphic methods of one form or another are now found in the courses in mathematics [in secondary schools] . . . in all countries, having gradually made their way from engineering, through thermodynamics and general physics, to pure mathematics. (p. 622.)

We need not dwell upon the causes of the movement for the promotion of technical education in Great Britain during the 1870s. Suffice to say that it is bound up with the reports of successive Royal Commissions, the question of maintaining British industrial supremacy or preventing its decline, and the professionalization of science itself by way of providing jobs for science and mathematics teachers. In addition to the older DSA, two of the principal bodies involved were the City of London Corporation and the ancient Guilds of London. These two latter bodies came together in 1878 as the City and Guilds of London Institute, followed a year later by their sponsorship of evening lectures in chemistry and electricity for artisans in what was to become Finsbury Technical College. Armstrong and W. E. Ayrton (1847–1908) were the first teachers and they were joined by Perry in 1882. Armstrong (1920) subsequently referred to the pioneering trio as the 'Finsbury Mohicans'. It was at Finsbury that Perry developed a syllabus of practical mathematics for engineers which, in 1899, was adopted by the DSA (soon to become the Board of Education) as an addition to the various stages of pure mathematics, and practical plane and solid geometry. The new elementary syllabus which crystallized out at Finsbury included an emphasis on decimals with approximation in arithmetic and the use of logarithmic tables, and the slide rule. In algebra the use of formulae was stressed, and the study of functions and graphs using squared paper, with an early introduction to the ideas of the calculus. Practical methods, including the use of squared paper, were encouraged in mensuration and geometry. Numerical trigonometry, also involving the use of tables, simple work in three dimensions and vectors were also included. All these elements were to be variously mixed, all the time keeping close links with science and laboratory work (Perry, 1901; pp. 44–56). As Perry (1912) later somewhat acidly remarked, this conception was at the time:

> . . . exceedingly different from what used to be the study of the mere mathematician on the same subjects. (p. 35.)

Unfortunately, Perry is ambiguous as to when he first used squared paper for teaching purposes. It may possibly have been at Clifton College between 1870 and 1873,[2] following which he joined William Thomson in Glasgow before going on to Japan as Professor of Mechanical Engineering at Yedo (Tokyo) in 1875. In 1871, soon after arriving at Clifton (one of the new middle-class proprietary schools which took science seriously), at, he claims, the headmaster's (John Percival) suggestion Perry established a physical laboratory and engineering workshop in which he put boys through a series of quantitative experiments in mechanics using a kind of standardized 'Meccano kit' apparatus which Robert Ball had devised for lecture demonstrations at the Royal College of Science in Dublin. (Perry had learned of the apparatus from his brother, who still lived in Ireland; Ball's apparatus was itself adapted from that devised by Robert Willis (1800–1875.) Perry (1926) may be quoted at this point:

There is a piece of apparatus which is generally employed to illustrate the laws of blocks and tackles by which a man's pulling force can lift a great weight. In this apparatus a small weight is shown to be able to balance a much larger weight. The student having been told about this in a lecture, he is taken to the laboratory, where he finds *printed instructions* on the piece of apparatus. A scale pan containing the larger weight is, say, A and that of the smaller is B. He puts a certain weight in A. He now adds to B until, on giving B a start downwards, a steady motion of the apparatus results. He now says that B balances A and friction. He takes another weight A and repeats, and so on until he has obtained a great number of observations. He now *plots the weights* B which balance weights A on squared paper, obtains the law for efficiency of the machine, and obtains the law connecting friction and load. It is impossible for a boy to perform this process without obtaining exact ideas of work, loss of energy by friction, co-ordinate geometry, the meaning of a physical law, and the algebraic expression of a physical law. (p. iv.)

Unfortunately, this is later Perry speaking – in 1887 – and we have to note that a few sentences later he qualifies this description by saying that:

. . . this *method* of instruction was not very completely carried out at Clifton College.

We have already referred to the prohibitive cost of squared paper while Perry was at Clifton and without further evidence we take Perry to mean, probably, that the graphical method was *not* developed there.

However, we do have Perry's (1926) explicit word, as well as corroborative evidence, that the use of squared paper was 'more fully elaborated and tested' in Japan. Again, in his lectures of 1899 for the DSA on practical mathematics for working men, Perry (1899) says that in 1876:

Professor Ayrton and I used [sheets of squared paper] extensively in Japan, and when we returned to London and introduced, at the Finsbury Technical College, our methods of teaching mechanical and electrical engineering and laboratory work . . . we saw that one essential thing was the manufacture of cheap squared paper.

William Ayrton (1887) implicitly confirms this in the first edition of his *Practical Electricity,* published in 1887. Here he remarks that before the foundation of Finsbury by the City and Guilds of London in 1879:

Squared paper was practically used in England only for the recording of results of original experiments. And as these results, rather than the *training* of the experimenter, were the most important part of the investigation, the paper was very accurately divided, and sold at a high price totally out of reach of students. It became, therefore, necessary to have squared paper specially made, cheap, and at the same time sufficiently accurately divided for students' purposes.

Final independent corroboration, for use in Japan, comes from an anonymous report on Ayrton's laboratory, in the *Japan Weekly Mail* (1878) in October 1878, soon after Ayrton (but not Perry) had returned to England. The writer goes upstairs and visits a drawing office:

. . . some of the students were making working drawings of instruments, but the majority were reducing observations and drawing curves on squared paper.

This drawing office was to be found duplicated on the second floor of the new Finsbury College building opened in 1884.

Given the very close collaboration and rapport that emerged between Ayrton and Perry in Japan – they published dozens of joint papers – it will probably be impossible to say which of them first developed the use of squared paper for technical students. All we can say with confidence is that the technique was introduced in Tokyo in 1876 or 1877 and was developing at Finsbury by 1882.

We have dwelt on what some may feel is a trivial matter because the introduction of squared paper is a vital symbol of a development which was of tremendous significance for both technical education and mathematical education in general. It is, we believe, in that Japanese experience of Ayrton and Perry, and its transculturation to the City of London, that we must look to find the early roots of the amazing transformation of mathematical curricula in Britain which had become inevitable by 1900, as well as one of the roots of technical education and, concomitantly, the tension which grew between advocates of the study of mathematics for its own sake and for its utility which came to a head following Perry's address at the Glasgow meeting of the British Association in 1901.

In the late nineteenth century, the engineering profession, and Perry in particular, was responsible for some of the earliest agitation concerning the general teaching of mathematics in this country and the content of journals such as *Nature, Electrician* and *Engineering* testifies to this. In as early as 1880, Perry outlined his conception of elementary mathematics in a paper read before the Society of Arts (Perry, 1900; MacGillivray, 1902). By the turn of the century agitation had spread to the columns of general educational periodicals such as the *Educational Times, Journal of Education* and *School World* (Findlay, 1902; Young, 1920, pp. 87–90).

It should be added that we are *not* here considering the early attempts to reform geometry teaching which can be traced to the 1860s. In this earlier period the movement came from *within* the mathematical community, it met powerful resistance from certain leading academics, and was not helped forward by the more general and potent forces to which we have drawn attention. In consequence, the efforts of the Association for the Improvement of Geometrical Teaching achieved comparatively little and as the Mathematical Association, from 1897, this group was initially conservative in outlook and was forced to react in the early years of the present century on a much changed general situation which was largely not of its own making, though we will not pursue the detail here (Brock, 1975). What is more important is Perry's campaign to extend the adoption of his 'practical mathematics' from within the education of engineers to the wider education of intending teachers in the colleges and pupils in the schools. As a 1903 'Correspondent' (1903) pointed out in *Engineering*:

> It is well known that Professor Perry holds the opinion that the method which advocates of teaching mathematics to engineers is the best method of teaching children, for whatever life intended. At the same time it will be easily realised that the new system would from its very nature prove less startling to those who are trained to work hand-in-hand with Nature, and would therefore be easier to introduce to the engineering world than to any other. Practical mathematics, in fact, is a system of teaching mathematics to all persons, of all kinds and all ages. (p. 803.)

Perry left Finsbury in 1896 to become Professor of Mathematics and Mechanics at the Royal College of Science, South Kensington, where it was understood that he would develop further his system of mathematical teaching. Significantly, in 1899, the DSA first offered 'elementary practical mathematics' within its scheme of examinations in science and art, with Perry appointed as an examiner, and in the same year the Department published Perry's (1899) six lectures for working men on the subject, which was soon to be followed by a crop of new textbooks on practical mathematics by writers such as Frank Castle (1901), who came under Perry's direct influence at South Kensington. The success of this new subject in the technical field was quite extraordinary over the first decade of the twentieth century. The Board of Education (1900) reported that in 1899, the first year, there were only 212 papers worked compared with around 12 500 for practical plane and solid geometry and 24 000 for mathematics (the traditional diet). By 1903 the Correspondent (1903, p. 803) in *Engineering* could refer to 'enormous' progress, and added:

> How else describe the effect at one college in these islands, at which preparation was made for ten or a dozen students in the new mathematics and to which several hundreds came?

Interest was spreading rapidly and in 1909 the new subject had stolen the lead, with 6500 candidates compared with 6000 for mathematics and 3500 for geometry. The numbers actually attending the evening courses would have been at least two or three times greater (Holmes, 1910, p. 300). P. Abbott (1912), in his paper of 1912 for the ICTM, stated:

> It is probably correct to say that nobody has done more to influence the teaching of Mathematics in this country during the last 15 or 20 years than Professor Perry; and in no branch of the work has he brought about greater changes than in the teaching of Mathematics to technical students . . . In 1910, in England alone, 6964 students presented themselves in Practical Mathematics while only 2841 presented themselves in the first three stages of Pure Mathematics.

By 1913, when he was forcibly retired, Perry (1899, 1913, pp. vii–viii) was able to claim that there were more students of science engaged in his subject than in any other and also that it was more popular with students than the traditional course, taking the level of attendance through the year as an indicator. In the same period up to the First World War Perry also contributed to the process of dissemination by giving short summer courses at South Kensington on the teaching of practical mathematics, as well as courses on mechanics, practical geometry and graphics (Board of Education, 1904). It should also be added that in 1903 it was the Institution of Civil Engineers which appointed a Committee to consider the general education of all classes of engineer. The Committee reported in 1906, following a survey, and supported a number of Perry's proposals for a more practical and broader approach in the schools, including trigonometry, logarithms, contracted methods and mental arithmetic in particular (Thompson, 1911, p. 287). What is remarkable is that the non-engineers were also carried forward on the wave of curriculum reform.

There is not space in this paper to pursue Perry's general campaign and the various responses of the British, Mathematical and Assistant Masters' Associations, the Headmasters' Conference, the Board of Education, the various important examining bodies, and the educational suppliers and publishers, all of whom became involved in what became known as the 'Perry movement' in Great Britain, 'Perryismus' on the Continent and the 'Laboratory Method' in America. Such was the range of Perry's influence (Young, 1920, pp. 87–121; Wolff, 1915; Perry, 1901; British Association, 1907). However, Perry's early hopes for the general adoption of his scheme in toto were to prove too sanguine. At the Belfast meeting of the British Association (1903), in 1902, he suggested:

> It seems probable that at the end of another five years no average boy of fifteen years of age will have been compelled to attempt any abstract reasoning about things of which he knows nothing; he will be versed in experimental mathematics, which he may or may not call mensuration; he will use logarithms, and mere multiplication and division will be a joy to him; he will have a working power with algebra and sines and cosines; he will be able to tackle at once any curious new problems which can be solved by squared paper; and he will have no fear of the symbols of the infinitesimal calculus . . . Five years hence it will be called 'elementary mathematics'. Four years ago it was an unorthodox subject called 'practical mathematics' . . .

However, as Findlay (1902) pertinently remarked, in the same year:

> I do not think Prof. Perry and his friends quite realize . . . what a great gulf separates their work with adults and artisans from that conducted by teachers of boys and girls in school classes. (p. 184.)

Certainly the mathematical requirements for intending army officers within the public schools, and at the Royal Naval Colleges of Osborne and Dartmouth, which admitted naval cadets at 13, moved quite rapidly at an early stage towards Perry's ideals and laboratory work involving the mathematical masters was therby stimulated (Boyt, 1906; Board of Education, 1912). As Godfrey (1912, p. 431) pointed out, in 1912, concerning the preliminary education of intending engineers, surveyors and army entrants:

> . . . the teaching . . . tends to be of a more intuitional and experimental character than that of other boys . . . The Army requirements during the last 10 years have no doubt had a powerful reaction on the character of the teaching throughout English public schools.

However, in elementary and secondary schools generally 'practical mathematics' variously interpreted as both a new approach and a wider conception of mathematical possibilities, also gradually gained a hold, with some support from the important university examining bodies within the chaotic English examination system. The new requirements were accompanied by a flood of new textbooks and stimulated the supply of mathematical tables, instruments and aids of various kinds for the classroom and, of course, squared paper in a variety of forms. In all this there was clearly much enthusiasm, experimentation and also confusion in the schools. In particular, 'graphic(al) algebra' might mean different things to different teachers, and working through a chapter on 'graphs' in an arithmetic, algebra, or geometry textbook (writers generally 'cashed in' on the craze) could itself become as tedious and mechanical as the rote learning of Euclid's demonstrations[3]. However, squared paper had entered the mathematical classroom to serve a variety of purposes, and its place has not been undermined to this day. It was one of the engineer's tools by which it was thought the mathematical heritage might best be made available to all.

NOTES

[1] This is probably also true of the continent, though Cajori (Cajori, 1917, p. 299) found Strehlke (1842) recommending "the graphic representation of functions . . . [as] an effective aid in instruction" as early as 1842.

[2] In his practical physics texts, Perry's successor at Clifton College, A. M. Worthington, still used ratios to test the proportionality between results. Thus in Worthington (1886, p. 131) the law of the simple pendulum is evaluated by recording observations of L/T^2 for different lengths and periods. Only at the end of the book and in only *one* experiment, on the calibration of the galvanometer, is 'a square foot of millimetre curve paper' used. (See Worthington 1886, pp. 294, 308.) We thank T. B. Akrill, Stone Science Librarian at Clifton College for this argument.

[3] These general remarks are partly based on a scrutiny of a complete run of the monthly *School World,* 1899–1918, which contains dozens of articles, reviews, reports, commentaries and letters on mathematical education. The *Journal of Education* and the

Educational Times are similarly illuminating for the period. The *Mathematical Gazette* on the other hand is not very helpful concerning general curricular tendencies until about 1907.

REFERENCES

Abbott, P.: 1912, 'The preliminary mathematical training of technical students' in Board of Education, *Special Reports on Educational Subjects,* Vol. 27, HMSO, London, p. 10.

Armstrong, H. E.: 1920, 'Obituary. Prof. John Perry, F.R.S.' *Nature* **105,** 751–752.

Ayrton, W. E.: 1887, *Practical Electricity,* Cassell, London, p. 30.

Bell, G. M.: 1912, 'Practical mathematics at Winchester College' in Board of Education, *Special Reports on Educational Subjects,* Vol. 26, HMSO, London, pp. 427–428.

Board of Education: 1900, *Report of the Board of Education 1899–1900,* Vol. II, HMSO, London, p. 35.

Board of Education: 1904, *Report of the Board of Education for the Year 1903–04,* HMSO, London, p. 18. See also *Report of the Board of Education for the Year 1907–08* (1909), p. 81 and *Report of the Board of Education for the Year 1911–12* (1913), p. 136.

Board of Education: 1909, Circular 884, *The place and Use of Graphs in Mathematical Teaching,* HMSO, London, p. 11; note also Board of Education: 1909, Circular 711, *Teaching of Geometry and Graphic Algebra in Secondary Schools,* HMSO, London.

Board of Education: 1912, *Suggestions for the Consideration of Teachers and Others Concerned in the Work of Public Elementary Schools. Instalment No. 3,* HMSO, London, rev. ed., p. 8.

Board of Education: 1912, *Special Reports on Educational Subjects.* Vol. 26, HMSO, London, pp. 183–219, 393–428.

Boyt, J. E.: 1906, 'Reviews. Practical (or experimental) mathematics', *Mathematical Gazette* **3,** 294–300.

Branford, B.: 1908, *A Study of Mathematical Education,* Clarendon Press, Oxford, p. viii.

British Association for the Advancement of Science: 1903, *Report of the Seventy-Second Meeting . . . Belfast 1902,* 719.

British Association for the advancement of Science: 1907, *Report of the Seventy-Sixth Meeting . . . York 1906,* 444–459.

Brock, W. H.: 1979, 'An experiment in technical education', *New Scientist* **84,** 622–623.

Brock, W. H.: 1975, 'Geometry and the universities: Euclid and his modern rivals 1860–1901', *History of Education* 4, 21–35.

Bryan, G. H.: 1902, 'The British Association discussion on the teaching of Mathematics', *School World* **4,** 88–91.

Cajori, F.: 1917, *A History of Elementary Mathematics,* Macmillan, New York, 2nd ed.

Castle, F.: 1901, *Practical Mathematics for Beginners,* Macmillan, London, pp. v–vi.

College of Preceptors: 1903, 'The teaching of elementary geometry', *Educational Times* **56,** 456–467.

A Correspondent: 19 June 1903, 'The reform in mathematical education', *Engineering,* 803–805.

Eggar, W. D.: 1903. *Practical Exercises in Geometry,* Macmillan, London, p. v; also Eggar, W. D.: 1901, 'The co-ordination of the teaching of elementary mathematics and physics', *School World* **3,** 361–363.

Fawdry, R. C.: 1912, 'Practical mathematics at Clifton College', in Board of Education, *Special Reports on Educational Subjects,* Vol. 26, HMSO London, pp. 400–402.

Findlay, J. J.: 1902, 'The teaching of mathematics: impending reforms', *Educational Times* **55**, 184–187.

Gay-Lussac, J. L.: 1819, *Annales de Chimie et de Physique* **11**, 181.

Godfrey, C: 1902, 'The teaching of mathematics at preparatory schools', *School World* **4**, 288–291. See also pp. 201–203, 243–245.

Godfrey, C. and Bell, G. M.: 1905, *Note-book of Experimental Mathematics,* Arnold, London.

Godfrey, C.: 1912, 'Methods of intuition and experiment in secondary schools', in Board of Education, *Special Reports on Educational Subjects,* Vol. 26, HMSO, London, p. 437.

Herschel, J. F. W.: 1833, 'On the investigation of the orbits of revolving double stars', *Memoirs of Royal Astronomical Society* **5**, 171–222, quoted in Tilling (1975), p. 110.

Holmes, H. T.: 1910, 'Elementary mathematics in evening schools', *Mathematical Gazette* **5**, 200–202.

Japan Weekly Main: 1878, **2**, 1129–31; quoted in Perry (1900), p. 102.

Jenkins, E. W.: 1979, *From Armstrong to Nuffield,* Murray, London, pp. 30–33.

Laisant, C. A.: 1904, *L'Education fondée sur la science,* Paris, quoted in Young (1920).

MacGillivray, D.: 1902, 'Correspondence. The teaching of mathematics', *School World* **4**, 38–39.

Montgomery, R. J.: 1965, *Examinations: An Account of their Evolution as Administrative Devices in England,* Longmans, London, pp. 29–30.

Nunn, T. P.: 1903, *William Ellis Endowed School. A Syllabus of the Courses in Mathematics and Science followed in the Upper School,* ESA, London.

Ostwald, W.: 1894, *Physico-Chemical Measurements,* Macmillan, London.

Perry, J.: 1899, *Practical Mathematics – Summary of Six Lectures Delivered to Working Men,* Department of Science and Art, London, p. 27. Revised in 1913 as *Elementary Practical Mathematics,* Macmillan, London.

Perry, J.: 1900 *England's Neglect of Science,* Fisher Unwin, London.

Perry, J. (ed): 1901, *British Association Meeting at Glasgow, 1901. Discussion on the Teaching of Mathematics,* Macmillan, London.

Perry, J.: 1902, 'Reform of mathematical education', *Engineer* **93**, 203.

Perry, J.: 1908, 'The correlation of the teaching of mathematics and science', *School World* **10**, 459–464.

Perry, J.: 1912, 'Practical Mathematics', *Nature* **90**, 34–35.

Perry, J.: 1926, Quoted in Perry's obituary notice, *Proceedings Royal Society* **IIIA**, i–vii.

Price, M. H.: 1976, 'Understanding mathematics a perennial problem?', *Mathematics in School* **5**, 18–20, 30–31.

Prout, W.: 1813, 'Observations on the quantity of carbonic acid emitted from the lungs during respiration', *Annals of Philosophy* **2**, 328–343.

Salmon, W. H.: 1903, 'Squared paper', *School World* **5**, 169–170.

Selleck, R. J. W.: 1968, *The New Education 1870–1914,* Pitman, London.

Smith, D. E.: 1913, 'Intuition and experiment in mathematical teaching in the secondary schools', in International Congress of Mathematicians, *Proceedings of the Fifth International Congress of Mathematicians,* Vol. 2, Cambridge University Press, Cambridge, pp. 611–632.

Strehlke, F.: 1842, 'Uebungsaufgaben für Schüler', *Grunert's Archiv der Mathematik und Physik* **2**, 111–112.

Thompson, S. P.: 1911, 'Mathematical and scientific subjects in the school education of engineers', *School World* **8**, 286–288.

Tilling, Laura: 1975, 'Early experimental graphs', *British Journal for the History of Science* **8**, 193–213.

Wolff, G.: 1915, *'Der Mathematische Unterricht der Höheren Knabenschulen Englands,* Teubner, Leipzig, pp. 67–84.

Wormell, R.: 1888, *Plotting or Graphic Mathematics,* Waterlow, London.

Wormell, R.: 1900, 'Unstable questions of method in the teaching of elementary science', *Educational Times* **53**, 240–243.

Worthington, A. M.: 1886, *A First Course of Physical Laboratory Practice,* Rivingtons, London.

Young, J. W. A.: 1920, *The Teaching of Mathematics in the Elementary and the Secondary School,* Longmans, New York, 2nd ed., p. 111.

XIV

Science, Technology and Education in *The English Mechanic*

The story of the *English Mechanic* is one of the strangest in science journalism — a cheap mass-circulation science journal run co-operatively with its largely working-class readers. Although not founded until 1865, its roots lay in the Mechanics' Institution movement of the 1820s.

As J.N. Hays has suggested in an essay on lecture courses in nineteenth-century London, during the nineteenth century 'curiosity about technical improvements was extensive in London across a wide social spectrum'.[1] Just as provincial industrialization is most associated with steam engines, textile machinery, iron founding, railways and coal mining, London was pre-eminent for ship-building, instrument- and watch-making, brewing and, as shown by patent registration, an extraordinary variety of 'improvements' and 'inventions' ranging from printing, carriages and musical instruments to umbrellas, sewing machines, typewriters (tachy typographs) and velocipedes (bicycles). However one interprets the mechanics' institutes movement,[2] it was for this reason no accident that the first institution to use this precise title was founded in London in 1824. Equally significantly, however, the foundation of the London Mechanics Institute (and those that followed rapidly in provincial centres

---

1 J. N. Hays, 'The London lecturing empire, 1800-50' in I. Inkster and Jack Morrell, eds., *Metropolis and Province. Science in British Culture* (London, 1983), pp. 91-119 (107). See also Thomas Kelly, *George Birkbeck* (Liverpool, 1957) and J. F. C. Harrison, *Learning and Living, 1790-1960* (London, 1961).

2 T. Kelly, *George Birkbeck* (Liverpool, 1957); S. Shapin and B. Barnes, 'Science, nature and control: interpreting Mechanics' Institutes', *Social Studies of Science,* 7 (1977), pp. 31-74; I. Inkster, 'The social context of an educational movement: a revisionist approach to English Mechanics' Institutes', *Oxford Review of Education,* 2(1976), pp. 277-307.

during the next few years) came after the establishment of a weekly journal in 1823 — the *Mechanic's Magazine*.

Although the *Mechanic's Magazine* was described by George Birkbeck as 'the most valuable gift which the hand of science has yet offered to the nation', the quality and significance of the journal declined in parallel with the fortunes of the mechanics' institutes themselves, becoming, by the 1850s, little more than a digest for patents for its owner, the patent agent R. A. Brooman. In July 1857 it was leased to a rising Cornish journalist and future philanthropist, John Passmore Edwards (1823-1911), who, following a large number of unsuccessful publishing ventures that had bankrupted him, was beginning his Smilesian 'rags-to-riches' climb through a policy of purchasing and rehabilitating ailing periodicals. However, since Edwards was not allowed a completely free hand with *Mechanic's Magazine*, he abandoned it in 1867, looking for a more amenable proprietor.[3] He found one in George Maddick (1824-81), the printer, publisher and owner of the *Court Circular* and *Sporting Life*, who had begun a new cheap weekly called *The English Mechanic* in 1865. It was Maddick who established the co-operative tone of the magazine which Passmore Edwards was to sustain and cultivate with phenomenal success.

By the 1860s, the railways had settled down to their period of steady expansion and development which, despite economic fluctuations, enabled Great Britain to retain its position as the workshop of the world. The whole country seemed to be changing over from agricultural to industrial pursuits. Engineering, in particular, began to assume considerable significance. Although *The Engineer* had been founded by Zerah Colburn as early as 1856, with such a rapid expansion of industries dependent upon mechanical progress, Maddick perceived that there was room for a cheaper weekly journal which dealt with engineering matters and the associated sciences in all their ramifications — in particular, a need for a journal which would appeal to the pride and skills of the employees of large and small manufacturing businesses as well as to self-employed inventors and improvers. Accordingly, on Friday 31 March 1865, Maddick launched the penny weekly *English Mechanic: A Record of Mechanical Invention, Scientific and Industrial Progress, Applied Chemistry, Arts, Manufacture, Engineering, Building, etc.* Its immediate success is best gauged from the fact that within a month the government's

---

3 *Mechanic's Magazine* struggled on until 1872 when it reappeared as *Iron*, finally becoming *Industry* in 1893.

Department of Science and Art began to advertise its examinations in the journal; and within a year it had expanded from 12 to 16, to 20, to 24, to 32 pages by September 1869, doubled its price to 2d in January 1866 (which it held until 1916), and achieved a claimed circulation of 80,000 copies (1868).[4]

In 1869, two years after having abandoned the ailing *Mechanic's Magazine*,[5] Passmore Edwards seized the chance of buying the *English Mechanic* from Maddick. For the next few years, under his ownership and the assistance of his political disciple, Ebeneezer Job Kibblewhite (1847-1928), the journal became 'a literary Aaron's rod', swallowing all opposition: *The Mechanic* (f. 1868), *World of Science* (f. 1867), *Penny Mechanic* (f. 1868), *Scientific Opinion* (1868-70). 'Though the individual dies the nation lives,' commented Edwards. 'Though scientific journals perish individually, the *English Mechanic*, by some means or other, absorbs the best portions of their animating spirit.'

The First World War caused serious damage to the journal by imposing cuts on its size, and by literally killing off its contributors and readers. Conscious of its outdated artisan's image, the title was altered to *English Mechanics* in 1923, but the expected readership of white-collar technical hobbyists did not materialise. (As early as September 1870 a reader had suggested a better title would be *The Amateur Philosopher* to prevent new readers thinking solely about machinery. The editor agreed it was to all intents and purposes a scientific journal and 'we have often regretted that a more generic name was not given to it at the commencement'.) The end came with the General Strike in 1926 when the paper was forced to amalgamate with *The Amateur Mechanic* — produced since 1890 by Edwards's old rival in popular journalism and education, Cassell & Co. Under the new name of *English and Amateur Mechanics* the journal did find a fresh audience intrigued by the wonders of wireless. In 1940 it was transformed again into a monthly *Home Mechanics* which, however, with the disappearance of interest in technical hobbies in the age of television, collapsed for ever in April 1959.[6]

---

4 Self-proclaimed circulations must be treated with caution. An 1869 editorial addresses 100,000 readers!

5 Between 1867 and 1869 Edwards had bought up another weekly, the *Building News* (f.1854), which he skilfully cultivated as a cheaper rival to Joseph Hansom's and William Godwin's *The Builder* (f.1842).

6 This account has been drawn freely from my 'Development of Commercial Science Journals in Victorian Britain' in A.J. Meadows, ed., *Development of*

To the modern eye, the early volumes are not particularly attractive to read. Until the First World War, each volume contained over 500 pages packed with brevier type set in three columns relieved only by woodcut engraving. The journal had originally appeared in small quarto (11" x 9"), but from vol. 11 (March 1870) the page size was expanded to foolscap (12" x 8") and the illustrations improved in quality and in number.[7]

Within a few years, much the most important feature of the journal had become its letters column. Queries were much encouraged by the advent of the penny postcard in October 1870. The editor repeatedly asked correspondents to sign their letters if they were merely queries so that they could be briefly answered within one of the columns; he pleaded that if pseudonyms were chosen, writers should not say ‘a subscriber’ or ‘a well-wisher’ since there were hundreds of these. Choose instead an ancient Greek name or an invention, he advised; consequently we find the paper riddled with velocipedarians, alphas, omegas and sigmas — the latter, an electrician named Sprague, soon became a regular columnist. By 1900 fewer pseudonyms were being used, showing, perhaps, that public education gave readers more confidence in the points they wished to make.

Longer letters containing information, comment and criticism were welcomed, and by 1870 had come to occupy 9-10 pages (25%) each week. Such letters were always prefaced by a quotation from Montaigne which was obviously dear to Passmore Edwards's philosophy:

> I would have every one write what he knows and as much as he knows, but no more; and that not in this only, but in all other subjects. For such a person may have some particular knowledge and experience of the nature of such a person or such a fountain that, as to other things, knows no more than what everybody does, and yet to keep a clutter with this little pittance of his, will undertake to write the whole body of physicks: a vice from whence great inconveniences derive their original.

Science Publishing in Europe (Amsterdam, New York, Oxford, 1980, pp. 95-122 (111-115).

7 During its early years, the engraver was Arthur Taylor, brother of the publisher, Richard Taylor. See W.H. Brock and A.J. Meadows, *The Lamp of Learning. Taylor & Francis and the Development of Scientific Publishing* (London, 1984).

Clearly this was a warning, 'stick to the point' and don't pontificate about things you don't know much about. But, of course, the great charm of the column was always that writers did not adhere to Montaigne.

There were many regular letter writers — especially 'Harmonious Blacksmith' who once (1869) had three letters in one issue occupying 10 columns. The eminent amateur astronomer, Richard Proctor, began to write regularly for the journal in the same year (1869) when answering an attack upon himself. He said later that he found the journal so interesting that he continued to read it and was thus sucked into its correspondence columns and from thence into the role of contributor. Proctor was a great controversialist, and had much fun in the 1870s in exposing 'paradoxists' who doubted Newton's laws of physics or who believed the earth was flat. One of his challengers was the notorious John Hampden whom Alfred Russel Wallace, the biologist, fought in a court case in 1871.[8] Proctor began a 'Column of Paradoxes' in 1873 in order to encourage right thinking, and the same year saw the establishment of regular mathematics, microscopy and medical columns.[9] The letters column demonstrates that *English Mechanic* was not all science, for there were earnest debates on Swedenborg, vegetarianism, spiritualism, the age of the earth, the Great Pyramid, and the patent laws.

Many other early features established by Maddick were continued and expanded by Passmore Edwards. The exception was an 'Employers' Wanted' column, introduced in the second issue for 1865, which was soon regularly filled with advertisements for jobs in architects' offices, for foremen bricklayers, civil engineers, draughtsmen and plumbers. But this *free* feature disappeared with Edwards who began to charge a shilling for 20 words for 'Situations vacant or wanted'. Apparently noticing a lack of take up, Edwards reintroduced an 'Employment' column at the rate of 6d. for 16 words in August 1870. He was rewarded immediately with advertisers who needed millwrights, asphalters, tar distillers and paperhangers.

Maddick had also introduced an 'Exchange Club' — a sort of Exchange and Mart — in January 1866 in which readers could freely advertise articles they wished to dispose of. There was a charge of a shilling per 20 words if they

---

8 A. R. Wallace, *My Life,* 2 vols. (London, 1905), vol.2, pp.365-76.

9 W.H. Brock, 'Medicine and the Victorian scientific press', in W. F. Bynum, S. Lock and R. Porter, eds., *Medical Journals and Medicial Knowledge* (London, 1992), pp.70-89.

were selling the article for money. By 1870 the charge had risen to 2s. per 30 words. A columnist of 1924 claimed, probably rightly, that these advertisements laid the foundation for many small businesses.

As to business advertising proper, the *English Mechanic* rapidly became a significant advertising medium. The early issues had only one, to one and a half, pages of advertisements, but by 1870 the advertising pages had risen to 10 (or 25% of the copy) and were of the order of 3s. for 5 lines, or £8. 8s. 0d. per page. A rough calculation suggests that the total advertising revenue per issue was between £60 to £80. With advertising revenue covering a substantial portion of the printer's bill, it is not difficult to understand how Edwards became a millionaire.

The advertisers cover a range of engineering and consumer products. Random inspection shows that the *English Mechanic* was a significant outlet for the philosophical instrument maker and for anyone selling what today would be called DIY equipment. For example, because of astronomical articles, which encouraged the making of lenses and mirrors for telescopes, we find advertisers offering telescopes for as little as £5. Such advertisements and associated articles also had other significant effects. Astronomy enthusiasts formed their own *English Mechanic* astronomy club in Liverpool in 1881, and its members became the nucleus of the British Astronomical Association founded in 1890.[10] One Liverpudlian, Jesse Nimkey, wrote proudly to the *English Mechanic* in 1889 to say that the chance purchase of the journal from a railway bookstall had led him to buy a 4 inch telescope advertised in the journal. Subsequently he had joined the Liverpool Astronomical Society, and had just been elected a Fellow of the Royal Astronomical Society.

Many other mutual improvement clubs and groups were encouraged by the journal. Readers in Edinburgh and Manchester formed *English Mechanic* clubs in 1870. The Edinburgh group became the Edinburgh Society of Arts. Some Londoners formed a Chemical Club with a laboratory in 1866, and there was much talk of Velocipede Clubs in the Midlands, and even of an *English Mechanic* colony in Tasmania (1871).

---

10 *English Mechanic*, 18 July 1890, pp 445-6; *Journal of the British Astronomical Association*, 1 (1890-91), p. 15. See also J. Gould, 'The *English Mechanic* and the £5 telescope', *Bulletin Scientific Instruments Society*, 10 (1986), pp. 11-12.

The *English Mechanic* was the perfect venue for inventors. To give but one example, Robert Paul (1869-1943), the son of a London shipowner, who trained at Finsbury Technical College as an electrical engineer in the 1880s, had begun his own electrical instrument-making company in London in 1891. In the same year he became interested in cinematography. Because the Americans refused to supply him with films, he began to make his own, and in 1896 he described in the *English Mechanic* his new Theatrograph, or projector. Paul stayed in the cinema or biograph business until the early 1900s, when he returned to scientific instruments, selling his firm to the Cambridge Scientific Instrument Company in 1919 to form The Cambridge and Paul Instrument Company.[11]

No doubt one could find similar connections with the journal of bicycle and car manufacturers. There were, for example, lots of motor car articles in the mid-1890s from F. C. Blake and T. Hyler White, both of whom became prominent automobile engineers in the early twentieth century.

There were female readers. That the journal had aimed at them from the start is clear from the way Maddick heavily advertised the *English Mechanic* in his new weekly newspaper, *The Lady's Own Paper* (founded November 1866, folded 1872), and that an occasional letter appears — as in August 1867 when a woman asked how to repair a broken sewing machine. More spectacularly, we know that the sheltered Mary Kingsley, who was forced to look after her invalid father, regularly found an outlet in the *English Mechanic*. As one biographer wrote in 1857:

> A burst pipe, or a failing stove, or a necessary job of carpentry gave her a chance to put its teaching to the test, and though she made mistakes, she learned how to use her hands — an invaluable preparation for the time when she would be in Africa, far from any help.[12]

A few editorial problems are detectable behind the paper's benign image. Early in the first volume, Maddick had to apologise to a firm of engineers (Messrs. Holtapfell) for plagiarising from their book *Mechanical Manipulation*.

---

11 M. J. G. Cattermole and A. F. Wolfe, *Horace Darwin's Shop. A History of the Cambridge Scientific Instrument Company 1878 to 1968* (Bristol, 1987), pp. 98-104.

12 Cecil Howard, *Mary Kingsley* (London, 1957), p.14. Another reader was the young physicist, Oliver Lodge. See his *Past Years. An Autobiography* (London, 1931), p. 67.

And, intriguingly, in the tenth volume there is a tantalisingly obscure reference to Edward Heuri Todé, editor of the *Fenian* and the 'obscene' *Illustrated Police News*, who is acidly described as no longer having *any* connection with the journal. A later fleeting reference in October 1869 implied Todé had been undermining the journal prior to Edwards's purchase. Todé seems to have founded the rival *British and Foreign Mechanic in 1869*; but as we have seen, he was soon bought out by Edwards.

What, then, does the *English Mechanic* tell the historian about education and, in particular, technical education?

Firstly, and most obviously, the journal was itself an experiment in technical education — a polytechnic offering a weekly course in a diversity of technologies. This policy, first established by Maddick, was continued and expanded by Edwards. When, following the inquest over Britain's relatively poor showing at the Parish Exhibition in 1867, readers asked 'What is this technical education?' which various commentators had identified as the reason for continental countries' greater success, the *English Mechanic* answered that it was what was to be found in the journal each week (26 July 1867). 'Our one great object,' it stated in August 1867, 'has been to educate, and so elevate the British workman — no matter what his class of labour may happen to be.' For example, the early volumes contain a long-running anonymous series on 'The Lathe and its Uses' (by James Luckin) which was later published in book form. This well-illustrated series gave clear and quite elaborate instructions for turning wood and metal and for making a large number of appliances and ornaments. Another author, Hermann Smith, who either was or became an organ builder and musicologist, serialised a 'Practical Treatise on the Harmonium', which provided instruction on how to build the instrument. Smith also regularly advertised 'The English Mechanic Harmonium', and says his articles have aroused the interest of organists and clergy, thus showing the *English Mechanic* is no mere class journal, but is 'household and national'. Indeed, Edwards noted rhetorically when he took over in 1869:

> our subscribers and correspondents are members of the largest mutual aid society that ever existed. The members of this society are composed of workmen, clergymen, inventors, students, chemists, botanists, astronomers, microscopists, mechanicians, engineers, curators of museums, travellers and others who are all more or less

inspired with a disposition to get from each other, and to give to each other, items of practical knowledge.

A letter from a Hull reader ('a man at the vice') had summed up readers' needs in April 1865. 'Don't make your paper too learned, but give us a little of all sorts of information.' And, in active collaboration with the readers, this was precisely what Maddick, and then Edwards and Kibblewhite, did. The *English Mechanic*'s sustained interest in the improvement of the velocipede is probably particularly significant. A long series of articles by Derwent in the 1880s gave precise instructions on how to build bicycles cheaply, and, given the known close economic and technical association between the manufacture of bicycles and the early motor car, it is not surprising, perhaps, that the *English Mechanic* appears to have been the first journal, in April 1896, to provide working instructions on building an automobile — *i.e.*, before the Act of Parliament allowed such machines on the road in the same year. Such articles, whether on the subject of bicycles, harmoniums, lathes, oil engines, phonographs, automobiles or, later, wireless sets, appear to have served a directly educational function and allowed many general mechanics, engineers and skilled workmen to specialize and start businesses in significant new technologies. The journal is also known to have determined at least one scientist, the physicist Oliver Lodge, the later Principal of the University of Birmingham, upon his career.[13] Yet others were significant contributors: Philip Gosse, the naturalist; Flanders Petrie, the future egyptologist, who wrote articles on chemistry; Oliver Heaviside, the mathematician; Lord Rayleigh, the physicist; the two astronomers, Richard Proctor and Andrew Noble (who used the pseudonym FRAS); Lord Grimthorpe (Sir Edward Beckett), who wrote about clocks; and the electrician, Cornelius Varley.

Secondly, but not surprisingly, each successive editor of the *English Mechanic* was explicitly interested in the promotion of technical education. Maddick's interest, in the early volumes, took the form of an active promotion of what might be called 'the post-1851 working men's exhibition movement' — a subject which deserves further study.[14] Thus, in reviewing the South London Exhibition in the very first issue in March 1865, Maddick felt that the event:

---

13 W. P. M Jolly, *Sir Oliver Lodge* (Cranbury, N.J., 1974), p.25

14 For pre-1851 see Toshio Kusamitsu, 'Great Exhibitions before 1851', *History Workshop*, No.9, Spring 1980, pp. 70-89.

> proved to the nobly descended and moneyed classes of England, that however high they may have graduated at Oxford or Cambridge, they have yet much to learn from the thinking and inventive mechinic. Greek, Latin and deep dippings into the classics and modern polite literature when backed by agreeable manners and the science of a Bond-street tailor, may enable the most soft-brained sprig of nobility to keep afloat in society so long as his tradesmen have faith. But the wonderful progress of science has even awoke something like a feeling of interest among the vapid votaries of fashion, and it is thought the thing in the higher circles of society to patronise the mechanic by visiting industrial exhibitions, to stare at and admire the products of thoughtful brains and skilled hands: and to return home with a vague notion that an inventive brain cannot be purchased like the freehold of a ten-acre field.

As this quotation suggests, there was never any doubt that the *English Mechanic* was on the side of the workman and always willing to encourage his aspirations to receive a fair wage and a fair shate of leisure for his efforts. Not surprisingly, therefore, the journal welcomed the news that the 'Brompton Boilers' at South Kensington (prefabricated buildings erected for the 1862 Exhibition) were to be sold off as re-erectable 'local museums of art'. (The most famous survives as the Bethnal Green Museum.) Similarly, its editors encouraged educational cooperation among working men by publicising their efforts to set up improvement societies, clubs, laboratories and workshops for the study of chemistry, lathe work and bicycle-making; and it was prepared, when necessary, to argue forcefully for a wage structure that would catalyse this sobriety and industriousness.

> [The 10 per cent wage increase being demanded in 1866] will pay only for that increase in rent which has taken place so recently and for the supply of those former luxuries of books and instruments which a true mechanic now looks upon as necessaries; or, if he prefers to live in cleanliness and health on the outskirts of the town, it will just pay for his 'workman's train' tickets... (9 March 1566).

Those who patronised working people by urging sobriety, cleanliness, industriousness and attendance at night school had to understand that these things were only possible if the workman was decently paid. Needless to say, the journal also strongly supported the Sunday opening of museums, galleries and lecture rooms.

Following the Paris Exhibition, the *English Mechanic* became explicitly concerned with the debates on secondary and technical education that ensued. 'We must meet the foreigner on the field of education,' it declared.

> We demand, then, a thorough reform in our educational system, or, rather, we require an educational system to be formed. However much the nation may have to pay for it, the cost will very soon be replaced.
> (23 August 1867)

Besides welcoming the deliberations of the Taunton and Samuelson Commissions,[15] for its own part the journal urged the reactivation of the decaying mechanics' institutes (and was always pleased to learn that some were still active) together with a vitalisation of the efforts of the Department of Science and Art to reach the artisans who really needed the system. Subsequently, it warmly welcomed Joseph Whitworth's endowment of thirty £100 scholarships and the renewed activities of the City Guilds. Indeed, through lively correspondence columns, the journal rapidly became the helpmate of young adults taking evening classes all over the country and, curiously, the chief vehicle for the dissemination of the new formulae and ideas of structural organic chemistry.[16]

During the 1870s, the journal became much more science oriented, the diffusion of Frankland's South Kensington chemistry becoming particularly prominent and significant. Astronomy was also brilliantly popularized by Noble, Proctor and others. In other words, the journal was catering for a new kind of 1870s science reader: the science student taking Department of Science and Art (and, later, City and Guilds) examinations or even London Matriculation, and the science and art teachers themselves. Thus we find abstracts of papers given at Chemical Society, Royal Astronomical Society, and even the Royal Society, and a new column of scientific intelligence. If we bear in mind that J. Norman Lockyer's *Nature*, with its lower circulation, was also seeking the same readership, it is not surprising, perhaps, that there was an anti-*Nature* tone to many *English Mechanic* articles. Suffice to say that certain of the *English Mechanic*'s regular correspondents in the 1870s and 1880s, particularly Noble and Proctor, were rabidly anti-South-Kensington, anti-

---

15 *I. e.*, the Taunton Commission on education for the middle classes, 1867-68; the Samuelson Commission on technical instruction, 1882-84.

16 C. A. Russell, *A History of Valency* (Leicester, 1971), chapter 4.

Department of Science and Art, anti-Lockyer and anti-the 'endowment of research' movement propagated by the Devonshire Commission.[17]

By scrutinising this technical journal, one can not only watch informal education at work, but begin to see some of the underlying tensions and suspicions that lay behind the emergence of the state's growing *rapport* with science and technical teaching.

Both editors and readers saw the journal as a venture in cooperation. As the journal's only historian remarked in 1924:

> How much this was appreciated may be judged by the fact that the Letters to the Editor, Replies and Queries frequently occupied over sixteen pages weekly, each number being a perfect mine of information on a variety of subjects... Men often give of their best in voluntary labour, and this is so with regard to the many hundreds of readers who have spent their time in helping others and exchanging views... Help and advice could be got gratis in the *English Mechanic* which a hundred guineas could not purchase ordinarily, and while the *nom-de-plume* of many of the late contributors must still be respected the names of the late Lords Grimthorpe (Sir Edmund Beckett (1816-1905), lawyer, architect, horologist) and Rayleigh can be taken as cases in point, while many of the leading scientists and engineers of today acknowledge their debt to the *English Mechanic* from an educational point of view.[18]

The *English Mechanic* was founded in 1865 at a time when British light engineering industries were making rapid progress. By appealing directly to the pride and skills of the employees of large and small manufacturing businesses, as well as to self-employed inventors and improvers, it rapidly achieved sales of 50,000 copies a week or more. As a 'record of mechanical invention, scientific and industrial progress, applied chemistry, arts, manufactures, engineering, building etc.' the journal was a literary Polytechnic, offering weekly serialised courses on a diversity of technologies, many of which (such as those on bicycle- and automobile-making) may have led readers into starting specialised businesses. Following the 1867 Paris Exhibition and the

---

17 R. M. MacLeod, 'Resources of science in Victorian England: the endowment of science movement, 1868-1900' in P. Mathias, ed., *Science and Society 1600-1900* (Cambridge, 1972), pp.111-66, esp. pp.155-58.

18 D. J. Smith, 'A History of *English Mechanics*', *English Mechanics*, 18 January 1924, pp. 5-6. See also 26 December and 25 January 1925.

fears of decline it evoked, the journal also became directly concerned with the promotion of technical education and its pages, particularly its lively correspondence columns, became the chief medium by which 'night school' pupils and their teachers aired their problems, grievances and enthusiasms.

XV

The Decline of Science

Charles Babbage's *Reflections on the Decline of Science in England* was first published in 1830, and reprinted in 1971. Once, when I was reading the smartly reprinted edition, a physicist colleague came into my room. He picked up a book from my desk, inspected the title, and commented: 'Is science declining in England? Another Doomsday Book!'

That colleague's reaction underlines an important point. Ever since Babbage's trenchant book was published, there have been scientists who have said that the scientific estate is declining — from being pleasantly satisfactory, things are now getting worse.

Here is Babbage in 1830:

> It cannot have escaped the attention of those, whose acquirements enable them to judge, and who have had the opportunity of examining the state of science in other countries, that in England, particularly with respect to the more difficult and abstract sciences, we are much below other nations, not merely of equal rank, but below several even of inferior power.[1]

Here is the Lancashire chemist Edward Frankland writing forty years later, in 1871:

> Not only are we behind in the aggregate of activity in discovery, but our individual productiveness is also markedly below that of Germany and France... It is highly remarkable that a country which, perhaps more than any other, owes its greatness to the discoveries of experimental

---

1 C. Babbage, *Reflections on the Decline of Science in England and on some of its Causes* (London, 1830; reprint Farnborough, 1969), p. 1.

> science, should be distinguished for its neglect of experimental research.[2]

Four decades later, during the Great War of 1914-18, we hear of a 'Neglect of Science Committee'[3], while in our own day we are familiar with that term from the 1960s — the 'Brain Drain' — and its supposed disastrous effects on our scientific and technological world leadership.

We must not think that the perennial argument about the decline of science was, or is, something peculiarly British. In the 1860s, for example, French scientists complained about the lack of research facilities in France compared with Germany;[4] and a century later the launching of the Soviet Sputnik in 1956 was utilised with singular effect by American scientists and educators.

It seems, then, that 'Declinism' has been a pretty standard response by the scientific community to certain historical situations from 1800 onwards. It is the task of the historian to understand how and why scientists as a group have found it *expedient* to use *threats* against society. *Threats*, however, because what the scientific community appears to say is, unless certain things are done there will be certain inevitable, and usually unpleasant, consequences. For example, Lyon Playfair, one of the organizers of the Great Exhibition in 1851, wiped away any feeling of British self-satisfaction at its success by saying:

> As surely as darkness follows the setting of the sun, so will England recede as a manufacturing nation, unless her industrial populations become much more conversant with science than they are now.[5]

However, historical analysis suggests that scientists can make imperfect diagnoses — that they may exaggerate certain symptoms, and can in fact exploit the patient for their own ends - that of the professionalization of science itself.

---

2 E. Frankland, 'Chemical research in England', *Nature*, 3 (1870-71), 445. Frankland reported this evidence before the Devonshire Commision, *Parliamentary Papers* 1874 [C.884] XXII, 4th Rep., p. 35.

3 See D. Layton, *Interpreters of Science* (London 1984), pp. 29, 193-8.

4 H. W. Paul, *From Knowledge to Power. The Rise of the Science Empire in France 1860-1939* (Cambridge, 1985); Mary Jo Nye, 'Scientific decline. Is quantitative evaluation enough?' *Isis*, 75 (1984), 697-708.

5 L. Playfair, in *Lectures on the Results of the Great Exhibition of 1851* (London, 1852), p. 202.

In order to demonstrate this we can examine Babbage's argument, and ask whether his analysis was a fair one. Was *British* science declining in 1830? There are six points we have to think about.

The first one arises from the title of Babbage's book which, while clearly embracing the whole of the United Kingdom, lazily uses the word 'England' not 'Britain'. In 1830, over a century after the Union between Scotland and England, the word England was still used indifferently for Britain. Not surprisingly, some Scotsmen found Babbage's book insulting; for, although he was supported by the Scots physicist David Brewster,[6] it was and is possible to argue that even if science was declining in geographical England, it was flourishing in Scotland.

The second, and perhaps most important, point concerns the difficulty in defining 'decline'. Did Babbage mean there was an absolute decline in British scientific activity; or merely a relative decline? That is, was scientific activity *really* decreasing; or was it increasing at a *lower rate* compared with that in another country? The problem can be clarified in terms of a definite commodity. The German chemist Liebig once suggested that the production of sulphuric acid is a barometer of a nation's industrial progress. Now, production of sulphuric acid in Britain increased during the period 1870 to 1910. But so did German production, which actually overtook British production about 1900; so that we can say Britain's share of world production of sulphuric acid declined. Of course, the fact that British production did increase will be hidden if we say that its percentage share of world production declined during this period. In any case, we should not forget the obvious — that Germany was a bigger country.

Sulphuric acid is a commodity whose production can be measured precisely. However, when it comes to scientific productivity it is much more difficult to know what to measure. There have been attempts to measure a country's scientific prosperity by counting its scientific publications. For example, in 1871 Frankland argued that British chemistry was declining because, if one took the one thousand or so chemistry papers that were published internationally in 1866, some 700 were by Germans, and some 200 by Frenchmen, but only about 100 (or roughly ten per cent) were by British chemists.

6 David Brewster, 'Decline of science in England and patent laws', *Quarterly Review*, 43 (1870), 305-42, and 'Observations on the decline of science in England', *Edinburgh Journal of Science*, 4 (1871), 261-3. On Brewster, see A. D. Morrison-Low and J. R. R. Christie, eds., *Martyr of Science. Sir David Brewster 1781-1868* (Royal Scottish Museum Studies, Edinburgh, 1984).

But what about the quality of those ten per cent? Let us suppose that those British papers were intrinsically more important than those from Germany or France. Clearly, bare statistics will not help us measure a decline or growth. They will only suggest a trend, or confirm a subjective impression. The same is true if we measure output in terms of scientific discoveries per annum, as some historians have tried to do. For example, it has been said that the British physiology declined between 1840 and 1870 in which year there were only 260 discoveries by British physiologists compared with 400 in France and 600 in Germany.[7] Here again we are left with the same problem of balancing the quality of discovery: and introduced to new difficulties over what constitutes a discovery and the historian's own national bias.

Babbage himself was gloriously vague and subjective about the decline he described. He cited no quantitative evidence whatsoever. What is clear, however, and this brings us to a third point, is that he was writing about two separate issues: the causes of the decline, or malaise, and its cure. He sees one of the principal causes of decline in the mismanagement and lack of professionalism in the Royal Society of London. (As Augustus Bozzi Granville aptly put it, Science was without a Head.[8]) A cure, then, was to reform the Royal Society - to return to the glorious days when Newton was President.

Now, attacks on the Royal Society were not new. We can find a particularly splendid and very funny example in the eighteenth-century attack by the naturalist and mountebank, Sir John Hill.[9] All historians agree that the Royal Society did need reforming in 1830. Babbage's book, and his own ‘behind-the-scenes’ activities did lead to attempted reforms, such as the election of a scientist, like John Herschel, as President. This plot in fact failed, and effective reforms, like the reduction in the numbers of Fellows elected each year, did not come until the 1840s.[10]

---

7 K. E. Rothschuh, *History of Physiology* (Huntingdon, N.Y., 1973), p. 348. See R. Hahn, *A Bibliography of Quantitative Studies on Science and its History* (Berkeley, 1980).

8 A. B. Granville, *Science without a Head* (London, 1830; reprint Farnborough, 1969).

9 C. Emery, 'Sir John Hill versus the Royal Society', *Isis*, 34 (1942), 16-42.

10 D. P. Miller, 'Behind hostile camps: Sir Humphry Davy's presidency of the Royal Society of London, 1820-1827', *British Journal for History of Science*, 16 (1983), 1-47; R. M. MacLeod, 'Whigs and savants: reflections on the reform movement in the Royal Society 1830-48',. in I. Inkster and J. B. Morrell, *Metropolis and Province. Science in British culture 1780-1850* London 1983), pp.55-90; Marie Boas Hall, *All Scientists Now. The Royal Society in the Nineteenth Century* (Cambridge, 1984).

It has been suggested that it was disillusionment with the Royal Society and the failure of reforms in the 1830s which led to the formation of the British Association for the Advancement of Science in 1831.[11] This is only partly true, however. The British Association would probably have been formed anyway in order to unite together provincial English and particularly Scottish scientists. Significantly, Babbage and Brewster were among the founders of the British Association, which they both saw as an organ for the advancement of science, that is for the professionalization of science and the arrest of its supposed declining fortunes.[12] Equally significantly, however, the opening speeches in 1831 show the historian that the new organization could only succeed providing it dropped any complaint of 'declining science'. The reason is to be found in a fourth point.

Many academic scientists found Babbage's argument and Brewster's support of it highly insulting. This was especially true of Oxford and Cambridge scientists like Adam Sedgwick and William Whewell, and scientists within the council of the Royal Society. They took it very personally. If science was really declining, then they were being blamed. Not unnaturally, they denied the decline and pointed to British talents like Thomas Young, Humphry Davy, John Herschel, and even Babbage and Brewster themselves. (Incidentally, we might well want to ask, if Britain was so backward, how did we produce a Darwin, a Maxwell, and a Kelvin in the following generation?)

However, if Babbage is read closely, it becomes clear that he was not suggesting that there was any decline in British experimentalism or applied science. But there was a failure to keep pace with other countries in mathematics and mathematical physics — the 'more difficult and abstract sciences'. We did not lack Davys and Faradays, but Laplaces and Poissons. It was pure theoretical science, not experimental applied science, that was in danger. Here Babbage was to be proved wrong. By the 1860s it had become clear to many that the danger lay in the widening gap between pure and applied science, and the failure to train experimentalists as well as theoreticians.

Babbage's complaint concerning Britain's failure to keep pace with other countries in mathematical science brings us to a fifth point: they way the 'declinist' always directs attention to another country as a model or standard of emulation. Babbage used France as his ideal; later in the century it was to be

---

11 L. P. Williams, 'The Royal Society and the founding of the British Association for the Advancement of Science', *Notes and Records of the Royal Society*, 16 (1961), 221-33.

12 J. B. Morrell and A. Thackray, *Gentemen of Science: Early Years of the British Association* (Oxford, 1978); R. M. MacLeod and P. Collins, eds., *The Parliament of Science* (London, 1981).

the turn of Germany; by about 1900 it was America and, curiously, Japan; today, it is America or Japan, Korea, Taiwan or Singapore. It should be obvious, though, that things may look very different from within the country that is being favourably painted. The Dutchman, Gerard Moll, answered Babbage in 1831. He praised the excellency of British experimental science, and British individualism.[13] By contrast, the French scientific scene that Babbage had so highly praised seemed to Moll to be riddled with jobbery, nepotism, as well as being pandering and inefficient in practice. Indeed, the historian can smile for, as Babbage wrote, the glorious days of French science had certainly passed, and by the 1860s French savants themselves were writing of decline.

This brings us to a final point about Babbage. The closer one looks at it, the more suspect does Babbage's argument become. It must also be admitted that his complaints were, on another level, extremely trivial compared with the real issues of the day: sanitation, the poor laws, child employment, and the slave trade. This triviality was increased by his complaint that scientific achievement was not rewarded by knighthoods, medals and profitable positions. Nor was his argument helped by the way he paid off old scores in the book — such as his attack on the then little-known Faraday for advising the Admiralty. According to Playfair: Babbage 'always considered himself a badly treated man, and this feeling at last produced an egotism which restricted the number of his friends.'[14] The son of a wealthy banker, he wanted financial independence from his father, and needed worldly success to support a wife and eight children. In 1827, his father, wife and two children died. He had a nervous breakdown. We must remember, then that *Decline* was written soon after a period of severe emotional stress. Brewster was an equally unreliable witness for *Decline*. He was entirely dependent on literary earnings for a living, and his continued failure to obtain a Professorship stimulated his desire for the reform of academic science.[15] Not until 1838 was he freed from financial worries when his political friendship with Lord Brougham was rewarded by the Principalship of St Andrew's University.

---

13 N. Reingold, 'Babbage and Moll on the state of science in Great Britain', *British Journal for History Science*, 4 (1968), 58-64

14 T. W. Reid, *Memoirs and Correspondence of Lyon Playfair* (London, 1899), p. 156. For Babbage's career, see M. Moseley, *Irascible Genius; A Life of Charles Babbage, Inventor* (London, 1964).

15 W. H. Brock, 'Brewster as a scientific journalist' in Morison-Low and Christie, *op.cit.* (6), pp. 37-44.

So, we must make careful note that both Babbage and Brewster had highly personal motives for their dissatisfaction with the state of science. To put it crudely, what was at stake was 'jobs for the boys'.

This brings us full circle to the question of expediency again. What have scientists got out of using declinist arguments? What usually lies behind the use of the decline argument? What really is usually proposed? The answer is scientific education.

The first chapter of Babbage's book was titled: 'On the Reciprocal Influence of Science and Education'. Says Babbage:

> It is not unreasonable to suppose that some portion of the neglect of science in England may be attributed to the system of education we pursue. A young man passes from our public schools to the Universities, ignorant almost of the elements of every branch of useful knowledge... It is in some measure to be attributed to the defects of our system of education, that scientific knowledge scarcely exists among the higher classes of society. The discussions in the Houses of Lords or of Commons, which arise on the occurrence of any subjects connected with science, sufficiently prove this fact.[16]

It will be noticed that Babbage saw the need for educational reform only in terms of the upper and upper-middle classes, and that he had nothing to say concerning the need to develop 'science for all'. Like Brewster, he also noted that, unlike France or Germany, most British science was being done outside the existing university system. Forty years later, with a government Department of Science and Art in operation, Edward Frankland saw the educational priorities very differently:

> The want of suitable buildings and apparatus for the prosecution of (experimental) researches.
>
> The non-recognition of experimental researches by any of our Universities.[17]

It is only too easy to show the defects and absurdities of Babbage's arguments. If the cause of decline was due to British educational institutions of higher learning, then since the system of schools and universities had existed

16 Babbage, *op.cit.* (1), p. 3.
17 Frankland, *op.cit.* (2), p. 445.

unaltered for centuries, how come science had ever flourished? Shouldn't it always have been declining, or, rather, never able to improve? And, insofar as Babbage was an absentee Cambridge Professor of Mathematics, he was hardly a fit person to criticise university standards of teaching! If he argued there was too little state support for science, this had always been true, and it is difficult to see as a cause of the decline of science. Again, insofar as Babbage had received £17,000 from the British government for his difference calculating engine, was he fair to speak uncharitably of the State? Was he not governed by self-interest? Babbage's analysis is also suspiciously defective because of what he ignores - the rich diversity of British scientific institutions ranging from the Scottish universities, the provincial mechanic's institutes and scientific societies, and the new scientific University of London.[18] Finally, his model state, France, when closely examined was by no means free from jobbery and nepotism and lack of state planning.

The truth is that British, indeed European, science was undergoing a dramatic shift in character during the first three decades of the nineteenth century. The work *scientist* was coined in the 1830s to express this shift from the amateur natural philosopher to the professional cultivator of science in full-time employment for science. [19] Whereas in the eighteenth century scientific activity had been the practice of individuals, and financed by individuals, it became so changed that government interference became necessary. Babbage certainly perceived this. His analysis was right that the pursuit of science in Britain was not a distinct profession like law. This was why it failed to recruit some of the best minds, or provide financial inducements or social prestige. As T.H. Huxley so succinctly put it: 'Science in England does everything but *pay*. You may earn praise but not pudding.'[20]

But Babbage's voice was hardly heard simply because individualism and amateurism still worked in 1830. By the 1860s, however, it had become obvious to many scientists that they worked no longer. And because action had not been taken, Britain was to be overtaken by other countries in many industrial areas — steel manufacture, synthetic dyestuffs and the developing electrical industry.

It was the threat of loss of industrial supremacy which was continually invoked in the 1860s, and its apparent fulfilment in Britain's poor performance

---

18 J. B. Morrell, 'Individualism and the structure of British science in 1830', *Historical Studies in the Physical Sciences*, 3(1971), 183-204.

19 W. H. Brock, 'Advancing science: The British Association and the professional practice of science' in MacLeod and Collins, *op.cit.* (12), 89-117. Reproduced in this volume as Essay II.

20 *Life and Letters of T. H. Huxley*, ed. L. Huxley (London, 1900), vol.1, p. 66.

at the Paris International Exhibition in 1867, that enabled the more militant 'Young Guard' of science to campaign for improvements in scientific and technical instruction, and to demand state support for science on a more lavish scale.[21]

The development of technical colleges, research laboratories in universities, and laboratories in British secondary schools, can all be traced back to the effective use by scientists and scientifically-minded industrialists to the deflation of national pride by comparison with the scientific output and facilities in Germany. To that extent, Babbage's argument for the decline of science has been of the greatest value in the professionalization of science, and the development of a technological society.[22]

21 R. M. MacLeod, 'Resources of science in Victorian England: the endowment of science movement, 1868-1900' in P. Mathias, ed., *Science and Society 1600-1900* (London, 1972), pp. 111-66; Peter Alter, *The Reluctant Patron. Science and the State in Britain, 1805-1920* (Oxford, 1987).

22 There is now a huge literature on this theme. See, for example, Martin Wiener, *English Culture and the Decline of the Industrial Spirit, 1850-1980* (Cambridge, 1982); B. Elbaum and W. Lazonick, eds., *The Decline of the British Economy* (Oxford, 1986); Donald C. Coleman and Christine MacLeod, 'Attitudes to new techniques: British business men, 1880-1950', *Economic History Review*, 39 (1986), 588-611; and David E.H. Edgerton, 'Science and technology in British business history', *Business History*, 29 (1987), 84-104.

XVI

THE JAPANESE CONNEXION: ENGINEERING IN TOKYO, LONDON, AND GLASGOW AT THE END OF THE NINETEENTH CENTURY

[The College of Engineering was] one of the most important factors in the development of Japan, since from it have come the majority of the engineers who are now working its resources and industries.

PRINCE ITŌ[1]

When I arrived in Japan in 1875, I found a marvellous laboratory, such as the world had not seen elsewhere. At Glasgow, at Cambridge, and at Berlin, there were three great personalities; the laboratories of Kelvin, and of Maxwell, and of Helmholtz, however, were not to be mentioned in comparison with that of Ayrton. Fine buildings, splendid apparatus, well-chosen, a never-resting keen-eyed chief of great originality and individuality: these are what I found in Japan.

JOHN PERRY[2]

This College is not altogether like what we have in our own country, but has gone beyond us and eclipsed all that we have ever done in that same direction . . . Might we not suggest that the experience of the Japanese would be useful to us, if some among us would take the matter in hand, project a technical school, in which French and German were common languages, where English was not neglected, and where theory and practice went hand in hand.

ANON, JAPAN WEEKLY MAIL[3]

Most of the newer technical colleges owe, directly or indirectly, in greater or lesser degree, something to the development of technical engineering and chemical training which was originated at Finsbury.

E. G. WALKER[4]

It is a rather curious fact in the history of education, that these courses [at the Glasgow and West of Scotland Technical College] are almost identical

This address was delivered at the Annual General Meeting of the British Society for the History of Science, at Chelsea College, London, on 24 May 1980.

In keeping with the conventions of such a lecture, documentation has been confined to basic references.

Acknowledgements: It is a pleasure to thank the archivists of the University of Glasgow (Mr M. Moss), the University of Strathclyde (Professor S. G. E. Lythe), Imperial College London (Mrs Jeanne Pingree), and the City & Guilds (Mr T. G. Bunce) for their assistance. Professor Tomasuka Terakawa (University of Hiroshima) kindly answered my Japanese queries, while Dr Olive Checkland, of the University of Glasgow, and Professor Sami Kita, of the University of Soka, Tokyo, enthusiastically shared with me their research findings on the social and economic connexions between Victorian Scotland and Japan.

with those which I had previously arranged for the Imperial College of Engineering, Japan.

HENRY DYER[5]

THAT the export of Scottish engineers and engineering teachers to Japan in the 1870s aided that country's astonishingly rapid process of modernization from a feudal to a capitalist, industrialized society will not occasion surprise or dissent. As the *Japan weekly mail* editorialized in 1878:

> In no direction has Japan symbolised her advance towards assimilation of the civilisation of the Western world more emphatically than in that of applied science.[6]

However, to suggest that Britain may have had something to learn from Japan may seem more questionable and warrant evidence, exegesis, and demonstration. But it will be my contention in this address that consequent upon the Japanese experience, British engineering education, notably in London and in Glasgow, underwent a significant alteration in design, pace, and execution; and that, as the epigraphs to this paper are intended to suggest, there was a 'connexion' between the practice of the engineering education of Glasgow, London, and Tokyo in the last quarter of the nineteenth century.

Japan, as E. J. Hobsbawm said of India and W. L. Burn said of Ireland, was a country 'to which *laissez-faire* never applied', and 'a social laboratory [where] the most conventional of Englishmen were willing to experiment . . . on lines which they were not prepared to contemplate or tolerate at home'.[7] Like the Royal Military Engineers in India who returned to England to organize and maintain the Department of Science and Art and who, in Crowther's words, urged 'the practical recognition that science must be organized',[8] the Japanese engineering experience enabled a number of British teachers to taste and test in the field at first hand the government-directed system of education hitherto associated with France, Germany, and Switzerland. It also enabled these same instructors to test the merits of a form of instruction, the sandwich course, which had been implicit in Scots and Irish university engineering education, but never formally designed and exploited by compulsion. Finally, faced by a young nation fresh to the ways of thought of Western mathematics and science, these educators were encouraged to experiment with the curriculum and their teaching methods, and to lay great stress upon laboratory and workshop skills.

The opening up of Japan to American and European trade has been amply documented by authorities like Grace Fox[9] and, more popularly, Pat Barr.[10] Until the mid-nineteenth century Japan remained isolated apart from sporadic and carefully controlled contacts with Portuguese and Dutch traders from the sixteenth century onwards during the regime of the hereditary Tokugawa Shoguns who left the Mikado or Emperor effectively a symbolic and powerless constitutional monarch: '. . . the

Shogun governed, but did not reign; while the Mikado reigned, but did not govern'.[11] Under the Shoguns, Japan deliberately isolated herself from the outside world until 1853, when the American Mathew Perry moved his gunboats into Yokohoma Bay and revealed, to the astonished Japanese, the wonders of a toy steam locomotive and of the electric telegraph.

By 1858 Lord Elgin (a Scot) had negotiated a trade treaty between Great Britain and the Shogunate, and the Scots firm of shipowners, merchants, and agents, Jardine & Matheson, who had been long established in Hong Kong and Shanghai, quickly put down bases at the treaty ports.[12] But by signing treaties, the Shogunate offended the unconsulted Mikado and territorial feudal barons; consequently Government and commerce remained in turmoil and in a state of civil war until 1868, when the young 15-year-old pro-Western Mikado, Mutsushito, abolished the Shogunate and restored supreme power to himself through an oligarchic group of pro-Western Samurai, one of whom was Hirobumi Itō (1841–1909). The restoration of the Meiji dynasty abolished Japanese feudalism. 'Knowledge', the emperor proclaimed, 'shall be sought throughout the world so as to strengthen the foundations of imperial rule',[13] or as Itō expressed it in 1886 when he had become Prime Minister:

> The only way to maintain the nation's strength and to guarantee the welfare of our people in perpetuity is through the results of science . . . Nations will only prosper by applying science . . . If we wish to place our own country on a secure foundation, insure its future prosperity, and to make it the equal of the advanced nations, the best way to do it is to increase our knowledge and to waste no time in developing scientific research.[14]

Japan, as James Bartholomew has suggested, did not make the mistake Russia had made in the eighteenth century. Since the main stress 'was placed on the systematic importation of existing knowledge, rather than the creation of new knowledge', Japan's universities were designed to be *teaching* institutions 'in which engineering and other applied sciences had a central place'.[15]

This deliberate decision to accept and to exploit the contagion of foreign culture as a cure for what Bartholomew and Inkster have described as a 'careful calculation of ends and means . . . an innovative response to the problem of technical and scientific backwardness',[16] undoubtedly owed something to the pre-Restoration experiences in England and Scotland of a small group of young Samurai.

Secretly, and in considerable danger, Itō and four other Samurai of the Choshu clan, including Yamao Yōzō, visited England in 1863 through the help of Jardine's shipping agent in Yokohoma, William Keswick. As Grace Fox has noted:

> Within the compass of British official policy and sometimes in defiance of it, it was the British merchants, in conjunction with their Western competitors, who introduced or abetted the most numerous changes in Japanese life and thought in this period.[17]

Hugh Matheson, the head of Matheson & Jardine in London, met the youths and boarded three of them (including Itō) with Alexander Williamson, the Professor of Chemistry at University College.[18] In 1865, Itō returned to Japan, where he played an important role in smoothing the way for restoration. Yamao went to Glasgow, where he worked in Napier's shipyard while attending evening classes at Anderson's College at the same time as a young Scots engineer Henry Dyer. Early in 1870 Yamao returned to Japan and joined the Ministry of Public Works, whose minister Itō had become.[19] In April 1871, Itō and Yamao drew up plans for a technical or engineering school (Kōgakkō-ryo) to train technicians for the execution of public works and the industrialization of Japan:

> It is . . . an urgent need to develop technological science in order to lay the foundation of welfare and utilization. The responsibility for the promotion of the study of science is laid upon this [Technology] department and a school of technology should be established—divided into two schools, at college and elementary levels, and foreign instructors should be invited in order to attain a desirable result in the future.[20]

This was one of several educational schemes drawn up at this time—for example, a naval college, which was founded in 1872 as the Imperial Naval College, an agricultural college, and the *Kasei*, or High School—and it was undoubtedly modelled on Yamao's experience at Anderson's College in Glasgow.

However, the technical school's intended opening date of May 1872 had to be postponed, since foreign experts were not forthcoming from the British Foreign Office 'owing to the demand for competent scholars in England and India'.[21] Hence, before his departure to America and Europe as a member of the Iwakura Mission in 1871, Itō was briefed to hire up to seven foreign teachers for the proposed college.[22] The Mission arrived in London in August 1872, a year after the British Government had been forced to set up a Royal Engineering College at Cooper's Hill to solve the problem of the alarming shortage of engineers in India.[23]

According to Hugh Matheson, speaking in 1882, as the agent for the Japanese Ministry of Public Works in London, he himself had had 'the honour to found' an engineering college in Japan.

> Several of those who are now prominent members of [the Japanese] Government had been placed under my care at an early period, and years after their return to their own country, and when the Revolution occurred

> which abolished the Daimios [i.e., feudal Barons], and established a partially constitutional Government, several of my friends became Ministers of State, and they requested me to assist them in founding, at Tokio, the Capital of Japan, an institution which would train young men for efficient service in the Public Works Department, which for the first time was felt to be all important in the development of the country. This was a very responsible Commission, but I did not shrink from it, and having taken counsel with an eminent friend . . ., Professor Lewis Gordon, I submitted the scheme of a College with Principal and half-a-dozen Professors, which I was in due time authorised to establish. I was fortunate, on the advice of the late Professor Macquorn Rankine of Glasgow, in choosing as Principal a young man of twenty-four, Mr. Henry Dyer. What they wanted was that he should be my representative with the Minister, a kind of alter ego, with whom they might confer on all matters connected with the work of the College and with the practical work of the Public Works Department.[24]

Dyer himself barely mentions Matheson's role, but Matheson was undoubtedly Itō's intermediary.[25]

Lewis Gordon (1815–76), the first Professor of Engineering at Glasgow from 1840 to 1855, who established extensive connexions in Glasgow and London following his return to full-time engineering consultancy,[26] passed Matheson's request to his former pupil and successor to the Glasgow chair, the thermodynamicist, W. J. M. Rankine. In turn, the latter had no difficulty in recommending his most brilliant recent engineering graduate, Henry Dyer. Dyer had served an apprenticeship, from 1863 to July 1868, under the Glasgow marine engineer, Alexander Kirk (*d.* 1891) in the firm of James Aitken & Co, foundrymen of Cranstonhill, Glasgow. During his apprenticeship he attended evening classes at Anderson's College, entering Glasgow University in 1868, where he won a prestigious Whitworth Scholarship in 1870.[27]

Several mysteries surround Dyer, not the least of which is how and why such an obviously talented man failed to establish any kind of national reputation in Britain when he returned in 1882. Today his name is scarcely known outside Japan and Glasgow, and he merits no entry in the *Dictionary of national biography*.[28] Why did the Institute of Civil Engineers ignore his Report on the Education of Civil and Mechanical Engineers which he submitted to them through Matheson in 1880?[29] Was it because the system of combined academic and practical training he advocated threatened the lucrative pupillage system? How was a man of 24, who seems never to have travelled outside Glasgow, able to draw up such perfect plans for the establishment of a foreign power's system of technical education?

Dyer himself tells us, by way of explanation, that he had been making 'a special study of all the chief methods of scientific and engineering study in the different countries of the world and of the organization of some of the most important institutions, with the intention of devoting myself to the advancement of engineering education in Britain, so that I had fairly definite ideas both as to what was desirable and what was possible'.[30] Such an interest would not have been unusual in the early 1870s. In 1869 the eminent naval engineer, John Scott Russell, had published a remarkable plea for state-sponsored technical education in *Systematic technical education,*[31] which he dramatically dedicated to Queen Victoria; the year before, the long-established Institution of Civil Engineers, conscious of the prestige which the profession of engineer had acquired in Victorian Britain, commissioned an extensive report on the current education of engineers at home and abroad.[32] This *Report* was published in 1870 and included important observations by Fleeming Jenkin, Professor of Engineering at Edinburgh and business partner of William Thomson in Glasgow. Jenkin also published at this time a number of speeches he made to Scottish audiences on technical education.[33] Russell and Jenkin, together with his own experience of Glasgow engineering training, were probably, therefore, Dyer's principal sources on the need for a more academic approach to offset the existing bias towards engineering pupillage.

From Jenkin, Dyer took the stress he laid upon mechanical drawing in the Japanese curriculum, and the importance of a broad, university-type education. From his own experience of the long Glasgow summer vacation which could be devoted to workshop pupillage, he coupled Jenkin's emphasis upon practical training in an engineering workshop separate from theoretical studies and Russell's strong recommendation of alternating periods of theory and practice, to fashion the idea of a factory or workshop in which Japanese students would be obliged to spend stated periods of time.[34] Again from Jenkin, he adopted the notion of project examinations for the final diploma. From either Russell, or the Institution of Civil Engineer's *Report,* he had access to the syllabus of the engineering courses of the Zurich Polytechnic. From Russell, and, no doubt, from Glasgow's splendid Hunterian Museum, he took the idea of departmental museums. From William Thomson's voluntary practical physics classes for more serious students[35] (and Glasgow's tradition of practical chemistry classes) he perhaps saw a painless way of introducing aristocratic Japanese students to using their hands without losing caste. Finally, from the Ecole Polytechnique or the Naval Academy which the British had established in Tokyo in 1870, either Dyer or his Japanese masters introduced uniform, a strict code of discipline (which was apparently never necessary for such diligent students) and the labelling of the students as cadets.[36]

As Grace Fox has noted, Dyer 'combined the best in the British and continental systems for training engineers'[37] in a six-year course. Itō's brief to Dyer, that the College was 'to train men who would be able to design and superintend the works which were necessary for Japan to carry on if she adopted Western methods',[38] was advanced during Dyer's voyage out to Japan. His design for a curriculum was immediately accepted by Yamao, both men recognizing each other from their evenings at Anderson's College. For the first two years, following an entrance examination, students took a broad science-biased secondary course, including English language and literature, geography, mathematics, mechanics, chemistry, physics, and drawing. In the following two years, six months of each year was spent in the works Dyer established and personally ran on behalf of the Ministry of Works at Akabane,[39] the other six months being spent on theoretical and laboratory college course work in any one of seven branches of engineering (civil, mechanical, chemical, telegraphy, mining, metallurgy, architecture). The College was thus a Polytechnic like Zurich. The final two years were spent entirely in the Akabane works, or any other public works under State control. Students' education was free providing they pledged themselves to work for the government for seven years after graduation (on penalty of returning the fees). Assessment was continuous by fortnightly class tests, with pass or fail examinations at the end of the second and fourth years.

The College began teaching in August 1873 with fifty-six students, twenty-five of whom only were admitted to the first year course, the remainder, at Dyer's insistence, entering an emergency preparatory school.[40] Temporary accommodation was used until the impressive brick buildings, designed by the Anglo-French architect, Charles Chastel de Boinville (1850-97),[41] were completed in 1877, at which time the College was renamed the Imperial College of Engineering (Kobu Daigakko).

W. G. Dixon, a Professor of English at the College, later described the College buildings as:

> The handsomest which the Government has yet erected in foreign style. After entering by an elegant gate, we pass a Gothic building with a clock tower. This was originally intended for the college proper, but was at once found too small, and is now used [1882] as a museum of objects in connection with the different technical branches taught in the college—civil, mechanical, and mining engineering, architecture, telegraphy, chemistry and metallurgy: the collection of engineering models is said to be the most complete in the world. Opposite this building, to our right, is a long row of dormitories. Proceeding further up the avenue, we come to the main building, a chaste French Renaissance erection forming two sides of a quadrangle.[42] A doorway between two towers leads into the library and common hall, a very handsome room capable of accommodating from 1000 to 1500 persons, while the galleries are lined with book-cases stocked with over 13000 volumes chiefly in the English and Japanese languages. The class-rooms are arranged exactly as they might be in a European college, while the various laboratories and drawing offices are furnished with the most approved appliances. Indeed, the arrangements as a whole speak volumes for the enlightened liberality of the officers of the Public Works Depart-

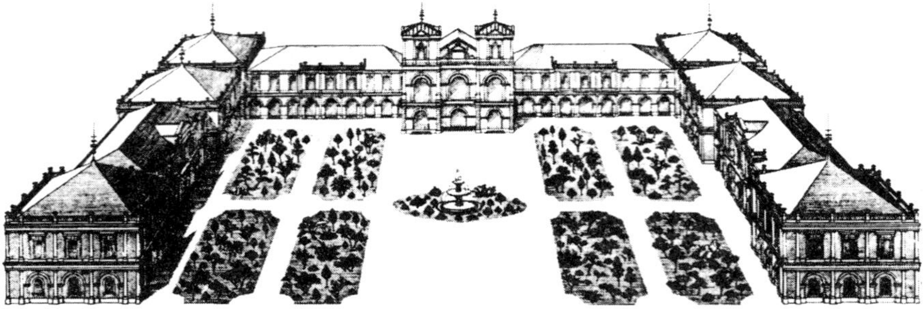

Figure 1. The Imperial College of Engineering, Tokyo, as conceived in 1877. The left wing was never built.
Source: *Imperial College of Engineering Tokei. General report by the Principal for the period 1873–77*. Courtesy Mitchell Library, Glasgow.

> ment, with which the college is connected. There can be no doubt that they are more complete than the arrangements usually found in the scientific colleges of England.[43]

If the College could boast that it had better provision for technical instruction than any comparable institution in England, it must be stressed that the facilities available to its teachers between 1873 and 1877 were extremely primitive. W. E. Ayrton,[44] the Professor of Natural Philosophy, reported to Dyer on 1 October 1877, the year before the new laboratories became available:

> Work is very difficult on account of the altogether inadequate laboratory accommodation only allowing a few students to participate in this course. It may here be mentioned that the Natural Philosophy Department has up to the present time occupied temporary quarters totally unsuited for the purpose as regards space and light, decidedly injurious to health from the total absence of ventilation . . . The laboratory is a mere room and unsupplied with either gas, water, or fire.[45]

Nevertheless, Ayrton and John Perry,[46] the Professor of Engineering, were somehow able to publish eleven joint research papers on electrical subjects before 1877.[47]

How far Dyer chose his fellow Professors is unclear. It was probably here that Matheson asserted his influence through his contacts at the

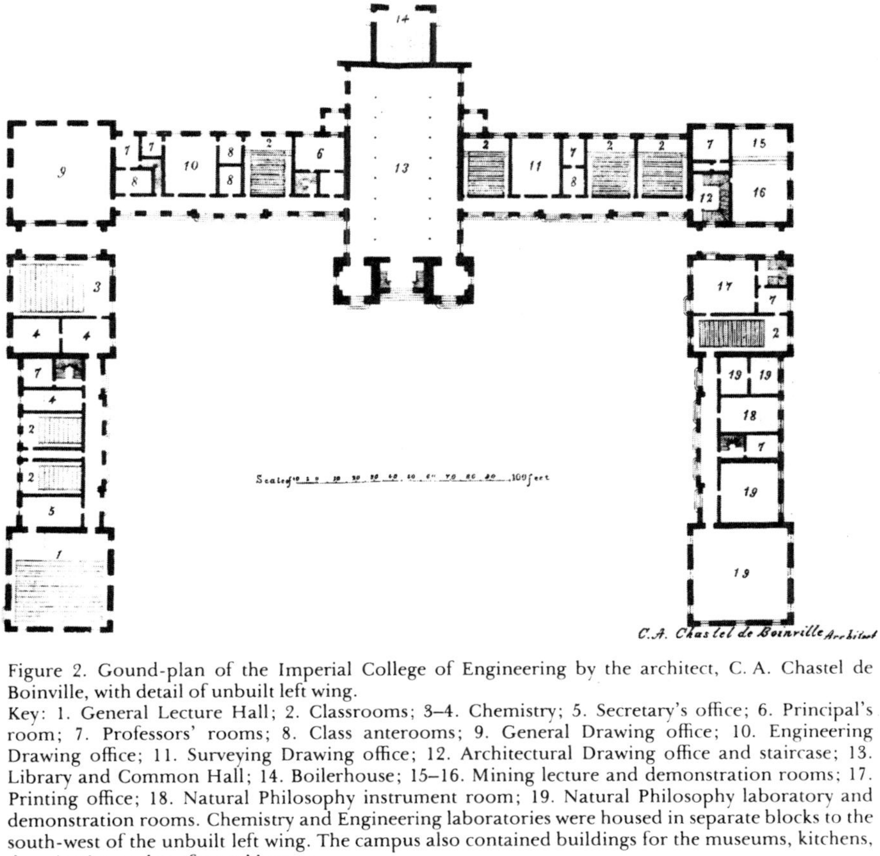

Figure 2. Gound-plan of the Imperial College of Engineering by the architect, C. A. Chastel de Boinville, with detail of unbuilt left wing.
Key: 1. General Lecture Hall; 2. Classrooms; 3–4. Chemistry; 5. Secretary's office; 6. Principal's room; 7. Professors' rooms; 8. Class anterooms; 9. General Drawing office; 10. Engineering Drawing office; 11. Surveying Drawing office; 12. Architectural Drawing office and staircase; 13. Library and Common Hall; 14. Boilerhouse; 15–16. Mining lecture and demonstration rooms; 17. Printing office; 18. Natural Philosophy instrument room; 19. Natural Philosophy laboratory and demonstration rooms. Chemistry and Engineering laboratories were housed in separate blocks to the south-west of the unbuilt left wing. The campus also contained buildings for the museums, kitchens, dormitories, and professors' houses.
Source: *Imperial College of Engineering Tokei, Calendar, Session MDCCCLXXVII-LXXVIII,* Tokyo, 1877. Courtesy Mitchell Library, Glasgow.

University of Glasgow and University College, London.[48] However the choices were made, they were all remarkably young men in their twenties and early thirties—indeed, even more remarkable, Dyer, their Principal, at 24, was the youngest. Since most of their students were only a few years younger, the College must have had the air of a Public School with masters fresh from Oxbridge—an observation reinforced by the following editorial in the *Japan weekly mail:*

> Among the qualifications of a teacher of Japanese youth that of a sound moral character should be the first. Among the people who put etiquette above morality [i.e., the Japanese], the foreign teacher, as a representative of a different and, as he believes, a better civilisation, should put morals before

> etiquette, and be himself a moral man. In truth, honour, devotion to duty, purity of life and personal chastity, temperance in eating and drinking, loyalty to his country, to her principles and best traditions, the spurning of all shams, the refusal of all bribes, detestation of all flattery, and disdain to pander to the vanity of his employers, in the foreign teacher should be firmly established.[49]

If Arnold, at Rugby, could influence and improve the young men of stable and fixed old England, what might not an Arnold accomplish in Japan where all was flux and its citizens looking for guidance? Certainly Dyer's printed speeches to College students read like a headmaster's homilies.

Whether because of the known policy of eventual Japanization,[50] or the financial retrenchment the Government was forced to make towards education in 1877 following the unsuccessful Satsuma Rebellion, Ayrton wrote to his former mathematics teacher, T. A. Hirst, in 1877, 'asking whether he thought that the time had come for starting [in England] . . . a course of applied physics somewhat on the lines of that given . . . in Japan. [Hirst] replied that England was not yet ripe for such a innovation'.[51] Ayrton nevertheless applied for the Chair of Technology which the newly-established City & Guilds of London Institute endowed at University College in 1878, but the post went to A. B. W. Kennedy, the sitting incumbent of the Chair of Civil Engineering.[52] Having left Japan at the end of the summer term in 1878 to work for Latimer Clark and to develop his own electrical consultancy practice, Ayrton was available for appointment in October 1878 as Professor of Applied Physics when, consequent upon the publication of the report on technical education by a Committee of the London Livery Companies,[53] the City & Guilds established evening classes in rooms hired from the Cowper Street Middle Class School in Finsbury. Here was the germ of both the Finsbury Technical College[54] which officially opened in a purpose-built building in February 1883, and the architecturally-splendid Central Insitution at South Kensington which opened in 1884.

In September 1878, his four-year service in Tokyo complete, John Perry rejoined Ayrton in London in electrical consultancy work, and was, therefore, free to join the Finsbury team when a Chair of Mechanical Engineering was added in 1882. Together with these 'Japanese twins', the Professor of Chemistry, Henry Armstrong,[55] completed a formidable teaching triplet. As Armstrong wrote later:

> Finsbury . . . was novel and remarkable in its arrangements. By accident, three men not only with special experience, but anxious to act as pioneers, in some strange way fell together as a team in complete unison . . . I have always regretted that Ayrton, Perry and I did not remain together at Finsbury during ten years or so.[56]

Finsbury Technical College has usually been discussed from the perspective of Armstrong, the stress being placed upon his successful application and development there of the heuristic method of laboratory instruction.[57] However, from the Japanese perspective it can be seen that Ayrton and Perry also made significant contributions to the 'Finsbury method'. Having experienced the common non-specializing two-year beginners' course at Tokyo, the notion of a common first-year course at Finsbury in which the bases of scientific, technical, and mathematical reasoning were analyzed and worked out, seemed quite natural; having experienced Dyer's stress on a workshop-laboratory based curriculum, Ayrton and Perry were happy to share Armstrong's Germanic zeal for 'research-like', 'problem-based' or 'heuristic' laboratory instruction, rather than didactic lectures. Moreover, Ayrton and Perry were able to establish, in London, a significant new teaching and recording device which they had tried successfully in Japan—the use of squared (graph) paper, which rendered practical work quantitative and helped to correlate mathematical and scientific work.[58] This technique rapidly spread to other teaching institutions. Having seen that, under a centralized government, it was possible to give industrial training, Finsbury avoided the mistake made by the British Government's Department of Science and Art (whose syllabuses avoided practice), and directly attempted the 'teaching of the practical applications of physics [and chemistry] to industry'[59] with day pupils and older evening class students from local firms. Having been untrammelled by either Government or private external examinations in Tokyo, Ayrton and Perry shared Armstrong's conviction (drawn from experience) that by making their own internal assessments it would be 'possible for . . . teaching to be directed solely to the *professional* needs of the students'.[60] Finally, since the minimum school leaving age was not raised to 14 in Britain until 1918 and there was no compulsion for local authorities to provide 'secondary' education until 1902, the College's policy of opening its doors as a day-school in 1882 (while continuing its evening classes) effectively introduced a new secondary grade of school—the technical school—into the English educational system. But such a technical school had already been established by Dyer, Ayrton, Perry and others in Tokyo almost a decade earlier.

In one thing only did Finsbury fail to reflect Dyer's Tokyo model. Despite Ayrton's strong recommendation, as well as that of his successor, Sylvanus P. Thompson (who had experienced the system at University College Bristol),[61] the City & Guilds refused to accept the principle of 'day-release' or 'sandwich' courses for London apprentices, or to establish further technical schools attached to factories on the Imperial College of Engineering's Akabane model. Indeed, apart from sporadic experiments in Bristol, and strong recommendations from the Institution of Civil Engineers in a *Report* of 1903, the sandwich principle remained undeveloped in English technical education until the 1920s.[62]

This was not, however, the case in Glasgow (and, possibly, Belfast[63]), where the long summer vacations and the proximity of the Clyde shipyards had allowed engineering students to intercalate industrial and engineering workshop experience with university education, or with evening courses at Anderson's Institution since the 1840s. But until the 1880s this was not a deliberately planned and compulsory arrangement.

Dyer left Japan in 1882, just three years before the abolition of the Department of Public Works and the absorption of the Engineering College into the newly-created Imperial University of Tokyo under the Department of Education. (Edward Divers succeeded him as Principal, a post he retained after 1886 when the College was absorbed into the Imperial University of Tokyo.[64]) Following some continental travel, Dyer resettled with his family in Glasgow where he continued to act as an agent of the Japanese Government and as a friend to the large numbers of Japanese students who came to study engineering in Glasgow. On the educational front he soon became actively involved in the Scottish Charity Commissioners' attempts to unite, in the common purpose of more efficient technical education, the motley educational institutions and charities of Anderson's College, the James Young Chair of Technical Chemistry, the Glasgow College of Science and Arts (i.e. the old mechanics institute), the Allen Glen's Institution (a quasi-technical school founded in 1853[65]) and the decaying Atkinson Institution. Politically, he became a Social Darwinist, much given to the message of health, sobriety, efficiency, and Empire.[66]

On 26 November 1886, under the 1882 Educational Endowments (Scotland) Act, the charitable educational institutions previously mentioned merged to become the Glasgow and West of Scotland Technical College (now Strathclyde University), with Dyer as one of its 30 Governors.[67] Dyer served on the 'Teaching and Staff' sub-committee and convened sub-committees on 'Libraries and Museum', and on the 'Calendar'. His appointment to the last no doubt explains the astonishing visual and material similarity between the Glasgow and Tokyo Calendars.

As in Japan, and effectively at Finsbury, the age of entry to the Technical College was 16 and was dependent upon a qualifying entrance test. Again, as in Japan and at Finsbury, there was a common first-year day course of mathematics, chemistry, physics, drawing, and one foreign language, followed by two years of courses chosen from a polytechnical range of special subjects, including civil, mechanical, chemical, electrical and marine engineering, naval architecture, metallurgy, and agriculture. But the important point is that, since courses ran from October to April, students were strongly advised to use the summer for employment in 'workshops, manufacturies, or drawing offices'.[68] Failing this they had to attend College classes in the pattern-making workshop.

Well could Dyer comment publicly in a lecture to the Glasgow Philosophical Society in 1893 on 'the curious fact' that Glasgow Technical College's 'courses are almost identical with those which I had previously arranged for the Imperial College of Engineering, Japan . . .'[69] Of course, only further research, perhaps that in progress by John Butt on the history of Strathclyde University and Sami Kita's on the Japanese presence in Victorian Glasgow, will confirm or modify Dyer's tongue-in-cheek observation.

It has not been my intention in this Address to add Japan to W. H. G. Armytage's remarkable quintet of essays *The American . . ., German . . ., Russian . . ., Swedish Influence on English Education,* since patently the transfer of a cultural package *from* Britain to Japan was far greater.[70] Nevertheless, in the limited field of technical education with which this essay had been concerned, the experience of reverse trans-culturation was significant. It was one thing to read reports of how continental countries trained their engineers; it was another to visit such countries, or to qualify oneself at a Zurich or Berlin Polytechnic; but it was surely quite another matter to have the opportunity of participating in a state-controlled technical education system, and to have the chance of developing the system *de nouveau.*

Although much more documentation is needed, and further study of the sociology of culture-transfer as well as the context of the British debates over engineering and technical education in the period 1867 to 1880 is necessary, there is little doubt that the shared Japanese experience of Ayrton and Perry was a necessary condition for the success of the Finsbury College venture. Since it is agreed that this College was the model for all the local technical colleges created in the 1880s and 1890s, including the Glasgow Technical College, it seems possible that the modernization of, and creation of a British technical education system geared to the second industrial revolution, owed a good deal to a rehearsal held in Japan in the 1870. As Henry Dyer said in 1906, in words which will strike the contemporary reader:

> The engineer is . . . [a] revolutionist, for his works bring about economic conditions which generate forces which cannot be resisted. Those nations only survive which adapt themselves to the altered conditions.[71]

NOTES

[1]Prince Itō, as quoted by W. H. Sharp, *The educational system of Japan,* Bombay, 1906, p. 206. Compare Itō's statement 'That Japan can boast today of being able to undertake such industrial works as the construction of railways, telegraphs, telephones, shipbuilding, working of mines, and other manufacturing works entirely by the hands of Japanese engineers is mainly attributable to the College so ably established and set in motion by [Dyer]', quoted in Henry Dyer, *Dai Nippon, the Britain of the East: A study in national evolution,* London, 1904, p. 7. For the College's significance as a training ground for Japanese electrical prowess, see Kenji Imazu, 'The beginning of electric engineers [*sic*] in Japan', *Japanese studies history of science,* 1978, *17,* 13–26.

[2]John Perry, 'W. E. Ayrton', *The central* [journal of the City and Guilds of London Institute, Central Institution, South Kensington], 1910, *7*, 70.

[3]*Japan weekly mail*, 24 August 1878.

[4]E. G. Walker, 'Finsbury Technical College' *The central*, 1933, *30*, 35–48 (35).

[5]H. Dyer, *Technical education in Glasgow and the West of Scotland*, Glasgow, 1893, p. 5 (reprinted from *Proceedings of the Philosophical Society of Glasgow*).

[6]*Japan weekly mail*, 1878, *2*, 52.

[7]E. J. Hobsbawm, *Industry and empire*, London, 1968, p. 123; W. L. Burn, 'Free trade in land: an aspect of the Irish question', *Transactions of the Royal Historical Society*, 1949, *31*, 61–74 (68). See also R. M. MacLeod, 'Scientific advice for British India: imperial perceptions and administrative goals, 1898–1923', *Modern asian studies*, 1975, *9*, 343–84 (343–4).

[8]J. G. Crowther, 'Alexander Strange', in *Statesmen of science*, London, 1965, 237–69 (242).

[9]Grace Fox, *Britain and Japan, 1858–1883*, Oxford, 1969.

[10]Patt Barr, *The coming of the barbarians*, London, 1967; *The deer cry pavilion. A story of Westerners in Japan, 1868–1905*, London, 1968. Other works which I have found useful, and which are not otherwise cited below, include: Stafford Ransome, *Japan in transition*, London & New York, 1899, which describes the later engineering faculty of Tokyo University in 1896; John R. Black, *Young Japan: Yokohoma and Yedo. A narrative of the settlement and the city from the signing of the treatises in 1858 to the close of the year 1879*, 2 vols., New York, London, & Yokohoma, 1883 (reprint ed. by Grace Fox, Tokyo, 1968); 'Les sciences physiques au Japon, 1868–92', *Archives internationales d'histoire des sciences*, 1956, *9*, 3–12; Masatoshi Hari, 'Introduction to the education of descriptive geometry and engineering drawing in the later Tokugawa regime and early Meiji period', *Proceedings of the XIV International Congress of the history of science*, 1975, iii, 142–5; Hiroshi Ishiyama, 'Henry Dyer (1848–1918) and his draft regulations for [the] proposed institution of engineers', ibid., iii, 211–14; Masao Watanabe, 'American science teachers in the early Meiji period', *Japanese studies in the history of science*, 1976, *15*, 127–44; Kenkichiro Koizumi, 'The emergence of Japan's first physicists, 1868–1900', *Historical studies in the physical sciences*, 1975, *6*, 3–108; G. C. Allen, 'Education, science and the economic development of Japan', *Oxford review of education*, 1978, *4*, 27–36.

[11]*Chambers' encyclopaedia*, London and Edinburgh, 1886, v, 687.

[12]Fox's study, op. cit. (9), is partly based upon the Jardine Matheson archives deposited at Cambridge.

[13]Fox, ibid., p. 261.

[14]From a speech to graduates of the Imperial University, quoted by J. R. Bartholomew, 'Japanese modernization and the Imperial universities, 1876–1920'. *Journal of Asian studies*, 1978, *37*, 251–71 (254).

[15]Ibid., p. 271.

[16]Ibid., p. 271. Ian Inkster, 'Meiji economic development in perspective. Revisionist comments upon the industrial revolution in Japan', *The developing economies* (Tokyo), 1979, *17*, 45–68, especially p. 54; I. Inkster, *Japan as a development model? Relative backwardness and technological transfer*, Bochum, 1980, chapter V.

[17]Fox, op. cit. (9), p. 548.

[18]Ibid., pp. 458–9. J. Harris and W. H. Brock, 'From Giessen to Gower Street: towards a biography of A. W. Williamson (1824-1904)', *Annals of science*, 1974, *31*, 95–130, especially pp. 122–6. Between 1863 and 1866 Jardine & Matheson connived to send several other students to England before the Shogunate officially permitted overseas study after 1866. According to Fox (p. 460) some twenty Japanese students' names are found in the *Calendars* of University College London, 1865–70. Fox comments, on the basis of diplomatic correspondence, that many students 'were in general without guidance and of some concern to the British Foreign Office, which wished the Japanese government to make an independent arrangement for their supervision'. Large numbers were recalled to Japan in 1874; see *Japan weekly mail*, 19 June 1874, 8 July 1874. The large influx of Japanese students at the University of Glasow did not begin until 1879 when the first graduates of the Imperial College of Engineering became eligible for study abroad. These students, and their later careers in Japan, are the subject of a study, 'Scotland and the industrialization of Japan, 1860–1914', being made jointly by Mrs Olive Checkland (University of Glasgow) and Dr Masami Kita (University of Soka, Tokyo).

[19]For Itō (1841–1909), see Kenji Hamada, *Prince Itō*, London, 1936, and J. Morris, *Makers of modern Japan*, London, 1906, pp. 119–53. In 1868 Itō was Secretary of the Bureau of Foreign Affairs and a provincial governor. In 1869 he became Vice-Minister of the Department of Finance and Home Affairs, spending 1870–71 in America studying financial institutions. For convenience in citing Japanese names, I have followed Western practice in placing the family name last, viz, Hirobumi Itō, rather than Itō Hirobumi. In transliteration, a macron denotes a long vowel.

[20]UNESCO, *History of industrial education in Japan*, Tokyo, 1959, p. 31. I have amended the syntax.

[21]Fox, op. cit. (9), p. 464.

[22]The Iwakura Mission was away from Japan from 23 December 1871 until 13 September 1973. See Marlene J. Mayo, 'The Iwakura Mission to the United States and Europe, 1871–1873', *Researches in the social sciences in Japan,* Columbia University East Asian Studies, No 6, 1959, pp. 28–47. Mayo implies that the English engineer Richard Brunton, then on leave of absence in Britain from building lighthouses, telegraphs, and railways in Japan, showed the Mission around the chief industrial sites of England and Scotland. On Brunton, see Fox, op. cit. (9), chapter XIII.

[23]Brendan P. Cuddy, 'The Royal Indian Engineering College, Cooper's Hill, 1871–1906. A case study of state involvement in professional civil engineering education', University of London PhD thesis, 1980.

[24]*The Ross-shire journal and general advertiser for the northern counties,* 15 September 1882, p. 3 (copy at Colindale Newspaper Library); partly quoted in *History of the Kobu-daigako,* Tokyo, 1931 (the text is otherwise in Japanese).

[25]Dyer, op. cit. (1), chapter I.

[26]*Proceedings of Royal Society of Edinburgh,* 1875–8, *9,* 212–16. See C. A. Oakley, *A history of a faculty: engineering at Glasgow university,* Glasgow, 1973.

[27]D. A. Low (ed.), *The Whitworth book,* London, 1926, p. 151. Dyer also won Glasgow University's Watt Prize for an essay on Newtonian science in the eighteenth century (copy Glasgow University archives).

[28]Ransome's comment, made in 1899, is relevant: 'The work accomplished by Mr. Dyer and his able staff of foreign professors will ever remain less noticed by the world in general than it ought to be, for the reason that it was all carried out so quietly; and although within twenty miles of a treaty port [Yokohoma], and in the capital of the country, it was out of the regular track of the tourist and the treaty-port resident. The professors, too, who did the work were leading a more or less retired life, as far as the rest of the European world in Japan was concerned'; Ransome, op. cit. (10), p. 96. For an obituary of Dyer, see *Engineering,* 1918, *126,* 291.

[29]H. Dyer, *The education of civil and mechanical engineers,* London, 1880.

[30]Dyer, op. cit. (1), p. 2.

[31]J. Scott Russell, *Systematic technical training for the English people,* London, 1869. See George S. Emmerson, *John Scott Russell, a great Victorian engineer and naval architect,* London, 1977, especially pp. 253–7; and, in more detail, George S. Emmerson, *Engineering education: a social history,* Newton Abbot & New York, 1973, chapter IX (Emmerson's chapter on Japan, pp. 225–30, is disappointing). Note also G. P. Mabon, 'Russell. A forgotten champion of technical education', *Vocational aspect,* 1965, *17,* 228–9.

[32]*The education and status of civil engineers in the United Kingdom and in foreign countries,* London, 1870. For a good recent account of the debate, see Cuddy, op. cit. (23), chapter I.

[33]F. Jenkin, 'Technical education' (11 January 1869), reprinted in *Papers, literary, scientific, etc, by the late Fleeming Jenkin, FRS, LLD,* 2 vols., London, 1887, ii, 157–82.

[34]The half-time system of children's education established by the Factory Acts of 1833 and 1844, was another possible model. See Edmund and Ruth Frow, *A survey of the half-time system of education,* Manchester, 1970. For comments on the development of sandwich courses, and the rival claims of Glasgow University, the University College of Bristol, and Glasgow Technical College to have originated the system, see Alan G. Smithers, *Sandwich courses: an integrated education?,* Windsor, Berks, 1976, pp. 15–16.

Colonel George Chesney, who had been professor of Surveying at the 'Thomasson' Civil Engineering College at Roorkee, which the British Government had established in 1848 in order to train Indians and Europeans for employment in the Indian Public Works Department (Cuddy, op. cit. (23), chapter II), also established the 'sandwich principle' at Cooper's Hill in 1871 (ibid., p. 91, and chapter IV, *passim*). Chesney's proposals, which are not mentioned by Dyer, could nevertheless have been known to him through 'A return to an address of the Honourable the House of Commons dated 15 March 1871, for a copy of all correspondence which has taken place up to the present time between the Secretary of State for India in Council and the Governor General of India in Council, in reference to the establishment of an engineering college at Cooper's Hill', *Parliamentary papers,* 1871, No 115. Since the Glasgow University Engineering Department feared that Cooper's Hill would undermine its near monopoly of placing its graduates in India, it seems very probable that Dyer would have been interested.

[35]For Thomson's physics laboratory, see S. P. Thompson, *The life of William Thomson, Baron Kelvin of Largs,* 2 vols., London, 1910, ii, chapter VII, and the letter from Thomson to the Trustees of the Ferguson Trust Fund, 4 February 1858, Glasgow University Library, Kelvin Papers, MS.T154.

[36]Dyer's 'sources' are discussed in his letter to Itō, 1 October 1877, printed in Imperial College of Engineering, *Reports by the principal and professors for the period 1873–1877,* Tokei, 1877, pp. 3–62 (copy Mitchell Library, Glasgow).

[37]Fox, op. cit. (9), p. 464.

[38]Dyer, op. cit. (1), p 2.

[39]Akabane was 12 km. north of the College and about 8 km. north-west from Hongō, the present site of the University of Tokyo. The Akabane works were given over to the Japanese navy in 1883. See Kozo Yamamura, 'Success illgotten? The role of Meiji militarism in Japan's technological progress', *Journal of economic history*, 1977, *37*, 113–33.

[40]A committee history of the *Kobu-daigakko* was published in Japanese in Tokyo in 1931 (n. 24, above), and there exists a Japanese thesis on the College by Somchai Priyawatanawit of Hitotsubashi University; I thank Ian Inkster for this reference. Information on the College in English is readily obtainable from its annually published Calendar (collections at British Library, and Universities of Glasgow and Toronto), and from contemporary accounts in English periodicals and newspapers. See also Dyer, op. cit. (1); William Gray Dixon, *The land of the morning: an account of Japan and its people based on four years' residence in that country*, Edinburgh, 1882; and the Report of Her Majesty's Secretary of the British Legation (Robert G. Watson), 'On the present education system of Japan', *Parliamentary papers*, 1874, *65*, 53–81. Watson's description is clearly drawn from the College's first *Calendar*. For contemporary views, see *Japan weekly mail*, 9 February 1878, pp. 132–5; 16 February 1878, pp. 152–6.

[41]Chastel de Boinville was the son of an Anglo-French clergyman who moved to Glasgow in 1871 and was posted to Japan as a draughtsman. He became head of the Public Works Building Department in 1874. See his obituary, *Journal of the Royal Institute of British Architects*, 1896–7, *4*, 359–60.

[42]The quadrangle, which is illustrated in *Kobu-daigakko* (n. 24, above) and *Reports* (n. 36, above), was never completed.

[43]Dixon, op. cit. (40), pp. 170–1. For some more details connected with earthquake precautions, see Dyer, *Minutes and Proceedings of the Institute of Civil Engineers*, 1885–6, *83*, 309–12.

[44]For Ayrton (1847–1908), see (i) *Proceedings of the Royal Society*, 1911, *85A*, i-viii, and (ii) the special issue of *The central*, April 1910, *7*.

[45]*Reports*, op. cit. (36), 29–30. Contrast Perry's description of Ayrton's 1878 laboratory in his 'The teaching of technical physics', *Journal of the Society of Arts*, 1880, *28*, 167–76; reprinted in J. Perry, *England's neglect of science*, London, 1900. His description comes from an anonymous article, 'A visit to Professor Ayrton's laboratory', *Japan weekly mail*, 26 October 1878, pp. 1129–31. Perry's lecture was illustrated with drawings of the College which do not appear to have survived, though some details of fixtures are to be found in E. C. Robins, *Technical school and college buildings*, London, 1887. It is clear to me that Ayrton's laboratory at Finsbury was modelled on its predecessor in Tokyo. For Finsbury see Robins, ibid., pp. 145–6.

[46]For Perry (1850–1920), who had studied engineering at Queen's University, Belfast (BE, 1870), taught physics at Clifton College, Bristol, and been a research assistant of W. Thomson's at Glasgow, 1874–5, see *Proceedings of the Royal Society*, 1926, *111A*, i-vii.

[47]For a list of these, see *Reports*, op. cit. (36), p. 30. In an obituary of Ayrton, op. cit. (44, i), p. ii, Perry implies that the research was done at night.

[48]Besides Ayrton and Perry, who had both been personal assistants to Thomson in Glasgow (Ayrton was also known to Williamson as a former graduate of University College London), the staff consisted of: Edward Divers (1837–1912), chemistry (a pupil of Williamson's); David F. Marshall, mathematics (an Edinburgh graduate); John Milne (1850–1913), mining geology and mineralogy; Edmund F. Mondy, metallurgy and technical drawing (a product of the School of Mines); Josiah Condor (1852–1920), architecture; R. Rymer Jones, surveying; and William Craigie, English (Craigie died in 1877 and was succeeded by Dixon). For other assistants, see the College Calendars.

[49]*Japan weekly mail* (1873), as quoted by Watson, op. cit. (40), p. 106.

[50]The number of foreigners in government employment reached its peak in 1874, with two percent of national expenditure going on their salaries which, individually, were often ten times those paid to native Japanese. See Shigeru Nakayama, *Characteristics of scientific development in Japan*, New Delhi, 1977, p. 44.

[51]W. E. Ayrton, 'Presidential address: electrotechnics', *Journal of the Institution of Electrical Engineers*, 1892, *21*, 5–36 (6).

[52]H. H. Bellot, *University College London, 1826–1926*, London, 1929, pp. 308–9.

[53]Livery Companies' Committee, *Report on technical education*, London, 1878. See also Tom Girtin, *The triple crown: a narrative history of the Drapers' Company, 1364–1964*, London, 1964; Jennifer Lang, *City and Guilds of London Institute centenary, 1878–1978*, London, 1978.

[54]Walker, op. cit. (4); H. E. Armstrong, 'The City and Guilds of London Institute: its origins and development', *The central*, 1938, *35*, 14–45; J. V. Eyre, *Henry Edward Armstrong, 1848–1937*, London, 1958; W. H. Brock, 'An experiment in technical education', *New scientist*, 22 November 1979, pp. 622–3. The surviving papers of the College are housed with the City and Guilds' archives which are at present being transferred to the Guildhall Library, London.

[55]See Eyre, op. cit. (54), and W. H. Brock (ed.), *H. E. Armstrong and the teaching of science, 1880–1930*, Cambridge, 1973.

[56]H. E. Armstrong, 'The beginnings of Finsbury and the Central', *The central*, 1934, *31*, 1–15 (7). The historian will concur with Armstrong's judgement. Ayrton and Armstrong both moved to the Central Institution established by the City and Guilds at South Kensington, but their teaching there was compartmentalized and divorced. Perry stayed at Finsbury until 1896, when he moved to the Royal College of Science at South Kensington, an institution then quite separate from the Central Institution.

[57]W. H. Brock, 'Observe, experiment and conclude. Finsbury College's new course of experimental philosophy in 1879–80', *History of Education Society bulletin*, Autumn 1980, *26*, 46–50.

[58]W. H. Brock and M. H. Price, 'Squared paper in the nineteenth century: instrument of science and engineering, and symbol of reform in mathematical education', *Educational studies in mathematics*, 1980, *11*, 365–81.

[59]Ayrton, op. cit. (51), p. 6. See also Philip Magnus, *Technical education. The introductory address . . . at the opening of the Finsbury technical college on February 19th, 1883, London, 1883*.

[60]Ayrton, op. cit. (51), p. 9.

[61]J. S. and H. G. Thompson, *Silvanus Phillips Thompson: his life and letters*, London, 1920. For Thompson's support of the sandwich principle, in which cites his experience of it when a lecturer at University College, Bristol, see *Report to the committee* [of the City and Guilds] *on the status and outlook of the City & Guilds Technical College*, Finsbury, n.d. [1901], p. 7 (copy City & Guilds archives).

[62]Smithers, op. cit. (34). See *Report of the committee appointed on 24 November 1903 . . . upon the subject of the best methods of education and training for all classes of engineers*, London, 1906.

[63]The evidence for Queen's University, Belfast, is equivocal. According to T. W. Moody and J. C. Beckett, *Queen's Belfast, 1845–1949: the history of a university*, 2 vols., Belfast and London, 1959, the two-year diploma course in engineering established in 1849 had to be followed by three years in an engineering workshop before a diploma examination could be sat (p. 146). The requirement of workshop training was abandoned in 1859 when the academic course was increased to three years. The BE degree was introduced in 1868, and it seems some candidates, most notably John Perry himself, sat the degree 'sandwich' style, see Perry's obituary, op. cit. (46), p. i.

[64]Following the creation of Western-style government under Itō in December 1885, the Ministry of Public Works was abolished and many government enterprises placed in the hands of private companies. The Kobu-Daigakko became the responsibility of a new Ministry of Education which made it part of a new Imperial University of Tokyo on 2 March 1886. Dyer's successor as Professor of Civil and Mechanical Engineering, Charles Dickinson West, remained in the Engineering Faculty well into the twentieth century. However, with the exception of Divers, who stayed as Professor of Chemistry until 1899, the other British professors were replaced by Japanese. Teaching continued in the original premises until new buildings on the Imperial University site at Hongō were opened in 1888. The original College buildings continued in use as museums and schools until destroyed in the great earthquake of 1 September 1923. The site is at present occupied by the Ministry of Education.

[65]J. A. Rae, ed., *History of Allan Glen's school, 1853–1953*, Glasgow, 1953.

[66]Dyer fits the interpretation of 'public science' offered by Frank M. Turner, 'Public science in Britain', *Isis*, 1980, *71*, 589–608. Dyer's later writings include *Valedictory address to the students of the Imperial College of Engineering*, Tokyo, 1882; *The foundation of social politics*, Glasgow, 1889; *A modern university with special reference to the requirements of science*, Perth, 1889; *Christianity and social problems*, Glasgow, 1890; *The influence of modern industry on social and economic conditions*, Manchester, 1892; *Technical education*, loc. cit. (5); *Science teaching in schools*, London, 1893; *Education and work*, Dunfermline, 1906; *Education and the industrial training of boys and girls*, Glasgow, 1913. Most of Dyer's library of about 4,000 volumes, including the works cited, were donated to the Mitchell [Public] Library in Glasgow, c 1920.

[67]See, for example, *The Glasgow and West of Scotland Technical College Calendar for the year 1888–89—et seq.* A set is kept in the Andersonian Library Archives, University of Strathclyde, Glasgow.

[68]Ibid., p. 43.

[69]Dyer, op. cit. (5), p. 5 and Dyer, op. cit. (1), p. 11 for the same claim.

[70]W. H. G. Armytage's essays are: *The American influence on English education*, London, 1967; *The French influence on English education*, London, 1968; *Looking north; influence and inference from Sweden in English education*, Arthur Mellows Memorial Trust Lecture, Peterborough, 1969; *The German influence on English education*, London, 1969; *The Russian influence of English education*, London, 1969.

[71]Dyer, *Education and work*, op. cit. (66), p. 20. Compare Dyer, op. cit. (1), 13: 'The engineer is the real revolutionist, for his work changes social and economic conditions and brings forces into action which are more powerful than anything which can be done by mere legislation'.

Queenwood College, Hampshire
(Courtesy Taylor & Francis Ltd)

XVII

Queenwood College Revisited

> The objects aimed at in Queenwood College are, to maintain for Pupils a HOME, in the true sense of the word; and, to surround that home by educational appliances, the best the times will afford.[1]

Midway between the Hampshire villages of East Tytherley and Broughton lies Queenwood Farm. Close to the farm cottages, and hard by the main road, the present-day visitor will find a large overgrown crater in the ground. Closer inspection will reveal the remains of a large brick foundation and water reservoir, and amidst the rubble, coping stones, remnants of bedsprings and gas-lights — the only signs that a thriving proprietorial boarding school, Queenwood College, once occupied the site - and before that, Robert Owen's socialist community, Harmony Hall. The once bustling building was destroyed by a great fire in June 1901.

The essence of teaching at Queenwood College was its emphasis upon the practical, as well as the theoretical, aspects of each subject. Its aim, as its prospectus advertised, was:

> to give an education in harmony with the age, an education that shall combine practice with theory and assimilate, as far as possible, the occupations of a college with those of real business, - and thus to fit a youth to fill and adorn the position in life he is to achieve and maintain.[2]

If not the first school in Britain to teach science through practical laboratory and workshop practice, it was undoubtedly the first to do so on such an extensive scale. The school also gave employment to a number of science teachers, several of whom became Fellows of the Royal Society and distinguished in research as well as teaching.

---

1 From the masthead of the school's magazine, *The Queenwood Reporter*. Surviving copies, for 1848, at Friend's House Library, Euston Road, London.

2 *The Queenwood Reporter*, 15 October 1848, p. 49.

Although the broad outlines of the history of Queenwood College and its activities are relatively well-known,[3] further details concerning its day-to-day activity during its heyday in the late 1840s and 1850s may be gleaned from the surviving journals kept by its three greatest teachers, the physicist John Tyndall (later Professor and Director of the Royal Institution in London), the chemist Edward Frankland (later Professor and Director of the Royal College of Chemistry in London), and the mathematician Thomas Archer Hirst (later Professor of Mathematics at University College, London and Director of the Royal Naval College at Greenwich). In particular, these journals reveal that life at the school was not altogether harmonious and that there were frequent rifts between staff, boys and the school's Quaker headmaster, George Edmondson, and his wife, Anne.

Queenwood had been founded in 1839 by Robert Owen and his followers as a socialist, albeit paternistically-governed, community called 'Harmony Hall'. The community was to have combined both industrial and agricultural activities for working men and women, together with a boarding school for middle-class parents of Owenite persuasion, on farm land leased from the landowner, Francis Baily.[4] In the event, financial considerations led to stress being laid exclusively on agricultural self-sufficiency and profitability. It was hoped that, in the fullness of time, Harmony Hall would become a model for the reform of the old moral world of class and capital outside its community boundaries. Unfortunately, although intended to expand to some 500 people, community numbers never rose above 60, and only 130 different residents explored its lifestyle during the community's six-year existence. For, despite the careful selection of residents for their trade skills and moral probity, flinty soil, bad harvests, under-investment, huge debts incurred by Owen in his leasing of the estate, internal dissensions between middle- and working-class members, and the failure to attract sufficient numbers of agricultural labourers who had, instead to be hired locally despite the communitarian principle, proved the whole scheme unsound. In 1845 the Hall was abandoned, although a few of 'the farmers' continued to live on the site as tenants.

Designed by the famous architect Joseph Hansom on the generous lines of a Victorian country house with cental heating and ventillation, gas lighting from

---

3 D. Thompson, 'Queenwood College, Hampshire. A mid-nineteenth century experiment in science teaching', *Annals of Science*, 11 (1955), pp. 246-254.

4 Full details of the complicated ownership and lease-holdership of the Queenwood site are given in R. G. Garnett, *Cooperation and the Owenite Communities* (Manchester, 1972), chapter 6. Garnett is closely followed by Dennis Hardy, *Alternative Communities in Nineteenth-Century England* (London, New York, 1979), pp. 53-58. Both sources give site plans.

its own gas works, mains water to every bedroom, costly windows, a mahagony-lined dining room, an expensively-fitted kitchen with a horizontal dumb waiter to the dining hall, and richly-moulded ceilings, the Hall, completed in 1842, seemed to George Jacob Holyoake, 'a monument of ill-timed munificence' more fitted to Peel's home at Drayton Manor than for a community of workers and scholars.[5] In March 1847 this palatial three-storey red-bricked building, which had housed the community and its school, together with 800 of the original 1,000 acres of land, was leased to to a Lancastrian Quaker, George Edmondson (1798-1863), who transformed it into a proprietory school and agricultural college.

The new name, Queenwood College, was derived from the chief farm on the estate, probably in an attempt to disguise its former identity. Some years later the College received bad publicity when it was revealed that he Home Colonisation Society was receiving rent from Edmondson. Although there is no evidence to suggest that Edmondson was an Owenite, like Owen, he believed that external circumstances played the major role in moulding men's characters and consequently that the education of children and young men was of fundamental importance in producing a moral, just and reverential society. Owen's well-known experimentation with factory children at his New Lanarck Mills had been perpetuated in the short-lived fee-paying school at Harmony Hall run by a German disciple of Pestalozzi and Fellenberg, Dr Oestreicher.[6] It would appear that although the community experiment had had to be abandoned in 1845, Owen (as well as his fellow Governors John Finch and John Pare) hoped that the New Moral Order could continue to be succoured through the Queenwood school. Its perpetuation as a school seemed to them a much more useful plan than the alternative - to turn the premises into a lunatic asylum currently being made obligatory for English counties through an Act of Parliament. To this end, Owen, Pare and Finch were attracted to Edmondson because of his help with the foundation of the College of Preceptors in 1846 and because of his Society of Friends' conviction that human society could be improved if children were taught through play and vocational practice in an egalitarian atmosphere in which teachers were moral models as well as instructors. Further continuity between Harmony Hall school and Queenwood

---

5 G. J. Holyoake, *A Visit to Harmony Hall* (1844), p. 8, reprinted in *Cooperative Communities Plans and Descriptions* (New York, 1972). Despite criticism, Holyoake approved the experiment. Note also Alexander Sommerville, *Notes from the Farming Districts. No. xvii. A Journey to Harmony Hall in Hampshire* (1842) which is also reprinted in *Cooperative Communities*. Although finally completed in 1842, members of the community had moved into the Hall in 1841.

6 Owen had educated two of his sons with Fellenberg at Hofwyl in Switzerland.

College may be seen in the continuation of leisure activities such as concerts and charades, evening lectures and classes in drawing, painting and phonography (shorthand), or in the teetotal dances to which local Hampshire farmers and villagers were invited.

Edmondson, the product of a large Quaker family, had been educated at the famous Friends' coeducational boarding school at Ackworth which, like (but quite independently of) the Swiss school of Pestalozzi and Fellenberg at Hofwyl then attracting attention, combined the learning of the 3Rs with manual instruction in useful trades such as tailoring, shoe-making and sewing. Ackworth was, in fact, following out the ideas suggested by George Fox, the founder of the Society of Friends a century and a half earlier. Edmondson was to maintain and extend this vocational principle in each of his schools.[7]

On leaving Ackworth at fourteen, Edmondson had been apprenticed to William Singleton (c. 1770-1832), a former Ackworth reading master who had opened a day- and boarding-school at Broome Hall, Sheffield. Here Edmondson learned to teach and acquired practical skill in book-binding, printing and surveying. He also came to share Singleton's belief that corporal punishment was morally wrong — though in practice he appears to have been unable to make this a firm rule at Queenwood. Edmondson was to marry Singleton's daughter, Anne, whom many came to see as the dominant partner in the marriage, in 1821. Before this, at the age of twenty, he travelled to Russia to tutor the children of the Quaker missionary, Daniel Wheeler, who was engaged in a land reclamation scheme for the Emperor Alexander I at St Petersburg. This experience awakened Edmondson's interest in agriculture when, with the assistance of the London Quaker pharmacist, William Allen, he began to analyse the reclaimed soils for their fertility.[8] Apart from a brief visit to England to marry and collect Anne, he stayed in Russia for seven years.[9] In 1825 the couple returned to Lancashire and set up a small day school in Blackburn. Not surprisingly, Edmondson offered Russian as part of the curriculum, and as the school prospered Edmondson added manual instruction in carpentry, printing and metalwork to the timetable. The school's printing press was used to print three of the texts Edmondson compiled on geography,

---

7 D. Thompson, 'George Edmondson (1798-1863)', *Friends' Quarterly*, 10 (1956), pp. 24-29.

8 Edmondson to Singleton, 9 February 1819, Friends' House Library. I am grateful to E. H. Milligan for help with Edmondson's career.

9 The Edmondsons' daughter, Mrs Jane Benson, later wrote a fictional account of her parents' early lives together, *From the Lune to the Neva. Sixty Years Ago; with Ackworth and 'Quaker' Life by the Way* (London, 1879). See also her *Quaker Pioneers in Russia* (London, 1902).

geometry and perspective drawing for his pupils, as well as a school magazine entitled *The Lower Bank Monthly Magazine*. A school magazine was to be the hallmark of Edmondson's successive schools.[10]

In 1841 Edmondson acquired larger school premises at Tulketh Hall, Preston. This enabled him to expand the curriculum to include classical and modern languages, the 'use of philosophical and mathematical apparatus', and to attract the sons of a more affluent group of Friends from all over the country. He was joined by Anna's brother, Josiah Singleton. Surviving copies of a monthly *Tulketh Hall Mercury* show that surveying (an important vocational skill because of the great railway and sanitary building programmes of the 1840s), book-keeping, shorthand and mathematics were prominent features of the curriculum, which was examined by two-hour examinations in each subject during the last days of each term.[11] There was also a Tulketh Hall Natural History and Philosophical Society which was supplied with weekly lectures by Edmondson, Singleton and the pupils themselves. Pupils also attended itinerant lecturers' classes in Preston - for example, in 1845, courses on chemistry by a Dr Robinson, and on astronomy by a Mr Holden. It is clear that by 1845 practical chemistry was being taught since there is a reference to lectures by W. H. Balmain on chemical analysis:

> in which all the experiments are all to be performed by its members. In accordance with this object several additional accommodations have been made to the laboratory. Each pupil having been [*sic*] furnished with a furnace, a retort and a stand, and several other chemical apparatus.[12]

Edward Frankland was to claim later that the chemistry teaching at Tulketh Hall was made 'under my direction'.[13] This claim is not altogether implausible since Frankland, a pharmacist and fellow Lancastrian, was then living in Preston and was to be invited by Edmondson to teach chemistry at

---

10 It would appear from *The Cambridge Bibliography of English Literature*, that school magazines originated at Eton College in the eighteenth century. Both Harrow and Rugby had magazines by 1835.

11 Volume 5, nos. 1-5, 7, and vol. 6 (1844-46) are contained in the Spencer Bequest P373, Harris Library, Preston. For examinations, *Tulketh Hall Mercury*, vol. 6, No. 19, 1 April 1845, p. 185.

12 *Ibid.*, vol. 6, no. 27, 15 October 1845, p. 265. Balmain became a well-known chemistry teacher for the Department of Science and Art and wrote *Lessons in Chemistry, for the Use of Pupils in Schools* (London, 1844).

13 In a notice concerning the life of John Tyndall in *Westminster Gazette 'Popular'*, no. 6, December 1893, p. 22. Copy Royal Institution Archives.

Queenwood.[14] Whatever the truth of the matter, there is no doubt that both the Lower Bank School and Tulketh Hall were very extensively equipped with apparatus for scientific demonstrations. A list of scientific exhibits at the Preston Exhibition in 1840, much of it identified as belonging personally to Edmondson, reads:

> quadrant, air pump, compass, sun dial, terrestrial globes, electric canon, speaking trumpet, electric jar and conductor, electromagnets, electrometer, goniometer, hydrostatic balance, clinometer, barometer, stereoscope, thermometers, oxihydrogen microscope.

And a descriptive passage reads:

> The electric machine and electromagnetic apparatus are in constant requisition. The circular table, upon which the philosophical apparatus is arranged, becomes in the evening the focus towards which all visitors tend, to witness the experiments by Mr Moses Holden and other gentlemen, who have kindly lent their able assistance.... Cloud of Olympus, or Fire Cloud, in which waves of fire will be seen rolling along the ceiling of the room, displaying many of the prismatic colours in great splendour.[15]

Such lists and descriptions are reminiscent of the activities of many eighteenth-century dissenting academies,[16] but to these Edmondson had added manual training and 'hands-on' chemical analysis.

By the mid-1840s Edmondson's school was becoming well-known outside the circle of Friends. Not surprisingly, therefore, when Harmony Hall collapsed he was approached directly by the surviving Governors and told 'how admirably the place was adapted to carry out the enlarged views of education they had heard he entertained.'[17] Following the establishment of the Royal Agricultural Society in 1838, the publication of Liebig's *Agricultural Chemistry* in 1840 and the creation of the Royal College of Chemistry in 1845,

---

14 C. A. Russell, *Lancastrian Chemist. The Early Years of Sir Edward Frankland* (Milton Keynes, 1986). Part 2 of this biography is in press.

15 *Preston Exhibition 1840.* Spencer Bequest P373, Harris Library.

16 See N. Hans, *New Trends in Education in the Eighteenth Century* (London, 1951); A. E. Musson and E. Robinson, *Science and Technology in the Industrial Revolution* (Manchester, 1969).

17 Obituary of Edmondson in *The Friend*, 1 July 1863, quoted in Thompson, *op. cit.* (7), p. 27.

Edmondson could not have avoided being aware of the importance of 'High Farming' or scientific agriculture and the need to educate the next generation of farmers in the principles of scientific estate management. Perhaps Frankland had also drawn his attention to the possibility of introducing agricultural chemistry into the vocational side of the Tulketh Hall curriculum. Edmondson must also have been aware of the foundation of the Civil Engineering College at Putney in 1839 by a group of the nobility who were interested in creating engineers to serve the new railway system in which they had heavily invested.[18] Therefore, despite the well-publicised failure of Harmony Hall as an agricultural cooperative, Edmondson perhaps saw Queenwood as a golden opportunity to create a truly vocational college in tune with society's needs for scientifically-trained farmers and engineering apprentices.

Such an enterprise would need a specialist staff - who were to called 'tutors', not 'masters' — as well as the normal complement of teachers of classics and mathematics. Frankland, who had moved to London to conduct temporary research on coal at the Museum of Economic Geology, was delighted to accept a more permanent appointment in a school whose basement was to be fitted out as a chemical laboratory. Similarly pleased was the 27-year-old Irish surveyor, John Tyndall, whom Edmondson had met while Tyndall worked for the Manchester firm of Nevins & Lawton, consultant engineers for divers railway lines through Lancashire and Yorkshire. Impressed by Edmondson's offer of £150 p.a. *plus* rooms and full board, Tyndall confided to his journal:

> [Edmondson] is a man of enlarged understanding and fully joins in the onward movement of the day. He intends to make QC 'no. 1'.[19]

Such confidence was not to last.

The Edmondsons, together with their fourteen year old daughter, Janes, and mother-in-law Singleton, moved to Queenwood in August 1847. When Tyndall arrived later that month to teach mathematics, surveying and elementary physics he found that Edmondson also expected him to act as the College's accountant and to prepare evening lectures on Liebig's *Animal*

---

18 The College found it difficult to recruit students and collapsed in 1859. See R. Sviedrys, 'The rise of physics laboratories in Britain', *Historical Studies in the Physical Sciences*, 7 (1976), pp. 405-36.

19 *Journals of John Tyndall*, 15 July 1847, typescripts Royal Institution Archives. Further quotations will be references by date in the text. According to Frankland, however, Tyndall suffered 'pecuniary loss' in abandoning surveying. See his anonymous obituary of Tyndall, *Minutes and Proceedings Institution of Civil Engineers*, 116 (1893-94), p. 341.

Chemistry and Erasmus Wilson's *Treatise on the Skin* (*Journal*, 21 August 1847). With the arrival of Frankland in September, however, Tyndall was saved from trying to lecture on Liebig, though he duly gave one on skin and soap (8 September 1847).

At the end of a successful first term, the French tutor, John Yeats, who had been a disciple of Fellenberg's at Hofwyl, gave a lecture to the school in which he made its structure and philosophy explicit. The staff were of two kinds, instructors and educators (or, as Tyndall put it nearly forty years later, informers and stimulators[20]), the former imparting instruction during school hours, the latter appealing to 'every faculty of mind and body' outside formal lessons. An analogy with the emerging system of teachers and house masters in the contemporary English public school is apparent. Pupils, Yeats continued, were divided by age group. The first, boys below the age of 15, followed an elementary course whose object was 'development, and the awakening of a love for learning'. The second older group pursued science-based vocational courses with a view to farming, engineering or commerce. The Principal (Edmondson), Yeats declared, was firmly against cramming.

> He would rather strengthen his pupils' minds by proper exercise, than strain them by over exertion; he would rather that a youth under his care, should learn to carry a fair burden in after life, than that he should make up, and take up, at school, a huge load, to be thrown down from distaste, or from weariness, on his first entrance into business.[21]

Clearly, Queenwood College was not to be a feeder school to Oxford and Cambridge, though one or two pupils, like Walter Flight (1841-85), were inspired by individual tutors to spend time at German universities.[22] Yeats' address was printed in the College's *Queenwood Reporter*, to which both boys and staff contributed:

---

20 J. Tyndall, 'Address delivered at the Birkbeck Institution on October 22, 1884' in his *New Fragments* (London, 1892), p. 231. For Tyndall. 'a power of character must underlie and enforce the work of the intellect.'

21 *Queenwood Reporter,* 15 February 1848, p. 14.

22 Flight attended Queenwood in the mid-1850s, and was attracted to chemistry by Heinrich Debus, who encouraged him to study at Halle, Heidleberg and Berlin before taking the London D.Sc. in 1867. A Fellow of the Royal Society (1883), Flight worked as a mineralogist at the British Museum, specialising in meteorites. See *Dictionary of National Biography*.

> as a means of increasing the ability of our pupils, and of testing their ability, — to facilitate and regulate our pupils' intercourse with their parents, — to open up to us a channel of communication with our friends and the public at large. (*Queenwood Reporter*, 15 February 1848)

Tyndall, Frankland and Josiah Singleton (Mrs Edmondson's brother who also taught at the College) struck up a warm friendship and spent their leisure hours discussing religion, philosophy, poetry, science and, inevitably, the government of the school. Frankland and Tyndall soon made a pact that each would attend the other's classes — Tyndall, Frankland's on botany and chemistry, as well as private practical classes in chemical analysis; and Frankland, Tyndall's on the steam engine, railway surveying and physics. The staff also attended each others' evening lectures: Davidson on astronomy, Singleton on perspective, Edmondson on agriculture, and Yeats on Pestalozzi and the Swiss. From this brief roll call it seems clear that Edmondson fleshed out the curricuum by relying on the past experiences of his masters to stimulate and pass on information and something of their own personal moral philosophies. Nor were Edmondson's family excluded, for his elder daughter, Jane, gave regular lessons on phonography. Her humanising presence, together with Anne's, was praised by Yeats who pointedly contrasted Queenwood with the Clarendon public schools in which the absence of feminine control led to recklessness and depravity.[23]

Singleton, Frankland and Tyndall would rise at 4.00 a.m. each morning to practise chemical analysis until the school day began at 6.45 a.m., and they would sometimes return there in the evenings. Allowing three hours for the preparation of each lesson, Frankland calculated his hours of duty as: Mondays 10 hours, Tuesdays 4, Wednesdays 10, Thursdays 4, Fridays 5 and Saturdays 7. However, the actual student contact hours were not onerous. On Mondays, for example, he took a one hour chemistry class from 9.00-10.00 a.m., a two-hour practical from 12.00-2.00 p.m. and another hour of chemistry from 4.00-5.00 p.m. The day concluded with a lecture to the whole school from 6.30-7.00 p.m.[24] Edmondson also allowed Frankland a young laboratory assistant after April 1848. Frankland's spare time was spent in the laboratory concluding important research which he had begun in London with Hermann Kolbe, the background to which he introduced senior boys. The latter also once assisted

---

23 Tyndall, *op. cit.* (20), p. 16.

24 *Journal of Edward Frankland 1 January 1848-12 August 1849*, typescripts at Royal Institution and Royal Society.

him in analysing the contents of a cadaver's stomach for the local coroner's office. (Frankland's *Journal*, 14 March 1848 *et seq*) The laboratory was frequently the source of merriment for the young teachers and boys who, forbidden beer by Edmondson, became 'teetotally drunk' by inhaling ether, chloroform or nitrous oxide.

There were other attractions in the school:

> On coming down stairs at dinner Tris [Frankland] who walked before me seemed impressed by a mechanical impulse which impelled him along the corridor with a fervid velocity. On reaching the stair bottom I discovered the cause of the attraction. Miss E, like a pure planet, had checked his gravitating tendencies and lo! he stood radiant with smiles dropping joysparkles from his eyes as he clasped her hand, his countenance became a transparency through which the full proportions of his soul shone manifest; his blood tingled from his eyebrows to his finger ends, and wealthy with rich emotions his face became the avenue of what he felt (Tyndall's *Journal*, 18 January 1848).

A born educator, and inspired by reading Fichte's 'On the Nature and Vocation of a Scholar', Tyndall 'joyed' over his science lessons 'with a good humour and heartiness which resembles a musical chant, and the little fellows are never weary of working with me'.[25] One of his first pupils, James Mansergh, was inspired by Tyndall to become an engineering apprentice. He subsequently became an internationally-famous railway and sanitary engineering consultant.[26] Frankland was also able to strike up a close friendship with the Rev. Richard Dawes, a Cambridge mathematicain who held the living of King's Somborne, a village four miles from Queenwood. Dawes had opened a National Society school in the village at which he had introduced elementary science to young children. Frankland exchanged books and apparatus with Dawes and gave chemistry lessons to the school as well as to the assistant Dawes employed. In return Dawes taught Frankland and Tyndall the rudiments of German.[27]

Unfortunately, although Frankland and Tyndall had both initially expressed strong admiration for Edmondson, by the second term they were beginning to

---

25 Tyndall to Hirst, 16 February 1853, Royal Institution Archives.

26 James Mansergh (1834-1905). See *Minutes and Proceedings of the Institute of Civil Engineers*, 51 (1904-5), 350-54.

27 D. Layton, *Science for the People. The Origins of the School Science Curriculum in England* (London, 1973). Frankland's *Journal*, 3 May, 12 August and 17 September 1848. Frankland's course notes survive; see C. A. Russell, *Edward Frankland* in press.

have reservations. The main problem seemed to be that his generous flexibility made a consistent code of discipline difficult, and matters were not helped by Mrs Edmondson's strong personality. Several of the initial staff only stayed a term after 'rows with Farmers', that is the senior boys whom Mrs Tyndall later described appropriately as:

> verging on manhood, some were failures from the universities, others had disappointed their parents in other vocations, and they combined to form an element of idling and unruly tendency which [called] for strong and judicious handling.[28]

It will be recallled that similar discipline problems, and even rebellions, had taken place in the older public schools until unruliness was tamed by religion, competitive games, and a prefectorial system.[29] In their absence, and admittedly aided by smaller numbers of pupils, Queenwood College could only work if there was mutual trust between boys and tutors and if its Principal laid down carefully defined rules of behaviour. But here Edmondson seems to have failed. Although Edmondson opposed corporal punishment in principle, he turned a blind eye to the fact that this view was not shared, or rather that it was not practised by all his staff. In May 1848 a boy ran away after being beaten by Yeats who then beat three other boys for swimming in the College's drinking-water tank. Two days later the senior boys mutinied and refused to attend lessons until Yeats was told to abandon such punishments. Worse still, the father of Henry Fawcett, one of the beaten boys, a future Postmaster general, threatened to expose the matter in the newspapers.[30] How this particular episode was resolved is unclear, though to the disgust of Tyndall, who himself thrashed three boys on 3 June 1852, despite his endangerment of the school's reputation over a harmless bit of fun, Yeats remained in Edmondson's favour.

Matters were not helped by Tyndall's and Frankland's lack of deference towards the Edmondsons, in contrast to Yeats's. The latter was detested for his hypercritical, fawning attitude towards the headmaster, his unsuccessful attempts to obtain more hours for language teaching over mathematics, and, unforgiveably, because he reported the fact that Tyndall, Frankland and

---

28 Louisa Tyndall, *Preliminary Drafts for a Life of John Tyndall*, p. 169, typescript Royal Institution Archives.

29 J. R. de Honey, *Tom Brown's Universe. The Development of the English Public School in the Nineteenth Century* (London, 1877) provides a good discussion.

30 Leslie Stephen, *The Life of Henry Fawcett* (London, 1885) does not allude to this episode. See pp. 8-15 for Fawcett's period at Queenwood. My account is drawn from Tyndall's and Frankland's *Journals*, 5 May 1848 *et seq.*

Singleton had been seen drinking in a village pub! Although happy to have accepted a reprimand from Edmondson, the feeling that he had become the instrument of Yeats diminished the congeniality of Queenwood for Frankland and Tyndall. Edmondson, Tyndall decided, was 'a reed shaken by the wind ... a well cultured cabbage plant ... a kind of government fiction - like the Queen of England - useful, perhaps, as a point of reference in certain cases.'[31]

They were also dismayed by what seemed to them Mrs Edmondson's ungenerous treatment of her brother. Despised for his Anglicanism in a family of Friends, Singleton was also proposing to marry 'an inferior person' and inspired by the romance of the situation, all three decided to leave at the end of the summer in 1848 - Singleton to found his own school in Lancashire,[32] and Tyndall and Frankland to further their scientific careers with Robert Bunsen at the University of Marburg. Despite Mrs Edmondson's earnest attempts to dissuade them (one suspects they would have left even if the atmosphere had been more congenial) the sacking of Singleton on 31 May 1848, for allowing an older boy to give a younger lad beer, proved the last straw and they duly left for Germany in October 1848. At this point Mrs Edmondson had put her pride in her pocket and implored them both to stay and the two friends eventually departed from Queenwood without rancour or bitterness on either side. Indeed, the loss of Tyndall must have been a great blow to the College. When Tyndall gave his Yorkshire friend, Thomas Hirst, a copy of a valedictory lecture he had delivered to the whole school in October 1848, Hirst noted:

> I think often if I had heard such lectures when at school [in Wakefield] they would have benefited me much. He speaks more to them as his equals - there are no symptoms of pedantry visible about him, he speaks to them as boys, and not men, and gives them his experience when at the same age, recollects his feelings then towards such subjects and acts accordingly.[33]

---

31 Tyndall to Hirst, 29 July 1851.

32 Singleton founded 'Spring Bank School' at Darwen. When Tyndall met him again on 27 June 1848 he had already been widowed after less than a month's marriage. This may suggest that the real source of his sister's attitude was an illegitimate child. Singleton remarried in January 1849.

33 W. H. Brock and R. M. MacLeod, eds., *Natural Knowledge in Social Context. The Journals of Thomas Archer Hirst FRS (1830-1892)* (London, 1980), entry for 15 October 1848. Future references to this source will be given in the text. The original typescripts of the journals are in the Royal Institution Archives.

At Marburg, Frankland and Tyndall became warm friends with Heinrich Debus, a German chemical assistant in Bunsen's laboratory. In April 1851, as a result of letters from Edmondson, Tyndall recommended that Debus should teach chemistry at Queenwood, and in the same month Edmondson offered both Tyndall and Debus positions, adding that Yeats had resigned. The College's fortunes had declined dramatically since Tyndall and Frankland's departure, even though practical chemistry teaching had continued under Robert Galloway (1822-96), another Lancastrian pharmacist colleague of Frankland's.[34] When Galloway left in February 1851, Charles Bloxam (1813-87), a recent graduate of the Royal College of Chemistry, was asked to replace him. He was not impressed.

> Pretty spot - large farm (800 acres) - splendid house - small family - Mr E. shrewd little quaker - committee of unity - Mrs E. very kind - Miss E. rather pretty - very young (17) - Mas[ter] E. pudding head - teachers so-so-men. No private room for teachers. Laboratory scrubby. Chem. teacher to take all boys to Church on Sunday evening - will not do for me.[35]

Student numbers had sunk to only 33 by August 1851. Tyndall duly returned, with Debus, in September 1851 for another two years of teaching.

Relations with the Edmondsons remained strained. Of Mrs Edmondson, Tyndall observed, 'Her heart is large and tender, but her pride and love of mastery are inordinate' (*Journal*, 3 December 1851). Tyndall finally resigned in July 1853 on being appointed a Professor of Natural Philosophy at the Royal Institution in London. He was replaced by one of his younger friends, Thomas Archer Hirst, whom Tyndall had met and encouraged while Hirst was an apprentice railway surveyor in Halifax. Hirst had followed Tyndall to Marburg where he had specialised in geometrical research and taken a doctorate before making further mathematical studies in Berlin. The next phase in Queenwood's development can be deduced from a journal kept by Hirst.

Hirst had first visited the College in 1850 in order to see Tyndall, and been immensely impressed.

---

34 In 1851 Robert Galloway became Playfair's assistant in London before moving to the Museum of Irish Industry in Dublin. His Queenwood course was published as *A Manual of Qualitative Analysis* (1850) and *First Steps in Chemistry* (1851).

35 Entries for 17 February and 7 March 1851 in Diaries of C. L. Bloxam, King's College, London, archives.

> Queenwood itself is a beautiful spot. It stands in a rich undulated chalk district, the small knolls and valleys are always graceful and smooth and the rich woods with their beautiful beeches, yews and elms have a soothing effect. The building itself is interesting on many accounts, first, its architecture which is in a novel and picturesque style, mostly Italian. Secondly, its inward arrangements, which are the most convenient and beautiful that I have seen, and thirdly, its associations, for this is the celebrated Harmony Hall, where the socialists first practically tried to live by the law of love, and of course miserably failed. On one end of the building the characters R. O. (Robert Owen) may just be distinguished, though they have been obliterated. At the other end, in flint squares is written C of M 1842 [Commencement of Millenium] (*Journal*, 21 September 1850, folio 657).

When Hirst joined the College in July 1853 - he had previously given several lectures there when visiting Tyndall - it had 76 pupils. His duties, somewhat less onerous than Tyndall's and Frankland's, involved thirteen hours of class teaching a week together with the preparation and delivery of two lectures a week to the whole school (*Journal*, 14 August 1853). There was also cricket and his cello to be played in a school band led by the music master, William Cornwall. His salary was fixed at £150 p.a., plus full board. Hirst, whose only teaching experience had been in the Mutual Improvement Society of Halifax during the 1840s, immediately found teaching children congenial, even though it robbed him of time for his geometrical research. There were also growing calls on his time to lecture on electricity and magnetism in several Hampshire towns - experiences which soon decided him that, unlike his mentor Tyndall whom he hero-worshipped, he lacked the talent ever to shine as a popular lecturer (*Journal*, 30 October 1853).

Hirst taught mathematics (along with Richard Wright), elementary physics and lots of electricity and magnetism. Like Tyndall, he used mathematics as a tool to solve practical problems with the boys, such as swinging, see-sawing, rebounding tennis balls and trigonometry. It took him, he records, five sessions (two and a quarter years) to work through a full course of physics (*Journal*, 17 February 1856). The practical bent, and success, of Hirst's teaching is clear from his account of an elementary physics class:

> On Thursday I taught them how to do square roots in a very short time, and how to find the hypotenuse of a right-angled triangle. As examples for practice we measured the sides of the room, table, etc., and first calculated the diagonal and then actually measured it as a test

> of our work. They took great interest in the lesson, and worked exceedingly well. Afterwards I happened to hear them talking about it to one another, and ignorant of my presence I heard such sentences as these: 'I say, I like this lesson, don't you?' 'I like this hour better than any other', and so on - all of which little remarks were both instructive and satisfactory to me (*Journal*, 10 September 1854).

Hirst's experience of teaching geometry and surveying from a practical and heuristic viewpoint was later to make him a formidable spokesman for the need to revise the teaching of Euclidian geometry.[36]

By 1856 it is clear that some parents, aware of the College's scientific reputation, used it as a crammer. For example, Tyndall's former survey employer, Colonel George Wynne (d. 1890), sent his seventeen-year-old son Edward to Queenwood for lessons with Hirst on physics and mathematics, Debus on chemistry, and John Haas for French, prior to his taking the entrance examinations for the Woolwich Military Academy (*Journal*, 2 March 1856). Wynne received private lessons from the three masters in addition to attending their day classes.[37] It is amusing to see how the fees were shared.

> At a meeting at Queenwood on Thursday evening last it was agreed that £50 a quarter should be charged [to Col. Wynne]. £5 of this goes to Mr Edmondson for the privilege of attending classes and for five dinners a week. I am to receive £10 for board and lodgings, and of the remainder [£35] I am to receive $^3/_5$ [£21] and Debus and Haas each $^1/_5$ [£7] for private lessons (*Journal*, 2 March 1856).

Hirst's relationship with the school's other staff was extremely cordial. His journals are filled with accounts of their walks and talks together and their merry evening parties. There was one exception. If John Yeats was the staff villain in Tyndall's day, Hirst could not abide the classics teacher, Mummery, who held a certificate of teaching proficiency from the College of Preceptors. Not satisfied with the letters M.C.P., Mummery purchased a German doctoral degree for £25 in November 1853. Hirst and Debus, whose Ph.D.'s from the University of Marburg had been earned by mental effort and a thesis, were disgusted, and Hirst made it plain that he thought the action dishonest. By threatening to expose the matter to parents, Hirst engineered Mummery's

---

36 See essay XII, 'Geometry and the Universities', above.

37 Edward Wynne served as a lieutenant in the Royal Engineers in Canada and India before his death in 1889.

dismissal at the end of the year. He was replaced by Walter Tribe, an Oxford classicist, who remained a lifelong friend.[38] The local doctor at the neighbouring village of Broughton, Dr Fox, who sent six sons to Queenwood, also became a regular member of the staff room in his capacity as the College's medical officer.

With Hirst at Queenwood, Tyndall did not give up his hearty interest in the school's well being, and he was a frequent visitor to talk to the boys. There was the threat of a financial crisis in 1853 when William Pare, the cooperator who had supervised the leasing of the site to Edmondson, threatened to sell the property for £27,000 'to a community of American Shakers' (*Journal*, 20 December 1853). Tyndall promptly intervened by writing a strongly-worded letter to Pare saying that since the College was doing such a good job educationally, commercial considerations ought not to enter Pare's mind.[39] The school continued, although possibly because the conditions of Edmondson's lease were altered, or because Debus disliked teaching agricultural chemistry, the 'farm class' was abandoned.[40]

During the 1850s discipline was based upon a system of monetary fines; but so as not to put poorer students at a disadvantage (or punish the parents who otherwise would have had to pay their sons' fines) there were jobs and errands for which the staff paid boys who were on the punishment roster. For example, before his marriage Hirst had boys on punishment duty to service his room. Nevertheless, as Hirst discovered, some boys got badly into debt and consequently practised lying and deceit in order to borrow further money from other boys, or even masters (*Journal*, 7 January 1854).

Like Tyndall before him, Hirst soon found himself, in the absence of strong leadership from Edmondson, acting as disciplinarian or umpire in school disputes. In February 1854. two boys ran away back to their homes in Manchester. On their return, Edmondson was at a loss how to punish them for absconding. Hirst soon ascertained that the older boy of the two had been a bad moral influence on the younger absconder and, indeed, on the whole school. Persuaded by Hirst, Edmondson agreed to expel the older boy. And in an episode which might have formed the basis for a late Victorian school story, Hirst investigated a spate of petty thefts in the school. Marked money was left to tempt the thief, and when this succeeded, male servants were ordered into one room, female servants into another and the boys into the dining room. A

---

38 Walter Tribe later became Rector of Stockbridge. His daughter, Mary, became a Duchess of Bedford. See *Dictionary of National Biography*.

39 In fact, because of legal wrangles, Harmony Hall's debts were not settled until 1863. See Garnett, *op. cit.* (4). Note that Pare sent two of his sons to Queenwood.

40 Mrs Tyndall, *op. cit.* (28), p. 545.

thorough search of everyone's pockets revealed the culprit to be a College servant.

A more serious indication of Edmondson's lack of leadership - he was now in his 56th year — took place at the College's Mutual Improvement Society in 1854. In this venue, equality between pupils and staff when they presented and discussed papers had always been the rule. In October 1854, however, Edmondson had thought it necessary to admonish some boys for talking during a staff presentation. This led Hirst to comment that:

> it is an excellent thing when masters can keep their position by natural and inherent superiority; but it is a dangerous institution ... when masters are merely propped up, and do not stand on their own intrinsic merits (*Journal*, 15 October 1854).

But if the master concerned had failed as an educator by being unable to keep his audience's attention, the College saw the episode as divisive. Seventeen boys resigned from the Society because Edmondson had punished talking; another ten resigned on the grounds that they no longer found the proceedings of interest; while another tested the rules further by proposing that attendance should be voluntary. Edmondson, as President of the Society, was left 'towering with rage', forced as the Society's Chairman to abide by its egalitarian constitution even though the school was rebelling against his authority as Principal of Queenwood (*Journal*, 15 October 1854). The matter was resolved by Hirst taking over the Society's Presidency, while Edmondson oversaw those who had resigned from the meetings. In a brilliant speech, Hirst made it clear to the pupils remaining in the Society that although the reins of authority were slackened at their meetings, boys were ultimately still under authority, and boys should not abuse their freedom by interfering with College orders. Otherwise, the logical consequence would be that boys could elect not to attend ordinary lessons. Hirst was left embittered by the fact that he, and not Edmondson, had had to resolve the potentially anarchic situation.

More drama followed in the same month when Wesley, the sixteen-year-old son of the famous Winchester Cathedral organist and composer, Samuel Sebastian Wesley, threw a firework into a classroom where teachig was in progress. After confessing, Wesley absconded, but was caught by Hirst after a furious chase. Brought back to College, Wesley again escaped over the roof from a locked room. He was duly expelled.

Although ultimately Edmondson's responsibilty, Hirst blamed the College's wave of unruliness upon a new teacher, McDougall, who dealt with boys as if they were his natural enemies. He was continually detaining them for minor misdemeanours.

> Between the lads and McDougall there has been continual strife, and no one but a schoolmaster can understand what an effect such a strife has on the whole school. The good lads are made worse, the bad get into open rupture. Such strife existing, there is always an escape there for the malicious propensities of the worst lads. The two who have been expelled this half year, Wesley and Sinclair, were the scum of the school. Apart from McDougall, they deserve expulsion. McDougall being there but accelerated it for which we may be thankful (*Journal*, 19 November 1854).

Not surprisingly, McDougall left at the end of 1854, as did another teacher, Iliff, whom Hirst thought showed considerable potential. Iliff had decided to quit because, expecting to find Queenwood a 'model place', he found instead that it was 'conducted by an imbecile, and that its success is not owing to, but in spite of, [Edmondson's] management' (*Journal*, 23 April 1854). Hirst himself was driven to calling his Principal 'a foolish little man, a thick-skinned, acrid blockhead' (*Journal*, 26 November 1854).

During 1855 Hirst, and some of the other staff, were outraged by the discovery of homosexual practices amongst senior boys. Although Hirst thrashed several for 'sensual practices' and caused two boys to be expelled, he was more outraged by the fact that Edmondson seemed unperturbed about the possible moral contamination of his own son who was a pupil in the school. To a modern reader it is Hirst, and not Edmondson, who seems to have been naive and heavy-handed. No doubt long experience had taught Edmondson that boys of a certain age would always masturbate, and that providing it did not lead to the corruption of junior boys, it was best ignored.

One may think that boys from urban backgrounds found Edmondson's old-fashioned Quaker speech and mannerisms, his sabbatarianism and teetotalism, false and pedantic. Appropriate to a small village school thirty or forty years before, it must have appeared unsuited to a large boarding school filled with boys and staff from very different religious backgrounds. In 1855, happy to help a former pupil, whose school had failed, Edmondson loaned the man £300, only to find his generosity abused. To Hirst this misplaced confidence was another sign of weakness (*Journal*, 11 March 1855), although he himself had benefited from Edmondson's generosity when his salary was raised from £150 to £200 p.a. on his marriage in December 1854. One of Hirst's colleagues, Richard Wright, gently pointed out that Edmondson might not have been able to afford this and that if every married master (as Wright was) were paid in such proportion, the school would be bankrupted. Hirst's 'professional' answer is an interesting comment on the rise of the specialist teacher and man of science in mid-Victorian Britain.

> I must confess that in fixing my salary at £200 I did not consider what other masters had. I satisfied myself that I earned it, and I thought the place could afford it. Yesterday and today I have deliberated long on the subject, and come to the conclusion which my conscience approves. I do not think I ought to ask less because other masters may be receiving too little; I do not think it my duty in fixing my salary to fix that of others — I could not do it, and if I could, it would be invidious in me; neither should I feel inclined to submit to Wright's apportioning our salaries, for however competent in money matters, and however honest, I can conceive that his views of the value of instruction might differ considerably from my own. I have come to the decision to demand what I think I justly earn, but at the same time I am prepared on account of the interest I take in Queenwood to share its fortunes to a certain extent; and should I see it to be necessary to take less I will do so (*Journal*, 3 September 1854).

In May 1856, after less than eighteen months of happy married life together at Queenwood, Hirst's wife, Anna, became seriously ill with tuberculosis and Hirst was forced to resign his appointment so as to nurse her in Ireland and on the Continent. Ninety-seven boys were then enrolled. There was much discussion over his successor and Hirst did persuade Edmondson that he needed 'a superior man' on the same salary he had been receiving (*Journal*, 20 July 1856). Unfortunately, his successor's name is not recorded. However, after his wife's death in May 1857, and following the exposures in Holyoake's *Reasoner* that Queenwood College was still, in fact, owned by the socialists, Richard Wright tried unofficially to persuade Hirst to return and help arrest the declining enrolment. But Hirst had irrevocably decided to devote himself to mathematics. In refusing the offer he made his lack of respect from Edmondson clear. When he did revisit the College in July 1859 for old times' sake, he noted laconically that 'my three years absence had produced no alteration in [the Edmondsons]' (*Journal*, 28 July 1859).

When Edmondson died from a heart attack on15 May 1863 his College was almost bankrupt. His widow saw out the school year before selling the lease to a Mancunian, Charles Wilmore, another progressive teacher who had been running a small school, Linden Grove, in the Cheshire countryside. One of the pupils he brought with him to Queenwood - the name was maintained - was John Hopkinson, who went on from Queenwood to Owens College, Manchester, to become a distinguished engineer.[41] By then Wright had quit Mrs Edmondson's sinking ship, but Debus stayed until 1867 when he became a chemistry master at Clifton College, Bristol — a new middle-class boarding school which quickly inherited Queenwood's reputation for outstanding science teaching and became a model for the older public schools. Wright replaced Hirst as a mathematics master at University College School, Hirst having been made a professor of Mathematics at University College itself. Debus continued with his chemical research and was awarded a Fellowship of the Royal Society in 1861; Wright published a geometrical textbook and played a leading role in the reform of geometrical teaching.[42]

Wilmore was able to maintain the reputation that Frankland, Galloway and Debus had given to the College for training in chemical analysis with teachers such as C. Leopold Field and Frank Clowes.[43] Like Galloway previously, Clowes published his Queenwood course as *An Elementary Treatise on Practical Chemistry and Qualitative Analysis* (1874). No doubt, therefore, the College continued to attract parents whose sons were aiming for careers in medicine or as public analysts. For, as Galloway observed in 1881 of his experiences in the 1850s:

> boys of 14 or 15 years of age become excellent analysts. Those of them who, on leaving the school became pupils, could make medical, food and other analyses more accurate in every respect than the surgeons to whom they were articled; and they found on attending the London Medical Schools, there or four years afterwards, that the chemical instruction they had received in Hampshire was much more perfect and extensive than that given in those Schools.[44]

Unfortunately, no Frankland, Tyndall or Hirst recorded their days' teaching at Queenwood during the Wilmore era, though a few paltry surviving memorabilia

---

41 Bertram Hopkinson, 'Memoir of John Hopkinson' in *Original Papers by the late John Hopkinson, DSc, FRS*, 2 vols. (Cambridge, 1901). For Queenwood, see vol. 1, pp. xiii-xvi. See also John Greig, *John Hopkinson, Electrical Engineer* (London, 1970).

42 Debus stayed at Clifton College until 1870 when he moved to London to teach chemistry at Guy's Hospital (1870-80) and, through Hirst's influence, at the Royal Naval College, Greenwich (1873-88). He died in 1915. See *Who Was Who*.

43 Frank Clowes (1848-1923) left Queenwood in 1881 to become first Professor of Chemistry at University College, Nottingham. See *Who Was Who*.

44 R. Galloway, *Education: Scientific and Technical* (London, 1881), p. 262.

show that he maintained the tradition of a school magazine and the egalitarian Mutual Improvement Society.[45] In a lecture to the Birkbeck Institution, Tyndall had praised the standard of teaching at Queenwood during his days there. When the lecture was published in his *New Fragments* in 1892, Thomas Owen, one of Wilmore's former teachers, wrote to Tyndall saying Queenwood was now 'decadent'. Tyndall expressed surprise that the College had survived so long.

> [Wilmore's] predecessor Mr Edmondson had the good fortune of being backed by a number of men specially fitted for the work they took in hand at Queenwood. When they withdrew from the school it would not have been surprising if it had collapsed. It is a striking proof of Mr Wilmore's energy and organising power that this collapse did not occur.[46]

Wilmore retired, bankrupt, in 1896 and old boys seem to have rallied together to save him from pauperism. Not until 1899 was he able to find a purchaser for the building. The new owner, an electrical engineer, W. P. Duncombe, transformed the grounds into a poultry and fruit farm; he also kept bees and bred show rabbits! The building itself was turned into a college for training electrical engineers. The gas works was demolished and the Chapel, which Wilmore had made from one wing of a master's cottage, was turned into a dynamo house.[47] The success of this venture is not known. That Wilmore continued to live at the College as an adviser is clear from the fact that he was burned to death in bed when the College was gutted by fire on 10 June 1902.

Frankland, Tyndall and Hirst were all dead by this date, and only Heinrich Debus remained to mourn its loss.[48] He would have appreciated Hirst's last memory of Queenwood in December 1891, a few months before his death. Hirst had stopped at Marble Arch to listen to a blind beggar recite Poe's *The Raven.*

---

45 Friends' House Library has magazines for the autumn and spring terms of 1887 and 1888; an undated school photograph; and a printed memorandum by Wilmore, dated 17 December 1899, concerning the sale of the school.

46 Tyndall to Owen, 5 August 1892, Royal Institution Archives.

47 For these slender details, see Garnett, *op. cit.* (4), p. 217. Sviedrys, *op. cit.* (18), p. 420 notes in passing the opening of private schools of electric telegraphy from the 1860s onwards.

48 Debus to Louise Tyndall, 11 and 14 June 1902, Royal Institution Archives.

> By doing so, my memory was thrown back forty years, to the time when I recited the same poem, for the first time to Tyndall. It was at Queenwood and we both reclined, the while, on the green sward of the broad avenue leading to Baron's Court (*Journal*, 6 December 1891).

The memory was not of science teaching, or of the tussles with Edmondson, but of the wonderful social life the young staff had experienced together talking about religion, politics and literature.

Thompson, the College's only previous historian,[49] rightly made the point that by the mid-1850s (about the time Hirst left), Queenwood College had begun to outgrow its strengths and it was being copied and emulated by other, more richly endowed, public and proprietory schools. As with any school lasting more than a few years, some of Queenwood's pupils, like Henry Fawcett (the politician), Walter Flight (the mineralogist) and the engineers James Mansergh and John Hopkinson, achieved greatness, as did Frankland, Tyndall and Hirst, all three of whom made very significant contributions to science education.[50] Through its practical emphasis the College had shown that it was possible to combine the values of a liberal education with vocational ends, and in many ways it anticipated the later civic technical colleges. Edmondson undoubtedly deserves credit for his vision in seeing that was possible, even though the combined evidence of Frankland's, Tyndall's and Hirst's journals reveal that he lacked the qualities necessary to be a great leader of boys. His school achieved distinction through the quality of his staff and despite, not because of, his own managerial and teaching abilities.

As Henry Fawcett's biographer, Leslie Stephen was naturally fascinated by Edmondson's school. In 1887, when selecting names for the *Dictionary of National Biography*, he chose to include Edmondson, and asked Holyoake to compose the entry. Holyoake sensibly consulted Yeats (who had become Headmaster of a middle-class school in Pimlico after leaving Queenwood),

---

49 Thompson, *op. cit.* (3). See also his 'Contributions to scientific education and the teaching of science' in W. H. Brock, N. D. McMillian and R. C. Mollan, eds., *John Tyndall. Essays on a Natural Philosopher* (Dublin, 1981), pp. 145-56.

50 Tyndall, Frankland, Hirst, Debus, Fawcett, Flight, Mansergh and Hopkinson were all made Fellows of the Royal Society. I have been unable to confirm that Tyndall's one time assistant at the Royal Institution, William F. Barrett (1844-1925), ever taught at Queenwood, as claimed by Thompson, *op. cit.* (3), p. 252. He was Tyndall's assistant from 1862 until 1866, when they separated after a disagreement. Barrett did not become a lecturer at the International College until 1867 and might, therefore, have spent the year 1866-67 at Queenwood.. This is not mentioned in *Who Was Who* or *The Times*, 26 May 1925, p. 16.

Wilmore and Tyndall. The latter was appalled, for he feared that between them, Stephen and Holyoake would put Edmondson on a pedestal as 'the great educator' when, in his view, he had been contemptible.[51] A book about Queenwood, he told Owen in 1892, was 'not to be heard of'.[52] Advised by Tyndall, and perhaps by Frankland and Hirst as well, Holyoake composed a curiously ambiguous notice containing a very Tyndallian sentence — 'He was largely assisted by his wife, who, in the opinion of many, had a superior intellect to his own.'

---

51 Tyndall to Hirst, 9 September 1887. 'I wish they would not disturb the dear man's memory by giving prominence to the point which was only worthy of contempt.' Royal Institution Archives.

52 Tyndall to Owen, 5 August 1892. Royal Institution Archives.

XVIII

School Science Examinations: Sacrifice or Stimulus?

Introduction

When Chesterfield junior advised his father in Dickens's *All the Year Round* in 1866, he pointed out that the Victorian world had seen so many changes

> that your experience . . . is really of no use whatsoever. . . I speak with the utmost respect, [but] you learned so little in what you are pleased to call the good old times. There were no examinations in those days. . . .[1]

Master Chesterfield was in no doubt of the stimulus and superiority that these examinations had endowed his generation:

> . . . Worthy sir, I think your common sense – with which, I confess, that you appear to be well endowed – will show you that among the many changes which mark this great and glorious age must be ranked a considerable alteration in the relative positions of father and son – of senior and junior. Consider how splendidly we have been educated. Consider how glibly we could answer all sorts of questions on scientific and other subjects, by which I believe your contemporaries would be instantly gravelled. Try us with anything you like; the distance between the planet Mercury and the moon; the manner of the formation of the old red sandstone; dodge us about with any number of teazers of this sort, and see if we are not ready with the answers.

At the first official meeting of the Association of Public School Science Masters in 1901, W. D. Eggar, a leading science teacher at Eton College,[2] summarised the development of British school science teaching in an elegant parody. Schools, he declared, could be classified 'as Shakspere classifies great men. Some schools are born scientific, some achieve science, and some have science thrust upon them'.[3] Even if it was not always reflected explicitly in school examinations, within a few years of this remark being made it was certainly the case that *all* secondary schools in Great Britain were teaching science – ranging, at one extreme, from the forty hours a week for prospective medical students at Kingswood School, Bath, through the eighteen hours a week for potential Oxbridge science Exhibitioners at Rugby,[4] to the compulsory minimum of three hours a week required in state schools under the Board of Education's Code of 1904.[5]

The situation in 1851, when the Natural Sciences Tripos examinations were instituted at the University of Cambridge,[6] had been very different, for there had then been virtually no science teaching in any of the schools of Great Britain. As E. C. Mack and others have argued, the transformation of the public school was brought about both by external public opinion and by opinion within these schools themselves:

> The translation of the ideas of the sixties into public school practice began as early as 1862, developed considerable momentum in the last years of the decade, and was virtually completed by the early seventies. In these dozen years the public schools, acting in response to public opinion, to the suggestions of the Commissioners, and to the commands of the Public School Act, put their houses in order, with almost unparalleled speed. They remodelled their governing bodies, made changes in their curricula, increased their educational facilities, and bettered their physical equipment. By 1875 or thereabouts they had adapted themselves to the spirit of the age and had become institutions closely answering to the demands of the late Victorian upper classes.[7]

As far as curricular change was concerned, the ultimate constituency, for or against, was the fee-paying parent. If we may lay some of the blame for parents' disinterest in science in the earlier nineteenth century on snobbery and on the lack of any obvious need for scientific knowlege in their childrens' careers, how do we explain the transformation of attitudes – the new key of the piper's tune – between 1865 and 1900? The answer lies in the use of examinations.

Most parents were no doubt quite indifferent to the niceties of the debate over liberal education and whether science was or was not a "humane" discipline; nor were they probably all that concerned with the supposed expense of science teaching, or its possible effects on time-tables.[8] Their principal concern was with their childrens' careers. The purpose of most education before the 1860s was the socialization and initiation of children into the rites of 'gentlemanliness' and to fit them for patronage. In the 1860s, however, families whose male children had traditionally gone into the Civil Service, or the army through patronage,[9] discovered that these careers, and increasingly most professions, were only open to their children through success in competitive examinations. By the same token, these careers were opened competitively to the children of the industrial middle classes who could afford to send them to the growing number of private secondary schools. Since the new examinations could include scientific elements, this factor, together with the growing number of science scholarships offered by Oxford and Cambridge Universities from the 1870s onwards, and the growth of respectable career opportunities in science and medicine, encouraged parents to send their children to those schools which were establishing the means and wherewithall to teach science.

Other schools were quick to follow, or new schools like Wellington (1859)[10] and Clifton (1862) were established to satisfy the middle-class parental market. In order to deal as economically as possible with the variety of vocations and examinations for which they were preparing their students, and to forestall the wholesale loss of sixth forms to 'crammers', the schools were forced to organize or reorganize their structures into specialized 'classical', 'modern', etc., sides. In compensation, however, headmasters (especially headmasters of hitherto minor schools) discovered that successes in examinations and entrance scholarships to the universities could be used as advertisements for their schools.[11]

As the [Acland] Consultative Committee of the Board of Education discovered between 1910 and 1911, the development of the school examinations system was incredibly complex.[12] Although the College of Preceptors had begun examinations for proprietory schools in 1853, and the Royal Society of Arts had initiated a system of 'General Examinations', principally for working adults, in 1856, external school examinations may be said to have originated in July 1857 when the politician, Sir Thomas Dyke Acland, drawing on the notions of the Third Earl Fortescue, ran an experimental examination in Exeter in order to provide a public test for parents of the quality of the local middle-class school education for which they were paying.[13] Acland's scheme was strongly supported by Frederick Temple (who had just moved to the headmastership of Rugby from his position as a government teacher training college inspector) and together they persuaded the University of Oxford to establish a Delegacy for two 'Examinations of Students not Members of the University'. One examination was intended for under 15s ('Junior Locals'), the other for students of 18 ('Senior Locals').

The scheme was launched by Oxford in June 1858, while Cambridge University, not to be outdone, launched a rival, but virtually identical system of local examinations in December 1858.[14] In 1865, with strong support from G. D. Liveing, the Professor of Chemistry, the Cambridge Syndicate admitted women to its Local examinations, but the Oxford Delegacy did not follow until 1870.

The term 'local' emphasized that these examinations 'were looked upon as extensions of the university into the outer world rather than as a part of the school educational system'.[15] Indeed, candidates were put forward directly by parents rather than the schools, and the examinations were held in public halls by visiting invigilators from the two universities. Soon, however, it was seen that schools were providing most of the candidates and it seemed sensible to provide a parallel system of 'School Examinations' as opposed to public 'Local Examinations'. For the former, schools submitted details of their courses and textbooks to Oxford or Cambridge, who appointed examiners to set questions on a school's syllabus. Papers were marked by the teachers themselves and moderated by the university examiners, who also submitted a report to the school.

It is important to notice that the Public Schools, or to use the Taunton Commission's convention, the First Grade Schools, virtually ignored these Local Examinations until the 1870s. As Roach points out, the examinations were not cheap (Cambridge charged a pound per candidate, plus fees to the local centre, and a week's board and lodgings if the candidate did not live in the same town), and for a long time they lacked prestige. 'The state did not demand them; business men were not interested in them and preferred to depend on testimonials of good character',[16] and, as we shall see, until the late 1870s they were of no value for matriculation purposes. On the other hand, although the Locals were primarily designed for the children of middle class parents, or Grade II schools which did not send pupils to universities, both the Oxford Delegecy and the Cambridge Syndicate intended that their Local Examinations should encourage parents, and the lower grade schools, 'to prepare able boys for Oxford and Cambridge who would not otherwise have had the opportunity to get there'.[17] The Oxford colleges showed more enthusiasm for this than did their Cambridge counterparts. From 1867 onwards several Oxford colleges offered awards to students on the basis of their results in the Oxford Locals. However, despite promptings from the Syndicate, Cambridge did not follow this practice until 1872 when St. John's College began to offer two sizarships each year to the boys heading the list of distinctions in mathematics and classics respectively. No other college followed this lead, and although Roach presents evidence that a few wranglers owed their presence at Cambridge to their earlier success in local examinations, there is no evidence that any candidate for the Natural Science Tripos benefited in this way.[18] Indeed, despite the encouragement which the Locals may have given to science studies at certain schools,[19] they evidently had a negligible effect on the pool of potential candidates for the NST – at least until the 1880s when, as we shall see, circumstances altered.

Although London University had introduced a difficult, but 'highly valued',[20] matriculation examination as early as 1838 (an examination which, incidentally, proved extremely popular with schools in the Home Counties as a 'leaving examination' or general certificate of school achievement), the Oxford and Cambridge Local, or School Examinations were never technically 'matriculation' devices. Until the 1870s entrance to both Oxford and Cambridge Universities lay in the hands of individual colleges, and the test of competence to graduate remained Responsions at Oxford and the 'Little Go', or Previous, at Cambridge – both examinations being taken when the student was already a university resident.

However, a side effect of the deliberations of the Taunton Commission (1867), which had among other things recommended the setting up of a Government Examination Council, had been to make First Grade Schools look to the ancient universities for protection and to establish the Headmasters' Conference. In 1872 the latter recommended that the Universities of Oxford and Cambridge should together establish a new examination for sixth forms.[21] On 8 November 1873 the Oxford and Cambridge Examination Board (Joint Board) was established. The Board's official historian has summarized the first meetings as follows:

> No time was wasted. Both the [Oxford] Delegacy and the [Cambridge] Syndicate met to discuss plans for the first examinations [in June] and to consider matters of policy. The Cambridge Syndicate in particular, took the view that there was a need for 'a more systematic and authoritative inspection of endowed schools' than the existing examinations offered. The importance attached by the two universities to the new institution was reflected in the readiness of distinguished and busy university teachers to visit schools, and the first Regulations – published within a month – emphasized this part of the Board's activities. Schools would be examined at the request of the headmasters on the internal work which boys were undertaking, and a report made to the Board and to the school authorities. . . If the school wished to have Certificates – and the distinction between internal examining and the awarding of Certificates was made clear – they would be awarded on the basis of written papers and on the examiners' *viva voce* report. . . . The Board at first offered examinations for Certificates three times in the year – June, July and December – allowing papers to be taken at schools or at centres.... Candidates for the certificate had to select four subjects from at least three groups out of a possible four – Languages (ancient and modern), Mathematics, English subjects and the Sciences. The examination was in fact, if not yet in name (until 1887) a 'Higher' Certificate.[22]

These school examinations, in which, unlike the not very popular Schools examinations of the Oxford Delegacy and Cambridge Syndicate, the Joint Board used externally set common papers for *all* the schools concerned, were undoubtedly effective in stimulating teachers, pupils and schools into a coordinated academic efficiency, in strengthening the ties between Oxbridge and the First Grade Schools, and in providing 'machinery for the universities to influence the pre-university education of undergraduates'.[23] More significantly, from the beginning, the Joint Board Certificate in appropriate subjects was accepted as giving exemption from the Previous and Responsions; *ipso facto* the Board became a matriculation agency. Following protests from the Oxford and Cambridge Locals Examination Delegacy and Syndicate, their Senior Certificates were also recognized, in 1877, for gaining exemptions from Responsions and the Previous. By this means the Locals gained an enhanced prestige for those schools which wanted to raise their status from Grade II to Grade I establishments.[24]

The examination statistics of the Joint Board from 1874, together with those from the separate Oxford, and Cambridge, Locals from 1877 are, therefore, a potential measure of the pool of Oxbridge entrants. Although not all candidates who took the group of science subjects necessarily entered either Oxford or Cambridge Universities, they may be regarded reasonably as the most likely pool of qualified candidates for Natural Science at Oxford or the Natural Sciences Tripos (NST) at Cambridge, notwithstanding the fact that students who read science had not necessarily taken science subjects at school.[25] Unfortunately, the statistical presentations in the annual reports of the Cambridge Local Examinations Syndicate and the Oxford and Cambridge Joint Examinations Board are incommensurate, and any conclusions about the pool of qualified talent available to the Cambridge colleges would be hazardous and hedged by qualifications. For example, if we assume that most Cambridge candidates sat Joint Board then, from Table II, a total of 306 candidates obtained passes in the three divisions of natural philosophy, botany, biology and geology in 1890. Ignoring the possibility that a few candidates offered more than one science subject, if these 306 candidates were shared equally between Oxford and Cambridge Universities, we might conclude that there were at least 153 *well-qualified* candidates for the NST in 1890. Since the Part I list for 1892 (*i.e.*, two years on) included 132 candidates (ignoring pass-men), this suggests that (however dubious our assumptions may seem[26]) by the 1890s the colleges had a well-stocked pool of scientifically-qualified boys from which to select their candidates for the NST. After a shaky beginning this pool increased in size between 1890 and 1901, as the following figures show (Table I).

Table I

	No. of candidates successful in Part I (Men only)	Total Joint Board Passes in Science divided by 2
1881	8	46
1885	45	46
1890	72	153
1895	79	203
1900	122	183

(*Sources: Annual Reports of the Oxford & Cambridge Schools Examination Board*; Roy MacLeod and Russell Moseley, *The Natural Sciences Tripos: Data and Commentary, 1851–1914*).

The average pass mark of over fifty per cent in most science subjects taken in the Higher Schools examination over the period 1874 to 1902 (Table II) suggests a basic soundness in the schools' science teaching. Although distinctions were much easier to obtain in chemistry (averaging a third of the candidates each year) it appears that the examinations were selecting out scientific ability in the schools.[27] Chemistry, not unexpectedly, was the most popular subject, though this was followed closely by the mechanical division of natural philosophy (*i.e.*, mechanics and applied mathematics; until a separate physics paper emerged in 1889, questions on heat, light and electricity appeared on the chemistry paper).[28] See Table III.

The paper on physical geography and geology was the least popular of the sciences. Botany was always a minor subject in boys' schools, despite its important role in the introduction of science into some of the schools.[30] Zoology was not offered as a separate subject until a newly-styled Biology paper appeared in 1885, the paper becoming increasingly popular by the turn of the century as pressure mounted to recognize schools as preliminary training grounds for the medical profession.

The relative popularity of the mechanics papers may have been affected by the fact that candidates for honours at Cambridge sat an additional paper for the Previous Examination in either mechanics, French or German.[31] Thus, a Higher Certificate which included a pass in the mechanical division of natural philosophy might provide a student with exemption from the whole of the Previous.

The Sacrifice of Specialization

Exemptions from matriculation type examinations soon proved a mixed blessing for the schools. By the 1900s there was widespread alarm that 'the universities and the professions [were] to some extent defeating their own ends by their demands upon the schools'[32] by forcing them to ignore matters of

Table II

Table II. Higher Schools Examinations. Science Papers from Group 4, 1874–1902
(Source: Annual Reports, Oxford and Cambridge Schools Examination Board)

	Nat. Philosophy Mechanical			Nat. Philosophy Physical			Nat. Philosophy Chemical			Botany			Biology			Physical Geography Elementary Geology		
	No	Pass	Dist.	No.	Pass	Dist.	No.	Pass	Dist.	No.	Pass	Dist.	No.	Pass	Dist.	No.	Pass	Dist.
1874	18	10	0				17	11	2	4	2	0				4	2	0
1875	21	10	3				28	16	3	6	4	0				15	7	0
1876	29	19	0				47	25	5	7	2	0				21	12	0
1877	52	30	5				52	41	15	9	4	2				37	16	3
1878	28	10	2				60	39	11	7	4	2				33	22	3
1879	29	18	3				67	40	18	8	7	1				44	40	14
1880	42	33	3				54	28	9	3	1	0				52	34	2
1881	39	26	3				49	29	9	2	2	0				45	35	6
1882	42	18	0				37	23	8	6	4	0				37	35	6
1883	32	24	3				59	43	23	8	4	0				40	31	8
1884	42	35	11				51	43	16	6	5	0				49	38	9
1885	30	22	10				62	39	32	4	4	0	4	4	4	31	24	11
1886	52	42	23				87	57	30	9	9	1	8	6	1	36	29	18
1887	57	43	19				63	37	16	15	12	4	3	2	1	61	47	11
1888	93	72	21				67	49	29	21	13	3	3	3	0	40	38	13
1889	102	96	47	46	42	21	83	69	24	23	17	7	4	3	2	46	44	16
1890	115	91	38	65	55	29	98	83	38	41	36	10	7	5	3	40	36	18
1891	133	108	45	66	59	24	111	97	33	34	30	9	7	5	3	76	61	8
1892	145	118	36	69	54	23	123	69	27	57	38	5	9	4	0	65	37	5
1893	148	110	24	69	43	9	119	58	13	57	48	8	16	6	1	35	28	3
1894	159	141	37	78	62	18	128	93	22	67	61	8	15	11	3	49	44	9
1895	179	110	33	107	71	29	153	92	15	72	62	9	18	11	3	68	60	18
1896	162	110	32	103	74	22	155	96	30		Abandoned		69	44	8	74	59	13
1897	183	108	22	95	64	20	120	63	25				56	54	15	83	61	18
1898	188	124	21	100	68	24	139	88	39				122	87	11	60	45	6
1899	168	122	36	111	59	19	159	83	31				131	102	19	38	31	5
1900	157	97	35	109	68	28	124	84	34				115	86	19	34	33	3
1901	168	98	22	106	61	23	142	71	19				112	93	24	54	32	3
1902	156	108	21	119	75	20	123	82	21				146	126	34	41	31	2

1874 papers were titled "Natural Philosophy", "Heat and Chemistry"
1895 Botany abandoned and new regulations for Biology

All boys had to satisfy the examiners in one sitting in at least four subjects from not less than three groups of subjects: Group 1, Classics and Modern Languages; Group 2, Elementary and Additional Mathematics; Group 3, Scripture, English, History and Geography; Group 4, the sciences as above. Girls were allowed two sittings, taking two subjects at a time.

Table III

Order of Popularity of Subjects Examined for Joint Board Higher Schools Certificate in 1909

(G = group)

Subject	Candidates	Subject	Candidates
Elementary Mathematics (G2)	1858	Greek (G1)	885
French (G1)	1505	Additional Mathematics (G2)	615
History (G3)	1408	Biology[29] (G4)	334 (14%)
Scripture (G3)	1166	Chemistry (G4)	242 (10%)
English (G3)	1143	German (G1)	241
Latin (G1)	1097	Physics (G4)	196 (8%)
English Essay (G3)	1042	Mechanics (G4)	179 (7%)
		(Total Candidates = 2361)	

Source: B.Ed., *Consultative Committee on Education,* 1911.

general education for premature specialization. As early as 1889 the Headmasters' Conference had memoed the Joint Board on science examinations in the following terms:

> The small number of boys who take up Science compared to those who learn it, and the very high percentage of honours gained [*i.e.,* distinctions] prove that it is an Examination only for specialists in Science. It does not touch the General Science work of the Schools. Further, it serves to hamper the higher teaching. If it is to serve the first purpose, it should not require in Chemistry, the General Paper and practical work, and Physics: the last two should be alternatives. And for the second, the harder qualitative analyses should be alternative with some volumetric (quantitative) analyses.[33]

Although this memorandum touched on a point which was to greatly exercise science teachers in the twentieth century, the Joint Board responded, not without reason, that a school could be examined without taking Higher Certificate papers (presumably either by special inspection and examination or through the Lower Certificate), and it refused to lower or alter the standard of the science examinations in any way.[34]

Specialization also had a pernicious effect on the ordinary pupil in the new state secondary schools after 1902 for although between, for example, 1907 and 1909 only 0.93 per cent of pupils from state schools went to Oxford and Cambridge, and only a further 1.76 per cent went to other universities,[35] sixth-form work became 'modelled on the needs of the few pupils who intend to proceed to the University, the school studies of the majority being regulated by the special requirements of the few'.[36]

Moreover, by the Edwardian period, written examinations had worked their effects further back into the schooling system. In the first place, the 'free-place' regulation of 1907 had introduced the category of 'Direct Grant' schools in which existing endowed or proprietary secondary schools were able to acquire grants from the state in exchange for offering poorer pupils a number of free places. Inevitably such places were decided by examination competition.[37] Secondly, at the Public Schools, in order to avoid the inconvenience of setting individual examinations for every potential entrant aged 13½ to 14, the Headmasters' Conference introduced a Common Entrance Examination in 1904.[38] By January 1911, 48 Public Schools were using this selection examination for which (in 1910) there were 1902 candidates. Competitors sat a General Paper (which included Scripture, History and Geography), and papers in Latin (translation, comprehension and grammar), Greek or German, French, mathematics (arithmetic, algebra and geometry), English (composition and grammar) and either Latin verse or Nature Study. Not all of these papers were necessarily taken, since the choice depended upon a particular school's requirements. Generally speaking, however, a boy intended for the classical side of the school was expected to offer Latin and Greek, but not German or Nature Study. J. J. Thomson found that Preparatory schools usually made no provision for teaching science, mainly because the Public Schools paid no attention to it in the Common Entrance. Nature Study was, in fact, dropped from the examination in 1915.[39]

Preparatory schools had made rapid developments during the Victorian period,[40] becoming forcing grounds for passing Public School entrance examinations. Accusations that these schools encouraged specialization[41] rather than intelligence were defended by the Rev. C. E. Williams and by Board of Education investigations.[42] Nevertheless, accusations arising from the findings of the Inter-Departmental Committee on Physical Deterioration, that the cramming of preparatory school boys produced symptoms of brain exhaustion in late adolescence, or at university, were prominent throughout the Edwardian period.[43]

Not surprisingly, over-specialization was one of the dominant issues discussed during the early years of the Association of Public School Science Masters (APSSM). At its first meeting in 1901, Ashford of Harrow complained that 'the examination system . . . has much to answer for . . . [because a boy's] examination in form and division are based on the model of those for Scholarship and [Higher] Certificate'.[44] The themes of waste of effort, and unfairness and inequality in teaching the future university student, whose needs dictated those of the majority, were powerful ones in the Association. But, despite its efforts to modify the system, it was not until 1919 that a rational system of combined leaving and matriculation examination – the School Certificate – was introduced. Until then schools coped as best they could with a situation in which, for example, a form of twenty boys 'might conceivably contain thirteen candidates for thirteen *different* public examinations, each and all working at one or more subjects not prescribed for the rest of his companions'.[45]

As far as College entrance scholarships to Oxford and Cambridge were concerned, the APSSM was greatly concerned to see their scope restricted. According to Falkner of Weymouth College, they turned boys into machines, so that they thought along set patterns and lost any powers of original thought they might have possessed.[46] Although Cambridge set test papers in Latin, Greek, Mathematics and either French, German or Mechanics, as well as the science papers, the former were, he thought, of a fairly low standard compared with the stiff nature of the latter. 'Science masters . . . being expected to score as many Scholarships as possible, naturally wish to get the boy into their hands as soon as possible, and this necessitates the withdrawal of the boy from the ordinary course of education before he has derived, and just at a time when he is able to derive, the greatest benefits therefrom'.[47]

However, although specialization, the lack of uniformity in matriculation requirements, the small emphasis placed by Colleges on boys' practical work in the schools, and the over emphasis on physics and chemistry in scholarship papers dominated the discussions of the APSSM between 1901 and 1919, they met with little success in catalysing change in their discussions with members of the two universities.[48] That so little was done no doubt has much to do with the professional aims of the Association, for it was the very increase of specialization in schools which was producing new openings for science teachers in the schools. We thus have the rather inconsistent picture of members of the APSSM lobbying against science specialization on the one hand, and urging the claims of the introduction of biology for prospective medical students on the other.[49]

Effects on the School Curriculum

Examinations, as protesters to *The Nineteenth Century* noted in 1888, forced a certain kind of uniformity on science teaching, making innovation comparatively difficult. Although the heuristic science syllabus based on the teachings of H. E. Armstrong, H. Gordon and W. M. Heller, which had been adopted by the Incorporated Association of Headmasters, was approved by the Delegacy and Syndicate of the Oxford and Cambridge Local Examinations in 1895, it is doubtful whether schools found heuristic methods compatible with any system of external examination.[50] Practical examinations were notoriously difficult to assess in Armstrong's opinion:

> However valuable examinations may be as a means of 'putting on the screw' on both teachers and taught, it is impossible to overrate the injury done under our present system by unduly 'forcing the pace' and neglecting the apparently unpromising material on behalf of those whose work is more likely to afford 'results'. The undue encouragement given to 'literary' methods, owing to the extreme difficulty of properly examining practically, is one of the greatest evils the system entails; the interference with freedom of action in schools, and the consequent check on the development of methods of training, is another; and further confusion is introduced in consequence of an entire absence of co-ordination in the requirements of the different examining bodies.[51]

In practice, teachers only used the more rigorous forms of heurism in the lower and middle schools, and little should be read into the fact that schools like St. Dunstan's College and Christ's Hospital, where heurism was strongly favoured, scarcely figure in the lists of NST candidates.[52]

It was almost inevitable, then, that school science curricula were restricted to those syllabuses laid down by the university-based examination Boards. Until 1902, for example, schools were forced, often against the inclination of their mathematics staff, to follow the Euclidean order of proof in geometry,[53] even for professional examinations which also followed the model demanded by the Oxbridge examiners. Similarly, the introduction of graphical methods, which were ardently pressed by John Perry, and which were destined to play an important role in teaching experimental physics, was retarded by the absence of any positive action by examination boards.[54]

To see precisely how the school science curriculum was affected by the examination boards would demand a detailed analysis of thousands of examination questions over a long period. Only a few observations will be made here.[55]

By acknowledging the Cambridge Higher Local Examination and the Joint Board's Higher Schools Examination as the equivalent of the Previous, the universities effectively pushed back first-year university work into the schools. Furthermore, several of the science textbooks recommended for Cambridge Higher examinations would have been equally relevant for a student preparing for Part 1 of the NST; among them were J. H. Balfour, *A Manual of Botany* (1851), C. Lyell, *The Student's Elements of Geology* (1871), T. H. Huxley, *An Introduction to the Classification of Animals* (1869), W. A. Miller, *Introduction to Inorganic Chemistry*, H. E. Armstrong, *Organic Chemistry* (1874), the *Natural Philosophy* texts by A. Ganot and A. Privat-Deschenel, and M. Foster, *Textbook of Physiology* (1877).[56]

Many of the NST examiners also examined for the Joint Board's school examinations or the Cambridge Locals – Liveing being the most ubiquitous of the examiners.[57] The university examiners demanded high standards and did not hesitate to draw pointed conclusions in their published reports. Comments such as 'weak', 'standard not high' and 'unsatisfactory' are to be found, though (as in modern reports) these expressions are carefully tempered by words of praise and encouragement to the teachers.[58] Like the Department of Science and Art examiners in the 1870s, chemistry examiners for the Cambridge Locals and the Cambridge side of the Joint Board were particularly upset by pupils who reported impossible experiments because their teaching had been mainly didactic and non-experimental. These same examiners also demanded a display of knowledge of organic chemistry before they were prepared to grant distinctions.

During the early years of the Joint Board's examinations, questions followed the syllabus fairly mechanically, so that if a teacher had followed it fairly rigorously his pupils could hardly have failed. For example, if the chemistry syllabus laid down 'chemistry of the blast furnace', pupils could expect regular questions such as: 'What are the materials with which a blast furnace is supplied? Show what changes successively take place as the materials sinking above meet those coming from below' (1876). Physics questions often had simple numerical riders – 'Enunciate Ohm's law', followed by a worked application. That the pass figures were not higher than fifty per cent must be both a reflection of the lack of experience of science teachers in the Public Schools and the difficulties teachers faced in finding adequate pupil-contact hours to cover the syllabuses. Standards did rise, however, and by the 1880s questions of a more sophisticated quality became more frequent. Thus a question on Avogadro's hypothesis in 1879 ended with the query: 'Shew... that a molecule of hydrogen contains two atoms at least'. Why 'at least'? – which demanded a deep understanding of the issue, and would undoubtedly have weeded out distinction candidates. As has been found in Chemistry Tripos papers, older terminology might reappear with changing examiners. Students for the Cambridge Higher Local in 1887 were asked to explain 'the doctrine of quantivalence' [valency], a term little used after the 1860s. It is of interest that evolutionary questions were freely asked in both geological and zoological papers.

As in the early unstructured tripos papers, no minimum number of answers was demanded of Joint Board candidates; though no candidate could possibly have worked through all the questions offered. In the Higher Certificate, Mechanical Division of Natural Philosophy, for example, candidates had to satisfy examiners on Section A (Mechanics, 6–8 questions), and either Section B (Hydrostatics, 4 questions) or Section C (Astronomy, 4 questions) within two hours. An overall impression is that questions were confined to what might be termed 'well established normal science', and the closest approach to knowledge of recent science appears to have been a reference to the Krakatoa eruption (1883) in an 1891 Physical Geography and Geology paper for the

Joint Board!

Despite the improvements in facilities for science teaching in schools, the number of candidates offering science papers in Joint Board Certificate examinations remained relatively low until after the First World War, though examiners noted a continual improvement in standards between 1874 and 1904.[59]

By 1904 the Joint Board demanded a practical examination from any candidate aiming for distinction; organic chemistry had become an essential part of the chemistry papers for all candidates; problems 'to test the candidates' powers of observation, description, and inference' had become *de rigueur*;[60] and a new biology paper which embraced both botany and zoology was attracting more candidates than for chemistry.

There were two major changes in the schools' examination system during the Edwardian period. The introduction of the school-leaving examination, or School Certificate, in 1905, was a device to overcome the 'dilemma of the double role' of leaving and matriculation examination into which the Higher Certificate Examination had been forced. Although the School Certificate led to 'some decline in the number and quality of candidates for the Higher Certificate Examination',[61] science subjects do not seem to have been unduly affected. The final rationalization of these two examinations into Junior and Higher School Certificate examinations did not occur until 1918.

Table IV

Higher Certificate Numbers 1903–13

	Mechanics			**Physics**			**Chemistry**		
	No.	Pass	Dist.	No.	Pass	Dist.	No.	Pass	Dist.
1903	157	88	16	132	77	19	161	76	13
1908	146	110	27	141	103	22	166	128	19
1913	183	113	19	201	154	29	200	143	21

School Certificate Numbers 1905–13

	Mechanics		**Physics**		**Chemistry**		**Physics & Chemistry**	
	No.	Pass	No.	Pass	No.	Pass	No.	Pass
1905	–	–	13	12	13	4	19	11
1909	74	44	92	47	105	64	134	117
1913	61	30	104	43	140	70	72	33

Source: Annual Reports, Oxford and Cambridge Schools Examination Board.

The other significant change was the increasing role which the Board of Education and Local Education authorities came to play in the examination system. Whereas before 1900 only the schools, their pupils and the university examiners were involved, from 1902 public authority began to play an arbitrating role through the developing system of financial incentives and awards to candidates who were successful in examination competitions. Thus the nineteenth-century school examination which had been introduced as a test and guarantee of liberal education and culture was transformed into the twentieth-century device of competition, selection and social mobility.[62]

Natural Sciences Tripos Failures and the Schools

Finally, it is of some interest to look, in general terms, at the secondary education of those who were successful, and those who failed, in Part I of the NST between 1881 and 1904. In the following tabulation (Table V) schools are listed in the order of production of the largest number of candidates.
In addition, 34 per cent and 39 per cent of Part I failures were from the Greater and Lesser Public Schools respectively.

On the face of it, it would appear that NST candidates from the Greater Public Schools (Charterhouse, Eton, Winchester, Harrow, Rugby, Shrewsbury, Merchant Taylor's and St. Paul's) were much less successful in both Parts I and II than other proprietory and endowed schools. Does the fact that over fifty per cent of the NST intake from each of these schools failed Part I have any significance? Were these candidates less well-prepared at school for the rigours of Cambridge science?

It may be fair to blame the schools, rather than Cambridge itself, for failure in Part I. In some cases students' failure may have arisen from inadequate science facilities or tuition in the schools.

It is true that those schools that were in the forefront of the introduction of science teaching (Clifton, Rugby, Sherborne, Tonbridge) produced the largest number of candidates for the NST. Also, however, it should not be surprising to find that the schools, like Clifton, which produced very large numbers of candidates, also furnished a relatively-large number of failures. Nevertheless, despite these factors (and in the absence of similar data for the same schools' Oxford natural scientists), it is hard to escape the suggestion that some public-school scientists found it difficult to adjust to university examinations. Were these the exhausted sacrificial victims of competition that critics of examinations spoke about? Until after the First World War students of only moderate ability were being admitted to Cambridge from the public schools. It may be significant that two schools, Manchester Grammar and Newcastle High, Staffs., which were to enter the state's embrace after 1902, had relatively few or no failures.

Table V

School	Total Pupils taking Part I	Success	Failure
Clifton	41	20	21(41%)
Charterhouse	32	9	23(72%)
Marlborough	30	11	19(63%)
St. Paul's	23	8	15(65%)
University C.	22	14	8(36%)
Rugby	21	7	14(66%)
Harrow	18	8	10(55%)
Manchester G.	17	13	4(23%)
Haileybury	17	2	15(88%)
Dulwich	16	8	8(50%)
Shrewsbury	16	7	9(56%)
Repton	16	3	13(81%)
Leys	15	8	7(46%)

School	Total Pupils taking Part I	Success	Failure
Uppingham	14	6	8(57%)
Winchester	13	3	10(76%)
Bradford G.	13	13	–
Tonbridge	13	1	12(92%)
Cheltenham	13	3	10(76%)
Eton	13	4	9(69%)
Merchant Taylors	12	5	7(77%)
City of London	12	12	–
Malvern	10	3	7(70%)
Sherborne	10	–	10(100%)
Newcastle	9	9	–
Oundle	9	4	5(55%)

Source: R. Moseley, *The Natural Sciences Tripos, Data and Commentary, 1851–1914.*

Conclusion

Although the routes taken by Victorian boys and girls to Cambridge Natural Science were extremely various and unsystematic, between the 1870s and the turn of the century a definite pattern and drill had emerged. Attending a moderately-expensive preparatory school where, unlike children of their own age in the state elementary schools who learned the three Rs, some history and some geography, the future Cambridge candidate also studied Latin, Greek, algebra and, possibly, French and German. At thirteen they sat a Common Entrance Examination to a Headmasters' Conference Public School in which little science might be taught until the pupils reached the age of sixteen, at which point they faced either the Junior Local examination of the Cambridge Syndicate or the Lower Certificate of the Joint Board. Passes in five subjects would be demanded, though whether one of the subjects was a science depended on the school's taste. If science was taken, however, and the candidate showed promise, he or she would pass on to a science-based sixth form and prepare intensively for Higher Local examinations, the Higher Certificate of the Joint Board, and/or a Cambridge College Entrance Scholarship.

Specialization in science could be intense, and although it was estimated in 1912 that only a tenth of Headmasters' Conference school pupils actually went to Oxbridge,[63] the sixth form curriculum was dominated by the university entrance requirements of these few. Moreover, the ancient universities' hold over the schools was felt not only through the schools' examination system, but through the stimulus of science scholarships, the representation by Oxbridge dons on the governing bodies of the schools, and by the continuing tradition that the teachers should be Oxbridge educated. Of these controlling elements, the latter was undoubtedly the most significant, for as the products of the NST filtered back to school teaching any enthusiasm for science created by their experience of the NST was passed to their pupils.

In the changed social, intellectual and economic circumstances of Edwardian England, the public school had altered its curriculum, transformed its buildings and multiplied its staff.[64] Although it may not have seen even one of its functions as 'the production of the nation's scientists', in its determination to fulfil its function of feeding as many of its products as possible to the universities, the schools – aided by qualified Cambridge teachers, investment in laboratories, and an examination system – were prepared to accept science as a human, if not humanising, activity. Science had, indeed, been thrust upon them.

Acknowledgements

My thanks to Dr. H. F. King, Secretary of the Oxford and Cambridge Schools Examination Board (Cambridge); and Mr. P. C. J. Laidlaw, Assistant Secretary of the University of Cambridge Local Examinations Syndicate, for allowing me generous access to archival material.

Notes

1 *All the Year Round*, 20 January 1866, pp. 45–6.

2 For science at Eton, see J. L. Heilbron, *H. G. J. Moseley. The Life ad Letters of an English Physicist, 1887–1915*, (Berkeley, Los Angeles and London, 1974).

3 *Conference of Public School Science Masters, Proceedings at the Meeting Held on Saturday, January 19th, 1901*, (priv. pr. 1902), p. 6. Eggar classified Eton in the third category.

4 For Kingswood, see A. L. Mansell, 'The Influence of Medicine on Science Education in England 1892–1911', *History of Education*, **5,** (1976), 155–68 (162); for Rugby, *Conference, op. cit.* note 2, p. 33; Rugby specialists also did four hours of mathematics. Note also, Rev. A. Irving, 'Science and the London Matriculation Examination', *Nature*, **66,** (1902), 320, and *idem*, 'Science in the Public Schools', *ibid.*, 459.

5 *Regulations for Secondary Schools* (1904), quoted in J. S. Maclure, *Educational Documents England and Wales 1816–1967* (London, 1968), p. 159. 'Not less than 4½ hours per week must be allotted to English, Geography and History; not less than 3½ hours to the language where only one is taken or less than 6 hours where 2 are taken; and not less than 7½ hours to Science and Mathematics, of which at least 3 must be for Science. The instruction in Science must be both theoretical and practical.' As Selleck and others have pointed out, this Code probably reduced the amount of science teaching in many of the older Department of Science and Art maintained schools, at a time when the Public Schools were, on the whole, increasing their commitment. See R. J. W. Selleck, *The New Education 1870–1914*. (London, 1968).

6 This paper originally formed part of a longer essay, 'Cambridge and School Science 1850–1914' commissioned by Dr. R. M. MacLeod and Dr. R. Moseley as part of their SSRC research project on the development of the Cambridge Natural Sciences Tripos. Since the Cambridge feeder schools during the nineteenth century were almost entirely the Greater and Lesser Public Schools (known after 1870 as Headmasters' Conference Schools), the essay (and therefore this paper) ignores state schools. For the same reason, no mention is made of Department of Science and Art examinations, for which see H. Butterworth, 'The Department of Science and Art (1853–1900) and the Development of Secondary Education', *History of Education Society Bulletin*, No. **6,** (1970), 35–45.

7 E. C. Mack, *Public Schools and British Opinion since 1800* (Westport, New York, 1941; reprint 1971), p. 91.

8 The arguments against introducing science into the public school curriculum are considered in A. J. Meadows and W. H. Brock, 'Topics fit for Gentlemen: The Problem of Science in the Public School Curriculum' in B. Simon and I. Bradley, (eds.), *The Victorian Public School* (Dublin, 1975), pp. 95–114.

9 According to T. Hearl, 'Military Education and the School Curriculum 1800–1870', *History of Education*, **5,** (1976), 251–64, 'starting in 1857 . . . the foundation curriculum formerly taught at the military colleges' was transferred to public schools through the actions of the Council of Military Education (p. 262). See also J. R. De S. Honey, *Tom Brown's Universe. The development of the Public School in the 19th Century*, (London, 1977), pp. 151–3.

10 D. Newsome, *A History of Wellington College 1859–1959*, (London, 1959); this otherwise model history does not unfortunately consider science teaching.

11 Publication of school examination successes in *The Times* led to protests from the Headmasters' Conference in 1874, but were ignored. Note J. R. de S. Honey in Simon and Bradley, *op. cit.*, note 8, p. 24, and his *Tom Brown's Universe, op. cit.* note 9, p. 245.

12 Board of Education, *Report of the Consultative Committee on Examinations in Secondary Schools* (HMSO, 1911). The Chairman was A. H. Dyke Acland, the son of Sir Thomas.

13 *Fifty Years of Progress in Education: A Review of the Work of the College of Preceptors* (London, 1897); D. Hudson and K. W. Luckhurst, *The Royal Society of Arts 1754–1954* (London, 1954), chap. 15; F. E. Foden, 'The Reverend James Booth and the Genesis of the Society's Examinations,' *Journal of The Royal Society Arts*, **118,** (1970), 58ff, 646ff; Honey, *Tom Brown's Universe, op. cit.*, note 9, chap. 2.

14 For the full story, see J. Roach, *Public Examination in England, 1850–1900*, (Cambridge, 1971), chaps. 2–4.

15 J. L. Brereton, *The Case for Examinations*, (Cambridge, 1944), p. 74.

16 J. Roach, *op. cit.*, note 14, p. 242.

17 *Ibid.*, p. 140. Roach cautions, however, that it would be wrong to assume that because a

have gone to the university otherwise' (p. 141).

18 *Ibid.*, pp. 141–2.

19 They also pioneered practical examinations in science before the Department of Science and Art, though candidates had to supply their own sets of apparatus. These were conveniently supplied by J. J. Griffin who charged 14s 6d for an Oxford chemistry set compared with 12s for Cambridge. See advertisements in J. J. Griffin, *Chemical Handicraft*, (London, 1877).

20 Roach, *op. cit.*, note 14, p. 261. The examination was 'harder' because most of its papers were compulsory and there was not the same range of options as in the Locals. It was notorious for setting catch questions on a syllabus that many regarded as limited in range. See C. Hawkins, 'Examinations from the School Point of View' in *Special Reports on Educational Subjects*, vol. 26, (*The Teaching of Mathematics in the United Kingdom*) (London, 1912), p. 448; William Dodds, *A Complete Guide to Matriculation at the University of London*, 4th ed. (Manchester, 1879) blamed the severity of the compulsory natural philosophy papers.

21 G. M. D. Howat, *Oxford and Cambridge Schools Examination Board, 1873–1973*, (Oxford, 1973).

22 *Ibid.*, pp. 3–4. A 'Lower' certificate for students leaving school at 16, which was introduced in 1884, was modelled on 'Lower Schools' examinations already separately conducted by the universities. Cambridge had admitted women to its local examinations in 1865, and Oxford in 1870. The Joint Board examined girls' schools from 1879.

23 Brereton, *op. cit.* note 15, p. 77.

24 Roach, *op. cit.* note 14, p. 233.

25 However, given the nature of specialization, those candidates who had taken science at school, and especially those who gained distinctions in their science papers, were undoubtedly more likely to have read science at a university.

26 That the assumptions are invalid is indicated by the numbers of candidates for the Higher School Examination (all subjects) who would have obtained exemption from Part I of the Previous. These numbers are less than the total number of undergraduates in Cambridge in any year. Clearly, the pool of admissions was not confined in those taking Joint Board papers, or for that matter, Joint Board and Cambridge Locals.

27 Candidates for distinction in any science were required to take a practical examination, and this may have been a limiting factor in the award of distinctions until better practical facilities were available in most schools.

28 *Consultative Committee on Examinations*, *op. cit.* note 12, p. 185.

29 The figure for biology is larger than might be expected because this subject could not be taken in the School Certificate examination which the Joint Board had introduced in 1905 for candidates who were leaving school at the age of 17. Such candidates had to pass in five subjects, some of which (but not science) were compulsory.

30 Jean Russell-Gebbert, *Henslow of Hitcham* (Lavenham, Suffolk, 1977), chap. 5.

31 The Previous examination consisted of three parts; Part I, which was basically a test of competency in Latin and Greek; Part II, a test on either Paley's *Evidences* or logic, or elementary heat and chemistry, together with papers on mathematics and an English essay. Candidates who passed these posts could proceed to an Ordinary degree. For Honours, an additional paper in either mechanics, French or German had to be taken.

32 *Consultative Committee on Examinations*, *op. cit.* note 12, p. 90.

33 *Oxford and Cambridge Schools Examination Board Minutes 1873–1904*, p. 164.

34 *Ibid.*, p. 69.

35 *Consultative Committee on Examinations*, *op. cit.* note 12, p. 67.

36 *Ibid.*, *Report of the Committee on the Position of Natural Science*, (1918). [Cd. 9011]. ix, pp. 2–4. (Hereafter, *Thomson Committee*).

37 As was the '100 Per Cent Special Place System' whereby Municipal Grammar School places were filled by a common entrance examination – later to be called 'The Eleven Plus'. See John Roach, 'Examinations and the Secondary Schools 1900–1945', *History of Education*, **8**, (1979), 45–58.

38 The examination was controlled by a Board of managers consisting of three representatives of the Headmasters' Conference and three representatives from the Association of Preparatory Schools. See *Consultative Committee on Examination, op. cit.* note 12, p. 305.

39 *Thomson Committee, op. cit.* note 36, p. 29.

40 D. Leinster-Mackay, 'The Evolution of t'other schools: an Examination of Nineteenth-century developments of the Private Preparatory School', *History of Education,* **5,** (1976), 241–9.

41 *Royal Commission on Secondary Education*, 1895. [C. 7862]. xliii, pp. 173–4.

42 Rev. C. Eccles Williams, 'Entrance Scholarships and Cram' in *The Public School Within. A Collection of Essays on Public School Education* (London, 1906); Board of Education, *Special Reports on Educational Subjects*, vols. 6 and 8 (London, 1900).

43 Notably expressed in H. B. Gray, *The Public Schools and the Empire* (London, 1913).

44 *Conference of Public Schools, op. cit.* note 3, p. 1.

45 Gray, *op. cit.* note 43, p. 137.

46 *Conference of Public Schools, op. cit.* note 3, p. 15.

47 *Ibid.*, p. 16

48 The matter is briefly discussed by N. H. Mikhail, *An Historical Review of the Development of the Main Associations Concerned with the Promotion of Science Teaching in Secondary Schools* (Unpublished M.A. thesis, University of London, 1961), and can be followed in the Minutes of the Association housed at the offices of the Association for Science Education, Hatfield.

49 See Mansell *op. cit.* note 4.

50 'The Sacrifice of Education to Examinations', *Nineteenth Century*, **24,** (1888), 617–37; W. H. Brock, (ed.), *H. E. Armstrong and the Teaching of Science 1880–1930*, (Cambridge, 1973), pp. 30–1.

51 Brock, *op. cit.* note 50, p. 100, from Armstrong's essay 'The Aims and Practice of Teaching Chemistry' (1897).

52 Both schools tended to send boys to Oxford or, because of their former City connections, to The University of London. See L. F. Morris, *A History of St. Dunstan's College* (priv. pr., London, 1970); *The Christ's Hospital Book* (London, 1953).

53 W. H. Brock, 'Geometry and the Universities: Euclid and his Modern Rivals, 1860–1901', *History of Education*, **4,** (1975), 21–35.

54 Cf. W. D. Eggar's support for Perry, *Conference of Public Schools, op. cit.* note 3, p. 8. Perry received some support from the Department of Science and Art and from the new Board of Education after 1899.

55 For some methodological possibilities and recent applications, see E. W. Jenkins, *From Armstrong to Nuffield*, (London, 1979), chap. 8.

56 See *University of Cambridge Higher Examinations, Papers and Regulations.*

57 Liveing sat on the Joint Board 1874–8, 1880–3; and on the Syndicate of the Cambridge Locals during the 1870s. He also examined for the Council of Military Education.

58 A chemistry reporter in 1883 divided schools into excellent, moderate, but uniform in standard, questions beyond the powers of boys, and hopeless. *Tenth Annual Report of OCSEB for the Year Ended October 31, 1883,* (Oxford and Cambridge, 1883), p. 13.

59 *Summary Report of the Work of the Board During the Thirty Years 1874–1903* bound in the *Board's Publications 1903–1904.*

60 *Ibid.*, p. 21.

61 *Decennial Report of the Board for the Years 1904–1913*, p. 6 (Bound in the *Board's Publications 1912–13*).

62 Roach, *op. cit.* notes 14 and 37; R. J. Montgomery, *Examinations*, (London, 1965).

63 Gray, *op. cit.* note 43, p. 263, who adds that it had dwindled from one fifth in 1872.

64 T. Parsons, *Essays in Sociological Theory*, 2nd edition, (New York, 1958), pp. 337–8.

XIX

Science for All

'Scientific education', wrote the young assayer and future economist, William Stanley Jevons, in 1856, 'is one of the best things possible for any man, and worth any amount of Latin and Greek. It tends to give your opinions and thoughts a sort of certainty, force and clearness which form an excellent foundation for other sorts of knowledge less precisely determined and established, provided you do not allow your mind to become completely formed to science.'[1] Although the rhetoric of the mid- and late-Victorian advocates of the introduction of science into the British school and university curriculum agreed with Jevons's view that science was mind-strengthening and mind-broadening — indeed, that better than Latin or Greek, it promoted a genuinely 'liberal education' — the long term English tendency (one must exclude the Scots) was toward Jevons's fear of specialization. By the turn of the twentieth century, with the introduction of a state secondary education for all that reflected and perpetuated the curricula and values of private fee-paying ('public') schools, instruction and examinations in, and laboratories for, the separate disciplines of chemistry and physics and, increasingly, biology, had become the norm. School science teachers, whose work was by the Edwardian era dominated by the external exigencies of the maddeningly different, highly specialized matriculation and entry requirements of universities, the army, civil service and dozens of professional bodies and qualifying associations, were therefore forced to 'form' their pupils in science whether or not their future careers lay in science.

The days of early specialization had begun. By the 1950s and 1960s, this had often reached ludicrous levels whereby pupils would, on the one hand, drop history or geography at age fourteen in order to concentrate on one or more of the sciences, or on the other hand, abandon science altogether because the highly specialized character of the chemistry or physics taught was beyond their needs and capabilities. Examples could be given of small grammar schools of 400 pre-sixteen pupils all studying chemistry and physics even though only one would go on to take a university degree in one of these subjects. The goal of university chemistry or physics or biology and medicine enormously influenced

---

1 R. D. Collison Black, ed., *Papers and Correspondence of William Stanley Jevons* (London, 1973), vol. 2, p. 244.

the school curriculum 'top down' instead of 'bottom up', with the teachers themselves devising a curriculum more appropriate for the other 399 pupils. Such specialization undermined the Victorians' achievements in science education, and given the broader curricula taught on the Continent, North America and, closer to home. in Scotland, it was bound to become a matter of public debate at a time of postwar bulging school enrolments and incipient industrial decline.

This essay will explore the tension between what came to be called 'general science' and specialized science education that was implicit within the 'science for all' movement. Since this tension was explicit within the heuristic campaign launched by Henry Armstrong in 1884, the question of how far science should be taught practically, rather than didactically, will necessarily loom large in what follows.

Henry Edward Armstrong and heurism

If the Royal Chemical Society were to name its Education Section after a distinguished chemical educationist in line with its Dalton, Perkin and Faraday Sections, the most serious contender would be Henry Edward Armstrong (1848-1937). Although he made no outstanding chemical discovery, Armstrong's contributions to the elucidation of benzene derivatives and the structure of terpenes were significant, and, as historians of science are beginning to realise, he and his students constituted a socially-significant school of British chemists. However, it is his contribution to science education which has had a most long-term effect.

In June 1884 an International Conference on Education was held at South Kensington in the newly-opened City & Guilds of London Institute. It was here, on 5 August, that Armstrong, the Institute's new Professor of Chemistry, first began to make public his critical, but constructive, views on science education. During the remainder of his long life he was continually critical of didactic teacher-centred methods of learning, and he became the principal exponent of *heurism* — the child-centred method of learning through experience and experiment.

How and why did an organic chemist become critical of his contemporaries' teaching?

In the eighteenth century, following the expansion of Newtonianism, there were several methods by which the literate could obtain scientific instruction — through itinerant lecturers, the new provincial philosophical societies, a plethora of periodicals, textbooks and encyclopaedias, and at the exciting

Universities of Leyden, Edinburgh and Glasgow. However, little science reached the lower orders of mankind except by way of honest individual effort. Industrialisation and the opening and growth of mechanics' institutes and mutual improvement societies in the early nineteenth century altered this slightly, as did the transplantation of the ideals of the democratic Scottish Enlightenment to London in the shape of the University of London in the 1820s. On the other hand, the teaching of natural science in endowed schools, and in the Universities of Oxford and Cambridge hardly existed (mathematics was a different story), and where it did, made little impression until the 1850s. That a definite movement, or campaign, for more science teaching occurred in the generation before Armstrong undoubtedly owed much to Liebig's army of British research students who returned to the United Kingdom to assume places of responsibility during the 1840s. Two of Liebig's pupils founded and directed the private Royal College of Chemistry in 1845 while another, Lyon Playfair, helped to establish the Department of Science and Art and its effective, but destructive, method of teacher payment by examination results, as well as the Great Exhibition whose profits were to create an educational city at South Kensington in the 1870s.

In the national schools there was little attempt to introduce the rudiments of science except through natural theology and object lessons. Isolated exceptions, such as the work of Edward Frankland's friend, Richard Dawes, at King's Somborne in Hampshire, were nevertheless very influential. Dawes's secular work on the science of common things excited the schools' inspector, Henry Moseley, and helped lead to the introduction of scientific instruction in teacher training colleges. Unfortunately, or so it seems in retrospect, the natural development of science teaching in elementary schools was stymied by bickering and demarcational jealousies between the Department of Science and Art and the Board of Education. In 1861 Lowe's Revised Code confined payment to elementary schools for the three Rs alone and placed responsibility for science teaching entirely on the shoulders of the Department of Science and Art.[2]

In the 1860s, following the deeper understanding of physics, chemistry and biology developed through the concepts of energy, atomism, valency and evolution, the emerging scientific community began to form a powerful pressure group for urging on the wider public a fuller appreciation of, and a more systematic teaching of, their interests. Although there followed much debate over the relative merits of classics, mathematics and science as

---

2 See D. Layton, *Science for the People* (London, 1973); D. W. Sylvester, *Robert Lowe and Education* (London, 1974).

formative elements in 'liberal education', by the 1870s there was little doubt that science had grown strong roots into the British educational system.

Yet, when Armstrong began teaching chemistry in 1870 it seemed to him that his students had been completely dulled from being subjected to the authoritariansim of didactic teachers and textbooks whose words related little to the practical things of life. His students were unable to think for themselves and were prepared only to memorise facts and formulae for examination questions. His predecessors had successfully brought science into schools, colleges and evening classes, but had stifled students in the process because no one had paused to ask how science ought to be taught, or how the science curriculum should be sequentially arranged. Armstrong's solution was the heuristic method.

Armstrong had received a very practical education at the Royal College of Chemistry (now part of Imperial College, London) and had performed his first piece of research, an improved method of water analysis, with Edward Frankland, a brilliant experimentalist who had developed the use of chemical formulae for educational purposes. It was Frankland who encouraged Armstrong to undertake further training at Leipzig, where he was plunged into the excitement of research on aromatic chemistry and forced to think out solutions to experimental problems under the benevolent, but scathingly-critical, eye of Hermann Kolbe. Both Frankland and Kolbe can be seen to have encouraged Armstrong to believe in the value of self education through laboratory research — a philosophy which had many affinities with (and possibly its roots in) the child-centred pedagogical innovations of Pestalozzi (1746-1827).

In 1880 Armstrong appeared as a scientific witness in a patent trial. The experience of the court atmosphere, the dialectic, the way plaintiff and defendant marshalled their respective cases, and the clear incisive manner in which judgement was delivered was, he felt, the coping stone on his education. Scientific training had taught him to examine evidence and to ask questions; reading R. C. Trench's *Study of Words* (1851) had made him 'critical and anxious to get behind meanings;' the patent action made him alive to the 'need of searching cross-examination and judicial consideration of every item for or against a proposition'. (Incidentally, it was this that also made him a formidable foe of Arrhenius's ionic theory.)

Armstrong was to label this 'methodological use of knowledge' *scientific method*, and he decided that it was best learned in a laboratory atmosphere by an autodidactic process he called *heurism*. (The word comes from the Greek meaning 'I have found out', as in Archimedes' reputed exclamation, 'Eureka!')

'Heuristic methods,' said Armstrong, 'are methods which involve our placing students as far as possible in the attitude of the discoverer.' What a child or a student can find out for himself he best remembers because he truly understands. Moreover, because the student-investigator is genuinely interested and involved in finding a solution to a problem, he will learn about it more efficiently than if he is given the solution didactically.

It seems clear that Armstrong developed these conclusions largely by intuition and his experience as a research chemist. Similarly, it seemed intuitively obvious to him that children would grasp certain ideas better at different ages, and that science syllabuses should be age-graded accordingly. The more thorough investigations which Piaget and other psychologists have made on children's cognitive developments have confirmed Armstrong's hunches, though the liberal application of heurism would be painful to both pupil and teacher.[3]

Until his death in 1937, Armstrong whole-heartedly committed himself to promulgating the notion that understanding came through doing. Precisely what was learned (information) was less important, at least in the early stages, than the method (process) involved in learning, thinking, or finding out about something. He belived, as the history of science seemed to suggest, that once a person had learned the techniques of experimentation, he could tirelessly acquire by his own efforts the mass of information which other teacher-oriented systems of education forced into children by rote, demonstrations, and the penalty of examinations.

Hindsight suggests that Armstrong did not sufficiently distinguish between the kind of 'discovery of understanding' ('I get it') which a tyro makes in a school or university laboratory, and the original discovery of innovation which a research scientist may make. That this point has caused confusion and led to objections to discovery methods is understandable. Nevertheless, until the First World War heurism was probably the chief catalyst in transforming school science teaching in Great Britain, Armstrong himself being energetic in its promotion.

Ever since its foundation in 1831 the British Association for the Advancement of Science had been a valuable annual public platform for the announcement of new ideas or experimental work, or for the discussion and report of scientific enterprises. Papers on educational subjects were sometimes

---

3 H. E. Armstrong, *The Teaching of Scientific Method and Other Papers on Education* (London, 1903; 2nd ed., 1910); J. Piaget and B. Inhelder, *The Growth of Logical Thinking* (London, 1958); L. S. Shulman and E. R. Keisler, eds., *Learning by Discovery: A Critical Appraisal* (Chicago, 1966).

presented, especially in Section F, Economic Science and Statistics.[4] A sound precedent for the discussion of science education had been created by F. W. Farrar's Committee of 1866. A second wave of discussions at the British Association took place during the 1880s, following the Elementary Education Act, the creation of the School Boards and the establishment of the New Revised Code. In 1879 the chemist and philanthropist John Hall Gladstone (1827-1902), who had been elected to the London School Board in 1873 (and effectively replaced Thomas Henry Huxley who had resigned in 1872), succeeded in creating a British Association watchdog committee on elementary schools under the auspices of Section F.

In 1885 Armstrong was President of Section B, Chemistry, at Aberdeen. Over half of his address was devoted to education, and it was a signal that he intended to use the British Association as a propaganda machine for his ideas. His theme was an echo of the accusations by Frankland, Perkin, Roscoe and other British chemists that British chemistry was poor in quality compared with the growing output from German industrial and university laboratories. The root of the problem, said Armstrong, was that science teachers had not been trained in methods of research themselves. 'No amount of instruction, such as is ordinarily given, in the mere theory and practice of chemical science will [ever] confer the habits of mind, the acuteness of vision and the resourcefulness required of an efficient chemist in any works, any more than the mere placing of the best tools in a workman's hands will make him a skilful operator.'[5]

The solution was to copy the German example and create an atmosphere of research within a school of higher education. Even this would not be sufficient unless students came to such centres better prepared. At the moment, he complained, their mathematical knowledge was unpractical, they could not draw or read foreign languages, and the scientific parts of their minds were 'deadened from want of exercise'. Compared with evening students or female day students, the bulk of students were drained of motivation, manifested by 'their marked inability, often amounting to downright refusal, either to take proper notice of what happens in an experiment or to draw any logical conclusion from an observation.' Only wholesale revolution of the curriculum by the introduction 'of a *rational system* of practical science teaching into all schools' would revive motivation and morale. Otherwise it were better that no

---

4 P. M. D. Collins, 'The origins of the British Association's education section', *British Journal of Educational Studies*, 27 (1974), pp. 232-244; D. Layton, 'The schooling of science in England, 1854-1939' in R. MacLeod and P. Collins, eds., *The Parliament of Science. The British Association for the Advancement of Science 1831-1981* (Northwood, 1981), pp. 188-210.

5 *British Association Reports*, 1885, p. 946.

science at all be taught in schools, to the loss of the faculties of observation and of reasoning from observation and experiment. It was not entirely the fault of teachers. Examinations were set by specialists who often lacked educational sense and who were 'prone to discourage science by marking too severely; and as their questions govern teaching, instruction is given in schools without due reference to educational requirements and in a purely technical style.'[6]

Armstrong ended the educational prologue to his main address on the state of chemistry by turning to the subject of specialization. It was not necessarily the introduction of more science subjects into the school curriculum that was needed. The aim should not be to produce junior specialists, but to develop faculties, to encourage the aspirations of the young 'by inculcating broad and liberal views of our science, not an infinite number of petty details'. This meant abandoning the obsession with qualitative analysis, for example. Chemistry had recently undergone striking internal changes through the knowledge of the periodic law and of structural organic chemistry; such changes provided another impetus for revolutionising its teaching.

In the summer of 1886, no doubt with the discussion of a 'rational system' of instruction in mind, Armstrong wrote to both Gladstone and the former schoolteacher, William A. Tilden to suggest that the subject of science teaching in schools be brought before Section F. Tilden replied that he had already tried to persuade William Crookes, President of the Chemical Society that year, to discuss the subject in Section B; but Crookes would 'not bite'.[7] Gladstone welcomed the suggestion, but pointed out that plenty of opportunity for discussion already existed whenever his elementary schools' committee reported to the British Association.[8] Significantly, at the Birmingham meeting, Armstrong was coopted onto Gladstone's committee. However, its range of inquiry was limited to elementary schools, whereas Armstrong wanted to debate the fate of science, especially chemistry, teaching over the whole range of school education. In 1887, at the Manchester meeting, he succeeded in creating a committee to consider and report on 'the present methods adopted for teaching chemistry'. The committee comprised the chemists Armstrong, R. Meldola, A. Smithells, Gladstone, W. J. Russell, A. G. Vernon Harcourt, Pattison Muir and W. R. Dunstan (a pharmacist), and the chemistry school teachers J. T. Dunn, Francis Jones and W. A Shenstone.

In December 1887 this committee circularised 500 headmasters and training college Principals asking them why they taught chemistry, what difficulties they

---

6 *ibid.*, p. 948.

7 Tilden to Armstrong, 2 June 1886, Imperial College Archives.

8 Gladstone to Armstrong, June 1886, Imperial College Archives.

experienced, and what teaching methods they found to be most efficient. Although there were only 86 replies, the committee was able to compose an interesting report for the Bath meeting in 1888.[9] Chemistry, they found, was being taught both as a form of mental training and exercise and because of its applications in various walks of life. Teaching difficulties included expense, the low value put upon the subject compared with other subjects in the curriculum, the examination syllabuses, and the absence of good textbooks and qualified teachers; but, above all, there was a general difficulty over the lack of good laboratories. On method there was a surprising unanimity that the subject should be taught experimentally, and a general feeling that there was too great an emphasis on qualitative analysis and manipulation in existing practical courses. The Committee itself concluded that something 'should be done in the direction of promoting a more uniform and satisfactory treatment' of chemistry teaching. To this end, it promised to suggest new methods of teaching the subject.[10]

In 1889, again using the powerful public platform of the British Association for the Advancement of Science, Armstrong developed a series of helpful 'suggestions for a course of instruction in elementary science' divided into six age groups or stages, most of which involved pupil experimentation. Probably the most important aspect of this heuristic syllabus was the way it abolished the traditional catalogue approach to the chemistry of 'elements, oxides and salts' that dated back to the time of Lavoisier, and replaced it with real problems (e.g., why does iron rust?).

This syllabus attracted widespread attention. Following its approval by the Headmasters' Association (f. 1890), it began to receive favourable attention in Germany and Japan.[11]

Further opportunities to promote heurism in Britain came through Armstrong's contacts with Christ's Hospital and St. Dunstan's College, the practical classes which he held for London science teachers in 1896, and through a long essay he contributed to the Board of Education's widely-read *Special Reports* in 1898.

He also practised what he preached on his own children: he sent three of his boys to St. Dunstan's, and held weekend experimental sessions with his three youngest children on questions suggested by their storybooks. In one tale a mischevious monkey had survived drowning, despite being attached to a heavy

---

9 *British Association Reports*, 1888, pp. 73-89.

10 For Armstrong's new course at Finsbury College, see essay XI above.

11 T. Terakawa and W. H. Brock, 'The introduction of heurism into Japan', *History of Education*, 7 (1978), pp. 35-44.

stone because (the author claimed) the stone weighed less in water. To test the story's credibility Armstrong encouraged the children, whose ages ranged from eight to twelve, to make simple tests with home apparatus, and to keep independent records of their results. Much of the charm of these notebooks lies in seeing how each child tackled the problem raised in a different way, and how, in deducing Archimedes' Principle, they also formed ideas about density, averages and percentages, besides realising the importance of cleanliness and accuracy in measurement.[12]

Thus, in 1902, when Armstrong reviewed the situation before the Education Section of the British Association (which he had promoted in 1900), he quite properly associated the increase in the number of school laboratories in Great Britain from probably less than 150 in the 1870s to over 1,000 in 1902, as due in large measure to his heuristic campaign.[13]

The role of faculty psychology

Entrenched in the argument of both the older 'instrumental' education of the 1840s and of the new 'practical' education of the 1870s was faculty psychology. For Henry Armstrong, teaching the physical sciences would 'develop a side of the human intellect which ... is left uncultivated even after the most careful mathematical and literary training; the faculty of observing and reasoning from observation and experiment'.[14] Faculty psychology — the idea that the mind could be analysed in terms of various powers or faculties — originated in Greek philosophy. It seemed to gain powerful empirical support by the clinical observations of Gall and Spurzheim at the beginning of the nineteenth century — observations whose social and educational implications were developed in the phrenological movement by George Combe. Although the complex multiplication of entities by the British phrenologists, and phrenology's debasement by mesmerism and occultism, led to its discredit, the phrenologists' belief that full development of the faculties, or organs, of imagination, judgement, reasoning, perception and memory (and a host of others) depended upon a person's environment and training, continued to be a

---

12 The notebooks are in Imperial College Archives.

13 *British Association Reports*, 1902, pp. 820-44.

14 H. E. Armstrong, 'Teaching natural science as a part of the ordinary school course', in *International Conference on Education Proceedings* (London, 1884), vol. 2, pp. 69-82, 104-106. Reprinted in W. H. Brock. ed., *H. E. Armstrong and the Teaching of Science 1880-1930* (Cambridge, 1973), pp. 74-89.

useful educational way of justifying a particular curriculum throughout the nineteenth century.

By formally training one faculty, or a part of it, it was claimed that the whole mind was improved for future use in any situation. Although the theoretical justification of the instrumental education's emphasis on the 3 Rs was faculty psychology, this was largely an afterthought. The need to teach reading, writing and arithmetic was self-evident without deduction from theoretical faculties.[15] Yet, so firmly entrenched was faculty psychology amongst mathematicians and classicists in the public schools 'that a substantial group of new educationists, concerned to introduce extensive reforms in both the methods and the curriculum, were forced to argue their case on the basis of faculty psychology.'[16] The unquestioned assumption made by reformers like Armstrong (on top of the unquestioned acceptance of faculties and their trainability) was that training could be transferred. Heuristic methods trained the faculties 'of thoughfulness and power of seeing; accuracy of thought, of word and of deed'.[17] Outside school these attuned faculties would respond not merely to special scientific stimuli, but to the general stimuli of everyday life and work.

Such a theoretical justification for heurism was fated to be overtaken by the development of Herbartianism on the one hand, and of experimental psychology on the other. To both Armstrong at first turned a blind eye. Since the Herbartians emphasised the role of the teacher in moulding the child, and stressed the significance of books and literary studies, conflict with practical educationists like Armstrong was inevitable. For the remainder of his life, from 1900 to 1937, Armstrong was to thunder against the revival of the literary cult.[18] Meanwhile, in the USA, the American 'progressive' educational programmes associated with the names of Colonel Francis W. Parker (1837-1902) and John Dewey (1859-1952) were moving in parallel directions. For although neither Parker nor Dewey were chemists (Parker was a secondary schoolteacher and Dewey a professional philosopher), their educational philosophies were strikingly similar to Armstrong's: the attack on a morphological curriculum whose primary concern was 'classification, in naming parts and describing forms, rather than in developing an understanding of

---

15 R. J. W. Selleck, *The New Education. The English Background 1870-1914* (London and Melbourne, 1968), chapter 2.

16 *ibid.*, pp. 45-6.

17 H. E. Armstrong, *The Teaching of Scientific Method* (London, 1903), p. 196.

18 H. E. Armstrong, *Journal and Proceedings of the Royal Institute of Chemistry*, 1924, p. 10.

function and functioning',[19] the attack on stereotyped recitation and grading methods, the Herbartian emphasis on teaching for interest and relevance to a student's mental stage of development, the emphasis on a socially-relevant education for an industrial society, the claims for manual training, for experience, for learning by doing.

Indeed, in Dewey's laboratory school at the University of Chicago between 1894 and 1904, and his book *The School and Society* (1900), there is much which is reminiscent of Armstrong's experience in England. Children's work with textiles led to investigations of a scientific and historical nature, while cookery was 'the natural avenue of approach to simple but fundamental chemical facts and principles and to a study of plants as articles of food'.[20] Consider the following nature lesson for 11-12 year-olds in Dewey's school:[21]

> The subject was the growth of sand dunes, which had been suggested by a visit to the shores of Lake Michigan some weeks before. The children were encouraged to ask questions about what they saw and to mark the point on a small tree up to which it had already been buried in sand...
> 1. Why do some places when trodden on make a different sound from others?
> 2. How did the sand dunes get there?
> 3. Did the water make the mounds?
> On a subsequent visit to the shore the growth of the mounds and the rise of the sand up the tree was observed and measured. Then it was pointed out that the rise could not possibly be due to the action of water, and the angle at which the sand rested was taken by a simple clinometer made by the children, who found they could not pile [sand] up a larger angle than 32 deg. On the return to school they drew in coloured chalks their recollection of the scene, and, after describing and discussing their experiences, concluded that the most probably explanation was the action of the wind. This hypothesis was then tested by experiment, and it was found that with a bellows, sand, and a few obstacles, the same formations could be produced in the schoolroom.

---

19 Harold Rigg, 'The school curriculum, 1825-1880', in *The 26th Yearbook of the National Society for the Study of Education, the Foundation and Techniques of Curriculum Construction* (Bloomington, 1926), p. 22.

20 F. W. Garforth, *John Dewey. Selected Education Writings* (London, 1966), p. 66.

21 *Mosely Education Report* (London, 1904), p. 279. Excerpts in Brock, *op. cit.* (14), pp. 121-132.

Curiously, neither Dewey nor Armstrong ever cited one another. Armstrong, unlike four other members of the Mosely Commission, did not visit Chicago in 1902, but since he edited its *Report* he must have been aware of their favourable account of Dewey's laboratory school.[22] However, as his diary of the tour reveals, he was annoyed by the way Mosely had organised the trip, and was convinced before he left England that there was nothing to be learned from America about education, and that the Uesanians (as he dubbed Americans) had made mistakes (like female eachers) which the British should avoid at all times.[23] He praised the standard of buildings, organisation and equipment of American schools, and the 'absolute belief in the value of education both to the community at large, and to agriculture, commerce, manufactures and the service of the State', but he dismissed as worthless all American teaching methods except in medicine and manual training. There was little evidence, he claimed, that John Perry's 'practical methods of teaching mathematics and geometry', which were coming into vogue in England, were appreciated. The old academic methods of teaching science seeemed to prevail almost exclusively.

Armstrong was wrong of course, both in claiming America was backward and in denying that Britain had anything to learn, though it is fair to say that Dewey had little impact on English education, which went its own independent way, until the 1920s.[24] With isolated exceptions, like Helen Parkhurst's Dewey-inspired 'Dalton Plan' of monthly project work to be completed at a student's individual pace which was enthusiastically adopted in England during the 1920s (arguably because Armstrong's heuristic schemes in chemistry had made teachers more receptive), English and American curricula went their independent ways. They are examples of parallel, not interconnected, curriculum innovation.[25] The 'virus heuristicum' was operating on both sides of the Atlantic simultaneously, attacking the often administratively-convenient formalism of earlier science education.

---

22 *ibid.*, pp. 123, 203, 279, 356.

23 H. E. Armstrong, 'Diary of a visit with the Mosely Commission, 1903', typescript Imperial College Archives.

24 R. J. W. Selleck, *English Primary Education and the Progressives, 1914-1939* (London, 1972), p. 113.

25 B. MacDonald and R. Walker, *Changing the Curriculum* (London, 1976). A number of English commentators, influenced by Armstrong, often described American science education as heuristic; e.g., R. E. Holmes, *The Making of Citizens. A Study of Comparative Education* (London and Newcastle, 1902), p. 280. More critically, H. T. Mark in a report on 'moral education in American schools', saw Dewey's method as 'the fallacy of a too rigid application of the "heuristic method"', *Special Reports on Educational Subjects*, vol. 10 (HMSO, London, 1902).

Two factors seem to have been at work.[26]
1. The accumulating influences of nineteenth-century European theorists of education such as Pestalozzi (1746-1827), Froebel (1782-1852), and Herbart (1776-1841) who had all stressed the natural development of children mentally, physically, morally and socially.
2. The influence of education psychologists like E. L. Thorndike in America and Cyril Burt in Britain, who developed procedures of measurement and fact-finding in education largely as a result of the work of Francis Galton and Karl Pearson and the British school of biometricians.[27]

These two movements, which were reinforced by evolutionary theory (the Darwinian in Britain and the Spencerian in America), viewed the mind as a unity which grows and evolves, and not as a conglomeration of phrenological faculties such as 'observation', 'classification', 'reasoning' which had to be attuned by various exercises to fit a child for the adult world. Ironically, it was this kind of faculty psychology which Armstrong and preceding members of the British Victorian scientific community had used successfully to justify a place for science teaching in the largely classically-oriented curriculum.

For Frankland, Huxley and Armstrong, what was called a liberal education could only produce a truly cultured person if science was included in the curriculum, for other subjects left several faculties dormant and untrained. However, given the growing specialization of the British educational system (largely caused by a obsession with examinations), it was easier to justify science teaching economically in terms of vocations and the industrial importance of science — especially as a result of German and American competition and the First World War. This utilitarian view of science, particularly of chemistry teaching, was reinforced by the Herbartian lobby, which not only undermined faculty psychology, but justified science teaching by the argument that the curriculum ought to be a simplified microcosm of the community's larger interest: economic survival and prosperity.[28]

This tension between the needs to train a scientific manpower and the aim of making science part of everyone's culture, was to be the hallmark of twentieth-century rhetoric concerning science teaching.

---

26 See, for England, Selleck, *op. cit.* (24); and for America, L. A. Cremin, *The Transformation of the School: Progressivism in American Education, 1876-1957* (New York, 1962).

27 Rigg, *op. cit.* (19), chapter 4.

28 Armstrong was more Pestalozzian than Herbartian. He distrusted the English Herbartians for their emphasis on literary culture since this threatened the place science had found in the curriculum. See his debate with F. H. Hayward, *School World*, 1908, pp. 417-419, 477-478.

Armstrong himself was not directly challenged until 1904. His critic was Frank Herbert Hayward, a Herbartian school inspector, who felt that heurism was a poor method because a child's greatest need was 'not that his mind should be exercised, but — that his mind should be fed with a rich repast of historical and biographical ideas.'[29] In 1908 he returned to the downfall of the 'formal or faculty training' in the widely-read journal, *The School World*,[30] and although Hayward was really attacking Armstrong's disciples for their 'purblind' advocation of training for accuracy and observation, Armstrong saw this as a direct confrontation. In his reply he poured scorn on the pretensions of psychological verbiage, and the logic and slipshod English of Hayward's attack.

> As to whether it be a good thing to be generally observant, so few are that it is difficult to give an answer based upon the tests of experience. The consequences of lack of observing power are so frequent and manifest, however, that it is clear that we could put up with some extension of the class of observant persons.[31]

Faculty training was not yet discredited in his eyes; it was still a working hypothesis. 'The extent to which it can be carried is yet to be ascertained — the ways of giving it successfully are to be worked out.'[32] Formal training was not to be equated with drudgery, as Hayward had implied. Yet,

> whilst every teacher will admit that it is necessary to create interest, we all know that it is not always possible to maintain this at bursting point and that in school, as in the world, uninteresting work must be done sometimes; that in point of fact, it is most important to acquire the art of doing uninteresting work in a serious and determined way.[33]

Armstrong did not return to the subject again, although for the remainder of his life he reserved the more vitriolic portions of his extensive vocabulary for psychologists.[34] In fact, formal training continued to be used in the practical

---

29 F. H. Hayward, *The Secret of Herbart: An Essay on the Science of Education* (London, 1904), p. 76.

30 *op. cit.* (28).

31 *ibid.*, p. 477.

32 *ibid.*, p. 478.

33 *Mosely Report* (21), p. 10.

34 For example, 'Sydney Smith has well remarked, "God has made nothing so curious as human creatures" — which is very true and perhaps the most curious of them all is

arguments of educationists and teachers, and in the end, during the last decades of Armstrong's life, psychologists grudgingly admitted that the doctrine of formal training held an element of truth which traditionally had been presented in a psychologically-invalid manner.[35] A person's capacities were improved by formal training, but the transfer of capacities only occurred in a limited way between related fields. This admission was to help the English revival of heurism in the 1930s.

The Association of Public School Science Masters

Many teachers' organisations and pressure groups emerged between 1870 and 1910. The Mathematical Association which was founded in 1897 to improve mathematics teaching in schools is a good example. It drew on the experience of its forerunner, the Association for the Improvement of Geometrical Teaching, which had been founded by Hirst and Wilson in 1871 to help abolish Euclid and teach geometry in a more practical manner. A Modern Languages Association and Geographical Association had similarly been founded in 1892 and 1893 to help teachers of these school subjects. In 1900, in order to counteract their feelings of loneliness in public schools, where classics were snobbishly thought to be superior to the sciences, the Association of Public School Science Masters (APSSM) was founded. Women teachers followed in 1912 with their own association. The Classical Association (1903), Historical Association (1906) and English Association (1907) completed the sextet of pre-First World War secondary school subject associations.

In 1919 the APSSM opened its ranks to other secondary school science teachers under the title Science Masters Association, and began the important journal *School Science Review* on the model of the *Mathematical Gazette* of the Mathematical Association. Since their foundations these professional organisations have played an effective role in pressurising for improved syllabuses and curricula, better and fairer examinations and laboratory facilities. In 1963, in the wake of Nuffield enthusiasm, men and women teachers

---

the psychologist. He assumes that he can deal with the human *psyche*, knowing nothing of his *corpus* ... [Section L] contributed nothing to the advance of knowledge in education — as usual it gave itself over to the tender mercies of freaks...', 'The British Association meeting, 1935', *The Chemical Age*, 1935.

35 Cyril Burt, 'Formal training', *School Science Review*, 20 (1938-9), pp. 653-66.

combined forces in the pressure group, the Association for Science Education.[36]

The APSSM was originally conceived by four Eton science masters, R. C. Porter, W. D. Eggar, M. D. Hill and H. de Havilland in 1900. They held their first meeting in London in 1901 under the chairmanship of Oswald H. Latter, a master at Charterhouse, with Sir Henry Roscoe (not Armstrong, who nevertheless welcomed the organisation) as their first President. One of Latter's pupils was Arthur Vassall, who taught at Harrow.[37] Latter, Hill and Vassall were the leaders of the APSSM and instrumental in getting biology taught in the public schools, the momentum of which was to lead to the 'General Science' movement in secondary schools.[38] It was Vassall, in a paper described by *Nature* as of 'remarkable lucidity', who made the APSSM aware of the defects of faculty psychology. Citing recent psychological opinion, he denied that powers of, say, observation, were increased through scientific studies, or that training in one scientific field 'overflowed' into others. From this he concluded that schools would have to cultivate wider areas of knowledge and interest and pay more attention to content and subject matter. However, Armstrong refused to accept the experimental finding that were the basis of Vassall's inference.[39]

Although Armstrong's discovery method aims were achieved positively in a few schools and, at a more advanced level, at his own Central Institution, his ideals were often debased into a heavy emphasis upon physical measurement and the physical sciences. The Ministerial code of 1904 had directed that instruction was to be given practically in two unspecified sciences. In practice, this came to mean chemistry and physics; astronomy was too esoteric, anatomy and physiology too indelicate or bound up with ethical problems of vivisection, and geology was still identified with religious controversy.[40] Armstrong had always urged syllabuses based upon a science of everyday life which would include elements of astronomy, geology and biology, besides physics and chemistry; but in many hands the enthusiastic probing for the inculcation of

---

36 David Layton, *Interpreters of Science. A History of the Association for Science Education* (London, 1984); Michael H. Price, *Mathematics for the Multitude? A History of the Mathematical Association* (Leicester, 1994).

37 *Special Reports on Educational Subjects* (HMSO, London, 1900), vol. 6, pp. 257-65.

38 Obituary of Latter, *School Science Review*, 30 (1948-9), p. 244.

39 *Nature*, 88 (1911-12), p. 394.

40 B. S. Cane, 'Scientific and technical subjects in the curriculum of English secondary schools at the turn of the century', *British Journal of Educational Studies*, 8 (1959), pp. 52-64.

'scientific method' left pupils with very little factual knowledge and a depressed scientific vocabulary. Dissatisfaction with this state of affairs found an outlet through the APSSM and the British Association's educational section.

At the British Association in 1902, C. W. Kimmins, Chief Inspector of the London County Council's Technical Education Board, praised an earlier British Association report on chemistry teaching and suggested that similar reports on other sciences were needed. 'The biological side should receive as much attention as that already given to the physical side' in schools;[41] too much time and money were wasted teaching sciences in separate compartments. In a paper on 'a rational curriculum' delivered in 1903, Armstrong largely agreed. There was no reason why biology should not be introduced and he hoped in future to find schools giving their pupils a broad view of nature, creating 'an interest in all that goes on around them'. However, 'physics and chemistry are the foundations ... of scientific belief,' he argued. 'They underlie all phenomena, all vital changes.'[42]

This theme of non-specialization, which had always been central to Armstrong's arguments, but overlooked in many circles, received further support in 1904 by J. H. Leonard, who again emphasised the absence of biology in most curricula,[43] and by Armstrong's friend and supporter Arthur Smithells (1860-1939), who presented a school syllabus in which 'scientific discipline and scientific method can be inculcated by simple experimental work, based entirely on matters of the household and of daily life; where the information acquired is truly useful knowledge.'[44] This syllabus owed much to Smithells's experience with and interest in Domestic Science for girls' schools. But it was T. E. Page of Charterhouse, the Secretary of the Incorporated Association of Assistant Masters who successfully launched a committee of the British Association to discuss the curriculum of secondary schools.[45]

Chaired by Oliver Lodge, with Stuart as its Secretary, the committee included Page, Sadler, Magnus and Armstrong. In 1907 it was joined by Richard Gregory, a former laboratory assistant of Worthington's, a textbook writer, staff writer for *Nature* and editor of *School World*.[46] It was Gregory who chaired a sub-committee, which included Latter (but not Armstrong) on the

41 *British Association Reports*, 1902, p. 844.

42 *ibid.*, 1903, p. 884.

43 *ibid.*, 1904, p. 844. For another perspective on the development of specialization, see A. J. Meadows, 'Specialization — the recurring debate', *Chemistry in Britain*, 9 (1973), pp. 504-6.

44 *British Association Reports*, 1906, p. 783.

45 *ibid.*, pp. 787-8.

46 W. H. G. Armytage, *Sir Richard Gregory* (London, 1957).

'sequence of studies in the science section of the curriculum'.[47] It found that although nature study was included in the first part of most secondary school curricula, this did not lead to a more detailed study of biology — except for intending medical stduents, as a post-matriculation subject. There was a lopsided emphasis on 'elementary physical measurements' and 'elementary heat', but little evident desire to tamper with this situation. Laboratory work was universal, though there were the usual complaints that examinations hindered or prevented heuristic methods from being used. The sub-committee found that opinion on the value of such 'problem' methods varied considerably, though their use was especially favoured in the early stages of teaching before a 'subject' approach was adopted.

Gregory's committee concluded that laboratory methods had been a great success, but there had to be both a greater freedom from examinations so that teachers could experiment with the curriculum, and a greater emphasis on the interdependence of the physical sciences. School work had to be brought into closer touch with everyday experience — a sentiment echoed again at the same meeting of the British Association by Armstrong and W. M. Heller (a former pupil) in their report on the curricula of elementary schools in which they tried to produce a better correlation between instruction in physical and biological science. So-called 'nature study', they argued, was usually an artificial subject and in its treatment of valueless objects 'a laborious elucidation of the obvious'. It was no substitute by itself for formal and systematic instrucion in practical physical science.[48] The physiology and biology of Heller's and Armstrong's syllabus appeared in the third and fourth years of their course, and rested on the foundation of physical sciences.[49]

Both Gregory's committee on secondary education and Armstrong's report on elementary schools effectively agreed that the sciences taught in schools must be correlated: the physical sciences with mathematics and biology. Most schools, however, continued to neglect biology. Through essays and the APSSM, Latter continually stressed the value of this science, and when London University tighened up its medical regulations and made biology a compulsory subject for their first MB examination (and so, effectively, for matriculation),

---

47 *British Association Reports*, 1908, p. 526.

48 *ibid.*, p. 509.

49 Armstrong and Heller were influenced by an experimental natural history course which had been developed by Marshall Ward in 1895. See E. W. Jenkins, 'Science, sentimentalism or social control. The nature study movement in England and Wales, 1899-1914', *History of Education*, 10 (1981), pp. 33-44; R. A. Baker and R. A. Bayliss, 'Louis Compton Miall, educator, 1842-1921', *Notes and Records of the Royal Society*, 37 (1983), pp. 201-34.

this was acknowledged as a triumph of APSSM policy — but disapproved of by Armstrong as a threat to non-specialization.[50]

When war came in 1914 Armstrong was overseas on a tour of the Far East. On his return he engaged immediately in offering unsolicited advice on how chemical products, especially dyestuffs which had hitherto been imported from Germany, could be obtained. He also played his part in the war mania by seeking out Fellows of German extraction within learned societies. Education was temporally forgotten.

Consequently, he played no role in the 'Science for All' campaign mounted in 1915 by a few teachers in the APSSM, led by M. D. Hill, which aimed to further the claims for a more general kind of science teaching. On 2 February 1916 *The Times* published a letter from the APSSM blaming the state of war on the neglect of science in British education. The latter soon led to the formation of 'The Neglect of Science Committee' whose most forceful members were the biologist E. R. Lankester, V. S. Bryant of Wellington School, and Richard Gregory.

Meanwhile, the APSSM had compiled a humanising syllabus entitled *Science for All* which effectively advocated what was soon afterwards called 'General Science.'[51] This, together with the resolutions on the scientific curriculum from the Neglect Committee, was presented to the 'Committee appointed by the Prime Minister to Enquire into the Position of Natural Science in the Educational System of Great Britain' which was appointed in August 1916 under the chairmanship of the physicist, J. J. Thomson. Simultaneously, Gregory persuaded the British Association to set up another committee on science in secondary schools, to be concerned with the essential place of science in general education. Armstrong, Sanderson, Vassall, and the dying Worthington, served on this committee which reported in 1917 in time to be noticed by the Thomson Committee.[52]

Science for all

The phrase 'science for all' was first coined in 1877 by the geographer and journalist, Robert Brown (1842-95), when he was compiling a monthly-parts encyclopaedia of science for the publishing house of Cassell, Petter and

50 A. L. Mansel, 'The influence of medicine on science education in England', *History of Education*, 5 (1976), pp. 155-68.

51 APSSM, *Report of General Meeting, 1916, with List of Members and Rules* (Harrow, 1917). The syllabus occupies pp. 32-43.

52 *British Association Reports*, 1917, pp. 123-207.

Galpin.[53] The latter had, for some years, found a ready market amongst British and Colonial readers for cheap illustrated works of useful knowledge and rational recreation concerned with popular science, geography, ethnology and exploration. Like most of their publications, *Science for All* appeared in monthly parts between 1877 and 1881. The parts, price 7d each, were to be bound up into five quarto volumes containing some 1,700 illustrations. Each issue consisted of a coloured frontispiece with 96 pages of illustrated text. An incentive to purchase the next issue was given by the fact that the last article in each part ended in the middle and was completed in the next number.

Brown was an old hand at this form of periodical journalism. An Edinburgh graduate with a doctorate from the University of Rostock, as a young man he had travelled around the world and been closely involved in the exploration of British Columbia during the 1860s.[54] Although he taught botany and natural history at the School of Arts, Heriot-Watt College in Edinburgh for some years, his disappointment in failing to obtain the Chair of Botany at Edinburgh University drove him to London in 1876. Here he took up full-time journalism as a staff-writer for *The Echo* and *The Standard*, and as a compiling editor for Cassell, Petter & Galpin, the market leaders for producing cheap works for the self-education of young readers and their families in an era of mass elementary education. By the time Brown launched *Science for All* he was already experienced in the editing of part works, having himself written *The Races of Mankind* (4 vols, 1873-76) and *The Countries of the World* (6 vols, 1876-81), both of which were in continuous production at the same time as *Science for All*.[55]

Brown's premise in editing such works for Cassell's was Johnson's aphorism that 'he who enlarges his curiosity after the works of nature demonstrably multiples the inlets to his happiness.' The science monographs of the day, Brown believed, were too technical for the average reader. Such an exclusive

---

53 *Science for All*. Edited by Robert Brown (Cassell, Petter and Galpin: London, Paris and New York), vol. 1 (1877); vol. 2 (1878); vol. 3 (1879); vol. 4 (1880); vol. 5 (1881). From internal evidence (British Library copies), the price per part was increased from 7d to 8½d as the work progressed. Reissued in 5 vols (1883-88) and in octavo (1890-94), with 90 pp. per part. A 'cheap edition' price 6d per part was issued in 5 vols, octavo by Cassell & Co. (London, Paris and Melbourne, 1895-97) and reissued 1899-1900. Note also a Swedish translation, *Vetenskap for Alla*, 5 vols, quarto (Stockholm, 1878).

54 See Brown's entry in *Dictionary of National Biography*.

55 Brown also serialised *The People of the World* (5 vols, 1882-85), *Our Earth and its Story* (a free translation of Kirchhoff's *Allgemeine Erdkunde*, 2 vols, 1887-88), and *The Story of Africa and its Explorers* (4 vols, 1892-95).

style was not to be adopted by the contributors Brown signed up for *Science for All.*

> We propose to provide food for our ... readers — food for the mind that feels an interest in every-day science, and science that can be understood by ALL... We shall not suppose that reader knows anything of science, and therefore will adopt the same method in explaining what is seen as the reader would do if he or she [*sic*] tried to find out all about it unaided. We shall lead our pupils on from the known to the less known, from facts which are familiar and patent to those which are less familiar and more abstruse. We shall show him the method of scientific reasoning by first pointing out the facts, and then asking him to agree with us or differ from us as to our interpretations of the information we have collected together. In other words, we shall adopt the natural method, taking our facts as we find them in nature, and reasoning from them, not commencing with theories, and then finding facts to support these foregone conclusions, which must first be taken for granted. Science is ever advancing, and our work, though in no sense a mere journal or newspaper, will be abreast of the latest discoveries and views.[56]

This inductivist rhetoric, which was responsible for giving so many laymen *and* scientists a totally false picture of the practice of science, was fortunately ignored by most of the contributors, some of whom, like R. A. Proctor, A. R. Wallace, W. F. Barrett and T. G. Bonney, were renowned for their skills in communicating astronomy, biology, physics and geology to the general public. However, the titles of the essays — 'Why the sea is salt', 'Rust', 'Voyages in Cloudland', 'Teeth', 'Moles and Mole-hills' — did tend to be object lessons and to reflect 'the science of common things' form of science teaching currently being replaced by problem-based laboratory science. For Brown, such essays were 'pegs on which to hang the scientific principles of which they were the embodiment'.[57]

Science for All was a phenomenal success. It was reissued in monthly parts, as well as annual volumes for the Christmas present market between 1883 and 1888; and again between 1890 and 1895. There was also a Swedish translation in 1878, while Brown, at the conclusion of the work, wrote of its favourable

56 *Science for All*, 'Introduction', vol. 1, pp. i-ii.

57 *ibid.*, 'Conclusion', vol. 5, pp. 374-6. Putting all the essays together, Brown concluded, formed an effective plan of 'the city of science'.

reception in America and the colonies. Cassells also issued a 'cheap edition' of the complete work on inferior paper, omitting the coloured illustrations, between 1895 and 1897, For this edition the price of the monthly parts was reduced to 6d. Some new articles were commissioned for the reissue where material had dated. For example, an 1877 essay by T. C. Hepworth entitled 'The light of the future' became in 1895 and essay by Lieut. Bradley A. Fiske on 'How electricity is produced': the future had become the present. There was a final issue of the cheap edition between 1899 and 1900.

The purchase of *Science for All* by school libraries and by science teachers undoubtedly ensured that the phrase was familiar to late Victorian and Edwardian readers. Although, unfortunately, there is no documentary evidence to make the link, it is not improbable that Brown's work was familiar to Arthur Vassall, the science master at Harrow School. Faced by classes of boys who, though good at classics, were dull at the kind of pure science he was then teaching, Vassall began to devise a more popular, less specialized and more general curriculum for boys below the age of sixteen. Possibly recalling the Cassell's series he had enjoyed as a boy himself, as we have seen, he began to call his syllabus 'Science for All'.

In passing, it is worth emphasising that, despite the development of specialization in science marked by publishers' catalogues of the 1870s, or by the scientific community's own International Science Series,[58] there was little decline in output of 'general science' books or of 'series' on the part of publishers. Cassells themselves went on to issue *Cassell's Popular Science* in 24 parts at 7d each between 1903 and 1904. Its editor, A. S. Galt, defended the popularisation of science as 'worthwhile' and attacked the assumption that 'popular science' involved 'inexact science'.[59] Havelock Ellis masterminded a 'Contemporary Science Series' with the Walter Scott Publishing Company in 1889. Ellis's strategem was to focus upon controversial growth points in science and the social sciences. By emphasising 'topical and controversial scientific subjects [the series] allowed its middlebrow audience to feel an intelligent part of scientific culture, and the key to this success was the series' assumption that the laymen's common sense enabled him to contribute meaningfully to popular scientific culture.'[60] Another significant series was

---

58 R. M. MacLeod, ' Evolutionism, internationalism and commercial enterprise in science: the International Science Series, 1871-1910', in A. J. Meadows, ed., *Development of Science Publishing in Europe* (Amsterdam, 1980), pp. 63-94.

59 *Cassell's Popular Science*, vol. 1, p. iii (1907). Cassells also issued multi-volumed, illustrated editions of a *Natural History* between 1870 and 1914.

60 Katherine Ring, *The Popularisation of Elementary Science through Popular Science Books c. 1870- c. 1939* (unpublished PhD, University of Kent, 1988), p. 76.

Popular Science edited by Arthur Mee in weekly parts between 1912 and 1913, and published by Alfred Harmsworth's Amalgamated Press. Each part, priced at 6d, contained essays covering cosmology, geology, biology, psychology, physiology, technology, sociology and eugenics — the latter being 'the new science that would control the future and create a nobler race'.[61]

The syllabus *Science for All*, which received wide distribution after the war through publication in C. L. Bryant's *School Science Review*,[62] was largely the work of W. D. Eggar of Eton, F. M. Oldham and Arthur Vassall. Later it was thought to have been written deliberately in opposition to heurism, but this was denied by Bryant.

> Indeed, we welcomed [Armstrong's] campaign in so far as it supported our claims for more time and equipment for practical work in the schools, but we thought that the dear old man was too insistent on one method of teaching; we did not see why every child should walk around Ireland before he could believe it was an island. It fell to my lot to tell him so. He didn't like it, not did I; but we remained friends all the same. *Science for All* was neither for or against Armstrong ... It just happened that the pamphlet was compiled at a time when relations between us were a bit strained.[63]

The strain was evidently over the APSSM's acceptance of the discreditability of faculty psychology, the Association's belief that heurism needed to be tempered with a more informative approach, and, above all, its view that biology should share equally with the physical science and not form a lame coda to a science syllabus. Specialized physics and chemistry lessons were not suited to youths whose *forte* lay in classics or the humanities, or who were simply average pupils. Hence, if one shared Jevons's belief that science was liberalising, humanising and useful knowledge in an increasingly science- and technological-based society (as, to their credit, most members of the APSSM did), it was clear that to make science available to all, a rather different, more general, form of science teaching would be necessary. Specialization was bad — science was to be taught as part of the general education of everyone and not as if it was to be the vocation of everyone but a few specialists: of course,

61 *Popular Science*, vol. 1, p. 7.

62 *School Science Review*, 2 (1920), pp. 197-212. See C. L. Bryant, 'Fifty years on'. *ibid*, 32 (1950-51), pp. 140-45. Vassall implied that he described his Harrow course as 'science for all' in *British Association Reports*, 1917, p. 149.

63 *School Science Review*, 32 (1950-51), pp. 143-44.

'everyone' here referred only to boys in the major public schools! Science had to appeal to the *imagination* as well as to the reason; this was something that 'the skilfully articulated skeleton of heurism' could not do.

> Does science fulfil the highest aims of education? Does it... rouse the imagination and kindle latent genius, or does it only develop talent? Does it cause young men to see visions... or only put studious boys in the way of earning a comfortable living? These are the alternatives on which depends the appropriate status of the subject.[64]

Science for All proved a successful strategy to give public school science masters status within their schools and science a firm place within the curriculum. Ironically, it also gave teachers a broader base upon which to build a sixth form of boys who would opt for scientific degrees at universities, thus aiding the very specialization *Science for All* had been designed to combat. But in achieving status, and despite a modicum of utilitarian information in the *Science for All* syllabus, its stress upon science as a form of high culture inevitable devalued technology. Indeed, when President of the Science Masters Association in 1920, Archer Vassall went so far as to deride students who 'desert science for industry' as interested only in commercial gain, not in the search for truth. As Mansell has commented, in this respect, 'a rank order had been established, sciences for the gentlemen, technology for the players'.[65] This would have mattered little but for the fact that after 1902 the public schools tended to be the models for the municipal grammar schools which also tended, therefore, to devalue the utilitarian aspects of science. *Science for All* was, in fact, science only for the upper and middle classes and the few working-class children who climbed the scholarship ladder.

Science for All included six tentative syllabus suggestions in which biology figured prominently, and which moved from cosmology to the gramophone. Suggestions for less able boys in the Lower School included an experimental investigation of chalk. Complete freedom was left to the teacher to decide on the precise content and method, but insofar as the syllabus was more extensive than any proposed before, proponents of General Science inevitably tended to favour teacher demonstrations rather than individual pupil experiments.

---

64 *ibid.*, 2 (1920-21), pp. 197-212.

65 A. L. Mansell, 'Science for all? Science education 1850-1920', *School Science Review*, 57 (1976), pp. 579-85; see p. 584.

Thomson's report on the teaching of natural science.

Gregory's British Association Committee reached similar conclusions which it saw as the original aims of F. W. Farrar's Committee of 1866. The most interesting part of their unanimous report concerned the method of science teaching, for it revealed that Armstrong accepted the fallibility of faculty psychology as well as the criticism of form and content in many heuristic syllabuses. The view had gained ground, they reported, that 'all subjects, in different ways and to different degrees, can be made to give a training in scientific method'. However, if it were admitted that the sole task of a science teacher was to train pupils in scientific method and that this method was common to all well-conducted intellectual inquiries, there would be no need for special science teachers!

> This paradoxical conclusion depends upon the assumption that the method of scientific investigation can be regarded as separable from the matter, which is not correct. ... No-one ought to expect a training in scientific method acquired in one field of inquiry to be transferable to — that is, to guarantee competence in — a field substantially different from the former.[66]

To Armstrong's disgust, Thomson was to take such an argument as evidence of the fallibility of heurism.

Despite the inflation of the cost of laboratory apparatus, the Committee stresses the unique value of laboratory work.[67] It brought a pupil 'into direct contact with reality through his own senses and his own manipulation', and tested his reasoning powers in ways that the best manual training could not. But should laboratory work be by the subject-method (demonstration exercise) or problem-method (heurism)? The latter method they stressed was not intended, 'as is sometimes supposed, to make pupils discover for themselves laws and principles previously unknown to them, though to some extent this can be done, but rather to provide a continuous thread of reasoning for the practical work and a definite purpose for whatever is undertaken.'[68] The method was acknowledged as difficult; freedom to use both methods independently of external examinations was needed. Unfortunately, however, there had been a tendency for practical work to limit the scope of science

66 *British Association Reports*, 1917, p. 134.
67 *ibid.*, p. 137.
68 *ibid.*, p. 139.

courses. 'Increased attention to laboratory exercises has ... often been associated with a very restricted acquaintance with the world of science. The tendency has been to make all teaching a matter of measurement, to the neglect of the human aspects of the pursuit of natural knowledge.' This ironic effect of Armstrong's success — making content 'narrow and special rather than broad and catholic' was not lost on Thomson.

Finally, Gregory's committee was concerned with the human aspects of science and, no doubt urged by Gregory and Sanderson, it recommended lessons in the history of science and the discussion of modern scientific issues in schools. At all times scientific instruction was to be related to the affairs of everyday life. Thomson also appreciated this point. Syllabuses recommended by Vassall, Sanderson (Oundle), T. Percy Nunn, and Armstrong (a course on practical food studies) were also reproduced. All the courses included elements from biology as the science which closely concerned human life.

Thomson's Committee, upon which school teachers were not represented, reported in 1918. All the evidence presented to it had suggested that school science courses had become too narrow, both because of the undue emphasis on physics and chemistry, and the undue restriction of subject matter. 'Principles are often taught without reference to the phenomena of nature which they explain; the course does not satisfy the natural curiosity of the pupils; it may give them some knowledge of laboratory methods, but little idea of wider generalizations, such as the principle of conservation of energy, which are quite within their powers of comprehension.'[69] Too many schools planned their courses for the specialized study of science by senior boys — to the disadvantage of those who left school early. The growth of laboratory teaching had been of the greatest value. Unfortunately, however, 'many teachers have become so dominated by the idea of the supreme value of experimental work that they have left on one side and neglected those sciences which do not lend themselves to experimental treatment in school; the tendency has been to restrict the work to parts of physics and chemistry in which the boys can do experiments for themselves.' Laboratory work was essential,

> but sometimes the performance of laboratory exercises has been considered too much as an end in itself — such an exercise loses the value of a real experiment when it becomes a piece of drill; often exercises succeed each other without forming part of a continuous or considered scheme for building up a boy's knowledge of his subject. ... Insistence on the view that experiments by the class must always be

69 J. J. Thomson, *Natural Science in Education* (HMSO, 1918), p. 41.

> preferred to demonstration experiments leads to great waste of time and provides an inferior substitute.[70]

So far, there was little for Armstrong to quarrel with; he had never denied the value of demonstration experiments, and he was aware of the limitations and enthusiasms of teachers. However, Thomson directed these limitations towards an attack on the heuristic method.

> Much of this waste of time is due to a conscientious desire of the teachers to encourage the spirit of inquiry by following the so-called heuristic method; the pupils are supposed to discover by their own experiments, with little or no suggestion from the teacher, the solutions of problems set to them or of problems which they themselves suggest. The spirit of inquiry should run through the whole of the scientific work, and everything should be done to encourage it, but it seems clear that the heuristic method can never be the main method by which the pupil acquires scientific training and knowledge. He cannot expect to rediscover in his school hours all that he may fairly be expected to know; to insist that he should try is to waste his time and his opportunities (*ibid*).

Here was a caricature of Armstrong's intentions. For him heurism meant 'directed inquiry', with the teacher playing a discrete, but positive, role. The method was not meant to be the means by which stores of information were acquired, nor was it to replace entirely other methods of instruction and learning. But even though Thomson did not include these castigations in his important official recommendations, the damage was done.

Not surprisingly, Armstrong was furious. When the report was discussed at the British Association in 1919 he dismissed it categorically as worthless (which it was not) and unlikely to influence educational opinion (which it did).[71]

Gregory, who is sometimes portrayed as the enemy of heurism, in fact strongly supported it provided there was not an over-concentration on method, and that laboratory work was complemented and supplemented by a broad general course of lessons. Nevertheless, it proved to be Gregory's, and not

---

70 *ibid*., p. 42.

71 'Education at the British Association', *Nature*, 104 (1919), pp. 521-2. Cf. the anti-heurism of Ida Freund, *The Experimental Basis of Chemistry. Suggestions for a Series of Experiments Illustrative of the Fundamental Principles of Chemistry* (Cambridge, 1920).

Armstrong's hour.[72] Training in experimental, scientific method, Gregory recommended, was to be separate from a 'general course of science' based upon descriptive lessons and reading. Supported by Thomson's major recommendation that 'natural science should be included in the general course of education of *all* up to the age of about sixteen', Gregory used the British Association, *Nature* (which he now edited), and the *School Science Review* as platforms to urge the teaching of 'general science' to all children who were unlikely to attend university.[73] Laboratory work was to be 'different in nature and intention from work that can be regarded as manual training, or from measurements which exemplify purely mathematical principles. It must be used to illustrate subjects dealt with in the classroom — the "Science for All".'[74]

Meanwhile, Vassall, the other arch-apostle of general science, attacked those who would 'limit the function of science in schools to one end ... a training in inductive reasoning from experimental data obtained by the pupils themselves'.[75] His three aims of science teaching were the inculcation of method and a critical faculty, knowledge of science, *and* the stimulation of imagination and the aesthetic senses. On another front, through the influence of scientists like Harold Hartley, physical chemistry, including the ionic theory that Armstrong despised, penetrated the schools and proved an elegant way of using theory to order and unify practical chemistry classes.

General science

In 1924 the Science Masters Association (which the APSSM had become) published its pamphlet 'General Science' which contained a revised syllabus of *Science for All*, and whose recommendations were accepted for the Lower School Certificate examination by most examining boards. Two years later, the publisher Ward Lock reused the old Cassell title of *Science for All* when they engaged the English neuro-physiologist, Sir Charles S. Sherrington, to edit an anthology of essays by leading scientists on the current state of science. This *Science for All. An Outline for Busy People* lacked the illustrative multi-volumed appeal of the Victorian title, and was frankly a pot-boiler. Nevertheless, published in both London and Melbourne, it is known to have

---

72 *British Association Reports*, 1919, p. 354.
73 D. Thompson, *School Science Review*, 40 (1958), pp. 109-22.
74 R. Gregory, *ibid.*, 1 (1919-20), pp. 93-4.
75 *ibid.*, 2 (1920-21), p. 241.

inspired one woman, Joan Freeman, to take up a career in physics.[76] *Science for All* (1926) was part of a popular series, 'Books for All', which included others on *Wireless, Railways* and *Ships*.[77] A fourth edition, revised completely in his own words by John T. Hogg, appeared 25 years' later under the title *Science for All. Beginner and Student*.[78] Despite its unexciting prose and format, the book kept the phrase 'Science for All' in the educational market place at a time when the Science Masters Association was looking forward to the post-war reconstruction of schools, curricula and examinations.

Inevitably, because the SMA had been originally involved in the republication of *Science for All* in 1920, it became an obvious generic title for a series of textbooks commissioned by George Henry Joseph Adlam, a science master at the City of London Boys' School and editor of the *School Science Review*. Adlam launched the 'Science for All' series in 1921 with his own *Chemistry*. Four years later he wrote a *Practical Companion to Chemistry* (1925). *Chemistry* had been followed by *Geology* (1923) by C. I. Gardiner of Cheltenham College, and a four volume *Physics* by W. J. R. Calvert of Harrow. Calvert's text, which is interesting for the attention it pays to drawing apparatus and describing experimental results, was probably closest in spirit to Vassall's Harrow 'Science for All' project. Calvert's first volume appeared in 1924; the others on *Sound, Light* and *Electricity and Magnetism* all appeared in 1929, by which the time the series had been transformed into a 'General Science Series'. This change of title was clearly mooted in 1924 when the original series issued the first volume of *A Textbook of General Science* by R. Acton, subtitled *Introduction to Physics*. This explicitly alluded to the Thomson Report's recommendation that more attention should be paid to the 97 per cent of boys who did not go on to university to read science. The second volume of this general science textbook, co-authored by G. H. J. Adlam and O. H. Latter, vividly illustrated the problem of cohesion and integration inherent in such an interdisciplinary subject — *Introduction to Chemistry, Geology, Biology* (1924). Latter's own contribution to this was expanded into a separate *Biology*

---

76 Joan Freeman, *A Passion for Physics. The Story of a Woman Physicist* (Bristol, 1991), p. 29. Cf. the case of the chemist, historian and museum curator, Frank Sherwood Taylor (1897-1956), cited by A. V. Simcock, *Ambix*, 34 (1987), p. 121.

77 Sherrington's 384 pp. volume lacked an essay on chemistry — something Brown would never have allowed.

78 John T. Hogg, *Science for All. Beginner and Student*, 4th ed. (London and Melbourne, 1950). Note also W. Mitchell, ed., *Science for All* (Adelaide, 1937), which included popular science articles form *The Advertiser* (Adelaide); and the National Book League, *Science for All. An Annotated Reading List for the Non-Specialist* (Cambridge, 1958).

in 1926. This volume completed the original set of titles appearing as part of the series 'Science for All'. All subsequent reprints appeared as part of the 'General Science Series'.

Another revised edition of *General Science* was published in 1932. Yet, when the British Association investigated in 1933 it found that, despite support for the syllabus by the Hadow Committee on the education of adolescents in 1926, the subject was only taught in 37 per cent of boys' schools, 40 per cent of girls', and 42 per cent of coeducational schools.[79] The perennial problem with general science syllabuses proved to be one of correlation; few teachers felt competent to teach the whole syllabus, and it was difficult to teach without compartmentalising into discrete blocks of chemistry, physics and biology.[80] Texts, such as E. N. da C. Andrade and J. S. Huxley's *An Introduction to Science* (1933), failed to solve the problem. Part of the trouble, perhaps, was that between 1920 and 1940 unemployed scientists went into teaching in large numbers. Trained as specialists, they rebelled against heurism and the general science movement and campaigned for their subject specialisms. They saw themselves as scientists *manqué*, not as science teachers. For these reasons, as examination statistics reveal, the subject was never popular in the schools, though it undoubtedly stimulated the development of biology teaching. Indeed, by 1960 campaigns for General Science were quietly dropped by the Science Masters Association.[81]

Heurism redivivus

In 1929, F. W. Westaway published his outstanding and influential book on *Science Teaching*. One of his aims was to increase the stock of biology teaching. Whilst he was appreciative of Armstrong's achievements, he was explicit in rejecting heurism. Westaway, who believed heurism too slow and unworkable, suspected that Armstrong had never revealed the true secret of his method, or that he was aware of its true nature himself: sheer personality; 'his rather tart impatience towards his students, his refusal to help them one iota more than is absolutely necessary, his amazingly clever and ever-ready questions to meet the needs of the moment, his resourcefulness under all experimental difficulties, his untiring help to his students, his rather "grumpy

---

79 W. M. Heller, 'The advancement of science in schools', *British Association Reports*, 1932, pp. 209-28; and 'General science in schools', *ibid.*, 1933, pp. 312-30.

80 *School Science Review*, 2 (1920-21), pp. 323-6.

81 'Report of the discussion on resolution on general science', *ibid.*, 40 (1958-59), pp. 523-8; E. W. Jenkins, *From Armstrong to Nuffield* (London, 1979), chapter 3.

manner", and his rare words of praise'.[82] The implication was, then, that only teachers with extraordinary personal magnetism could use heurism effectively, and that demonstrations were probably better for class discipline anyway.

Such curt dismissal hurt Armstrong. To him heurism was a practical system which worked. It was pedagogically sound and tested by his experiences. His own syllabuses did not have all the faults found by Westaway or the general science followers. He covered subjects broadly and systematically and at the level appropriate to the child. Good teachers could be found or trained, as his experience with the Board Schools showed. Complaints about incompatibility with examinations were invalid; he would have changed the whole system of examinations anyway. Objections as to cost were exaggerated and compensated by the need for fewer textbooks. A seminar on heurism published in the same year as Westaway's book suggested that its spirit was not entirely crushed.[83] Yet, not surprisingly, by 1937 the heuristic method, as originally conceived, had vanished, killed by the examination system, and the collective criticisms of the new psychology, the general science movement, and brilliant writers like Westaway; while as another war and the need for fresh stringent economies closed the 1930s, some teachers again questioned the British emphasis on, or obsession with, laboratory teaching.[84] But small cells of heuristic teaching continued, notable at Christ's Hospital with Gordon van Pragh and John Bradley, who began to construct a continuous syllabus of practical problems linked together by the continuity of materials or by the comparison of substances. Bradley's philosophically and pedagogically stimulating writings in the *School Science Review* were influential in chemistry teaching after Armstrong's death.[85]

---

82 F. W. Westaway, *Science Teaching. What it Was - What it Is - and What it Might Be* (London, 1929), p. 27.

83 'The heuristic method — a symposium', *School Science Review*, 11 (1929-30), pp. 65-70.

84 See the debate in *Nature*, 145 (1940), p. 863 (A. G. Lowndes), p. 1023 (V. J. Chapman); 146 (1940), p. 133 (Lowndes); also, L. Conell, 'Demonstrations and individual practical work in science teaching: a review of opinions', *School Science Review*, 52 (1971), pp. 692-701; and compare Jack Kerr's conclusion that the British emphasis on practical work had made it 'inflexible, repetitive, outmoded, and often inadequately integrated with the theory', in his *Practical Work in School Science* (Leicester, 1963), p. 95.

85 J. Bradley, 'Bread or stones of Nuffield chemistry', in *Black Paper Three* (London, 1970), pp. 87-90; *Education in Chemistry*, 4 (1967), pp. 58-64. Bradley's interesting heuristic scheme was published in nine parts: *School Science Review*, 44 (1963), p. 549 ff; 45 (1964), p. 365 ff; 46 (1964), p. 126 ff; 47 (1965-66), p. 65 ff and 72 ff; 48 (1966-67), p. 467 ff; 49 (1967-68), p. 467 ff; and 50 (1968-69), p. 87 ff. See also Gordon van

Armstrong first collected together his educational writings for publication by Macmillan in 1903. He saw his *Teaching of Scientific Method* as a critical contribution to the movement to modernise education following the 1902 Education Act. Arranged in non-chronological order, with the constructive essays on heurism and its syllabuses following didactic essays which castigated the educational system, the work's impact was weakened by its lack of coordination. Smithells rightly thought it a pity that Armstrong had not distilled his thoughts into a monograph.[86] A second edition appeared in 1910 and was reprinted in 1925 when Armstrong opened the book with a new prefatory essay reviewing the changes which had taken place a quarter of a century after 'the winter of our discontent'. It had been a period of searching criticism and reappraisal, but the rooted individualism of the British character had insured that nothing was organised and that little use had been made of past experience. Consequently, although we spent more on education than any other country, there was little to show for it. 'But worse than this — by allowing the academic party to control the work, we have suffered the foundations of our old pre-eminence to be sapped and undermined in every direction by constant dogmatic teaching and a lavish system of enforced lesson learning supplemented by inelastic and inappropriate examinations.'[87]

The outlook for the future was not bright. The chances of our changing our ways was slim. We were living in a new technological society, but it was difficult for people to recognise the value of the language of science 'when those who are called on to introduce it and supervise the teaching have no true knowledge of the subject in which they are to give instruction and little understanding of its beauties and value.'[88] This situation would continue until a far broader training programme was introduced into schools, until the phrase *experentia docet* was adopted in all schools, until the faculty of criticism was cultivated rather than acquisitiveness, and until, paradoxically, we learnt to rate theoretical knowledge much higher, as other nations did.

In 1938, a year after Armstrong's death, the *Spens Report* which launched the idea of the tripartite organisation of secondary school education that was implemented in 1944, gave only a lukewarm appraisal of science in the

---

Praagh, *The Teaching of Science at Christ's Hospital*, privately printed, the author, Crawley, 1992).

86 H. E. Armstrong, *op. cit.* (3). See A. Smithells, *Nature*, 69 (1904), pp. 289-90.

87 Armstrong, *op. cit.* (3), p. xviii.

88 *ibid.*, p. xix.

curriculum.[89] Harking back to the Thomson Report, it strongly advocated the general teaching of science (p. 177) to the 85 per cent of teenagers who left school before the age of sixteen and urged that, by analogy, algebra, geometry and trigonometry should cease to be separate compartments of mathematics (p. 235). As it transpired, the Spens Report's idea of treating 'mathematics as a science in which the topics should be suggested and introduced by the examination of practical questions which in their day have been of urgent interest and utility to man in his affairs' (p. 238) had little impact, though it was to become the hallmark of the 'new mathematics' in the 1960s. For the Spens Committee, the chief aims of science teaching were threefold: knowledge of the laws of nature, awareness of the methods of scientific thought and investigation of their relative value in different circumstances; and, complementing historical studies, a science teacher could demonstrate the influence that scientific thought and achievement had had, and would have, in the evaluation of civilisation (p. 245). This emphasis upon 'the qualities, both moral and intellectual, which go to make up the devoted scientist' had the effect, like heurism itself, of stressing the methodological quality of the scientific endeavour rather than content.

Such an image of science and the scientist as logical, objective and bias-free was explicit in the sociology of science that was developed by Robert K. Merton in the 1940s: scientists displayed universalism, communitarianism, disinterestedness and organised scepticism.[90] Practical lessons or learning by discovery, it was argued in the Nuffield science curriculum projects of the 1960s, would inculcate these attributes in pupils. Even pupils who were not bent upon scientific careers would benefit from exposure to these attitudes — for, once again, it was assumed that such traits would be transferred by pupils to other contexts.[91]

As Mary Waring and others have pointed out, the Nuffield Foundation Teaching Project, announced by the Minister of Education on 4 April 1962, was 'a major landmark in English education practice' — the first attempt on such a large scale to obtain widespread reform of *both* content *and* teaching approach in science and mathematics by encouraging teachers to develop an 'articulated and comprehensive set of tested teaching materials'. These were,

---

89 *Report of the Consultative Committee on Secondary Education with special reference to grammar schools and technical high schools* (HMSO, 1938), pp. 39-41. The chairman was Will Spens.

90 R. K. Merton, *The Sociology of Science. Theoretical and Empirical Investigations* (Chicago and London, 1973).

91 Robin Millar, 'Training the memory; continuity and change in the rhetoric of school science', *Journal of Curriculum Studies*, 17 (1985), 369-82.

so the argument ran, to be resource materials, not pre-packaged courses; resources from which classroom teachers in both grammar and secondary modern schools, as well as the growing number of comprehensive schools, could draw relevant and up-to-date materials in order to develop 'some insight into scientific thought and method' for both future science specialists and non-scientifically minded early school leavers.[92]

The two most remarkable features of the Nuffield Ordinary Level examination syllabuses for chemistry, physics and biology were their emphasis upon theoretical models and learning by experimental discovery.[93] For example, the Ordinary-level chemistry syllabus devised between 1962 and 1967 by H. F. Halliwell, with advice form the inorganic chemist, Ronald Nyholm, the chemistry textbook writer and examiner, E. H. Coulson, and Armstrong's Christ's Hospital School disciple, Gordon van Praagh, interpreted chemistry as 'process and product'. Pupils' powers of imagination, detective work and hunch-making were to be stimulated and encouraged by heuristic open-ended experimental work which was to be explained by the use of the concepts of energy and the gram-atom (mole). Complementing this process of investigation and explanation were the inculcation of the necessary degree of mathematical competence and the introduction of any relevant history of chemistry, and it relations with industry and the social improvement of mankind.

Although some of the roots of the Nuffield revolution can be traced back to Armstrong's *virus heuristicum*, it also borrowed freely from the post-Sputnik curriculum innovations of American science teachers who had, in their turn, found inspiration in the work of Dewey, Piaget and others, as well as in the rhetoric of 'learing by doing'. Chemistry, if not science, was for all.[94]

---

92 This paragraph summarises Mary Waring, *Social Pressures and Curriculum Innovation* (London, 1979), p. 3. The influence of the 'two cultures' debate instigated by C. P. Snow in 1959 is also apparent in the Nuffield movement. See C. P. Snow, *The Two Cultures; and a Second Look* (Cambridge, 1964).

93 Waring, *op. cit.* (92), chapter 5. For example, Nuffield O-level *Physics. Teachers' Guides* (London, 1966). The schemes for Combined Science, Secondary Science and Junior Science, together with the A-level syllabuses in Chemistry, Physics, Biological Sciences and Physical Science developed between 1964 and 1972, were similarly transformed.

94 See Susan Isaacs, *Intellectual Growth in Young Children* (London, 1930); R. J. Merrill, *The CHEM Study Story* (San Francisco, 1969), which describes an American curriculum project for High School chemistry students of 16-17. The Nuffield Foundation's *Physics. Teachers' Guide III* (London, 1966), pp. 1-2, referred to 'physics for all', rather than chemistry. For general orientation, see W. H. Brock, *Fontana History of Chemistry* (London, 1992), chapter 11.

In practice, teachers rapidly found that Nuffield heurism only worked amongst brighter pupils. Among the majority of pupils, on the other hand, apart from the perennial problem of discipline in a large laboratory class, pupils either got the 'wrong' results or wormed the 'right' answer by deft questioning of the teacher. Closer scrutiny of classroom learning also showed that children brought their own 'world view' preconceptions and intellectual baggage into the laboratory.[95] Two other factors undermined the credentials of discovery methods in the late 1960s: new ideas on the psychology of learning, and a new critical philosophy and sociology of science. Both had the effect of aiding the construction of syllabuses and classroom/laboratory practice which would create 'science for all' rather than 'science for the few'.

In October 1976, James Callaghan's Labour government launched a 'great educational debate' in an attempt to promote a greater awareness of the nation's economic dependence upon industry.[96] Callaghan specifically urged that science teachers should pay greater attention to technology and encourage more children towards industry. The implication was clear: the teaching of secondary school science needed revision again.[97] Although the Department of Education and Science did not use the phrase 'science for all', it did recommend that '*science* should find a secure place *for all* pupils at least to the age of 16'. But since much secondary school science was too specialised for all pupils, so that many opted out of science classes at the age of 14, it followed that new, more general, science courses would have to be devised.

These conclusions were echoed by the Association for Science Education's newly-formed Education (Research) Committee in the same year when its working party formulated a policy for science education in the 1980s.[98] In evidence on 'The Attainment of the School Leaver' to the cumbersomely-named Education, Arts and Home Office Sub-Committee of the House of Commons Expenditure Committee, the Association for Science Education drew attention to how the changing nature of society was affecting science teaching.

---

95 J. L. Wellington, 'What's supposed to happen, sir? Some problems with discovery learning', *School Science Review*, 63 (1981), 167-73. Rosalind Driver, *The Pupil as Scientist?* (Milton Keynes, 1983); Robin Millar and Rosalind Driver, 'Beyond processes', *Studies in Science Education*, 14 (1987), pp. 33-62.

96 Department of Education and Science, *Education in Schools. A Consultative Document*, Cmnd 6869 (HMSO, 1977).

97 Layton, *op. cit.* (36), chapter 10.

98 Brian E. Woolnought, *Physics Teaching in Schools 1960-85. Of People, Policy and Power* (London, 1988), p. 146.

As society became more egalitarian (they were referring to comprehensive schooling and to the abolition of the 11-plus examination system), the methods of teaching and the curricula were changing too because more mixed ability teaching was being demanded. There was now, in 1977, experimentation with combining science subjects and a greater emphasis on 'inquiry into and critical thought about science rather than the mere acquisition of factual knowledge'.[99] Science, the Committee emphasised, was an essential component of the curriculum up to the majority school-leaving age of sixteen because of 'three distinctive, but related, contributions to general education [made by science]: science for the enquiring mind;[100] science for action; and science for citizenship'.[101] Over-specialization in secondary schools had always obscured their contributions to general education, and to remedy this, there ought to be a balanced core curriculum for all pupils until sixteen. '*Science* should be an essential component of the education of *all* people.'[102]

It was in this context that the old public school science masters' phrase 'Science for All' was refurbished for the 1980s and 1990s. In November 1977, for example, F. A. Bollen, a science adviser in Newcastle-upon-Tyne, queried whether the Green Paper's recommendation of 'science for all' necessarily had to mean 'unified science;'[103] and when the Association for Science Education's working party, led by the Oxford educationist, J. J. Thompson, issued its own radical consultative document, *Alternatives for Science Education* in December 1979, the phrase was adopted unilaterally and without comment.[104]

Following hard on its controversial document, and continuing to operate from the premise that science must form part of the core curriculum, the Association for Science Education issued a definitive policy statement,

99 *Education in Science*, 74 (September 1977), p. 16.

100 This phrase came from Eric M. Rogers' widely-read textbook for non-scientists, *Physics for the Enquiring Mind* (Princeton and Oxford, 1960), which formed the basis for much Nuffield O-level physics. Rogers, who had taught at Charterhouse, became Professor of Physics at Princeton University where, in 1956, he became involved in the Physical Science Study Committee (PSSC) reforms of physics education in American High Schools.

101 *Education in Science*, 74 (September 1977), p. 16. Cf. L. Hogben, *Science for the Citizen* (London, 1938).

102 *Education in Science*, 74 (September 1977), p. 18, my stress.

103 *ibid*., 75 (November 1977), pp. 25-6.

104 Association for Science Education, *Alternatives for Science Education. A Consultative Document* (Hatfield, 1979), pp. 50, 53. Cf. Diana Manuel, 'Qualified to teach?' *Education in Science*, 86 (June 1980), pp. 26-8, where she argued that the phrase had yet to be subjected to serious analysis.

Education through Science in May 1981.[105] Here the pedagogic implications of 'science for all' were finally articulated: the levels of self-evaluation of science teachers would have to be raised as they were confronted 'by highly divergent ability groups' (p. 28), including recalcitrant fourteen-years olds to whom science had been anathema or impossibly difficult; consequently, teachers would have 'to adopt a wider and more flexible range of teaching styles' (pp. 28-9); more attention would need to be paid 'to the role of written and spoken languages in affective science teaching' (pp. 29-30) to make science less abstract and intimidating and closer to pupils' everyday personal experiences; and, as the APSSM had discovered seventy years' before, but now within the context of the computer age, closer attention would need to be paid to the correlation between mathematics and science in, and across, the curriculum (pp. 30-1).

Despite, and indeed because of, the new National Curriuclum which, with its battery of attainment targets, began to operate in September 1989, and which insists upon a core of scientific subjects up to the age of sixteen, there remain tensions and antagonisms within the English, Welsh and Northern Ireland state secondary systems between advocates of separate science discipline teaching and integarted science for pupils under the age of sixteen.

It will be interesting to see how in the remaining years of this century the now dominant integrated 'investigative, applied and problem solving' form of science teaching set in a personal, social and technological context, will work out in British classrooms. And how 'science for all' first mooted by Victorian popular educationists and Edwardian science teachers will affect the independent schools which are clearly destined to survive amidst the more numerous, but less prosperous, state comprehensive system which together aim to provide secondary eductaion for all in our stratified nation.

---

105 *Education through Science. Policy Statement* (Hatfield, 1981).

ADDENDA ET CORRIGENDA

I See in addition, Peter Alter, *The Reluctant Patron. Science and the State in Britain 1850-1920* (Berg: Oxford, Hamburg, New York, 1987)

II The definitive account of the early years of the British Association appeared at the same time as this essay: J. B. Morrell and A. Thackray, *Gentlemen of Science. Early Years of the British Association for the Advancement of Science* (Clarendon Press: Oxford, 1981).

V Compare John M. Robson, 'The fiat and finger of God: The Bridgewater Treatises' in R. J. Helmstadter and B. Lightman, eds., *Victorian Faith in Crisis* (Macmillan, 1990), pp. 71-125; and the literature cited in the Bibliography accompanying John H. Brooke, *Science and Religion. Some Historical Perspectives* (Cambridge University Press: Cambridge, 1991).

VII For a more general context, see D. E. Allen, *The Naturalist in Britain. A Social History* (Allen Lane: London, 1976; reprinted Princeton University Press: Princeton, NJ, 1994), together with my review, 'A taste for naturalists', *History of Science*, 15(1977), pp. 287-294.

XI For further context, see W. H. Brock, ''Building England's first technical college: the laboratories of Finsbury Technical College, 1878-1926' in F. A. J. L. James, ed., *The Development of the Laboratory* (Macmillan: Basingstoke, 1989), pp. 155-170.

XII p. 24, line 7, *for* work *read* word.
p. 24, line 41 *for* prossessed *read* possessed.
p. 25, line 11 *for* discouragemnt *read* discouragement.
p. 25, line 40, *for* 1878 *read* 1873.
p. 26, paragraph 4, the British Association Committee was established in 1869, not 1871 as stated.
p. 27, line 15, *for* syllabus *read* text.

See also W. H. Brock and R. M. MacLeod, eds., *Natural Knowledge in Social Context. The Journals of Thomas Archer Hirst FRS*, microfiches edition (Mansell: London, 1980); A. G. Howson, *A History of Mathematics Education in England* (Cambridge University Press: Cambridge, 1982), chapter 7 on James Wilson; Joan L. Richards, *Mathematical Visions: the Pursuit of Geometry in Victorian England* (Academic Press: Boston, 1988); M. H. Price, *Mathematics for the Multitude? A History of the Mathematical Association* (The Mathematical Association: Leicester, 1994).

XIII See M. H. Price, 'The Perry movement in school mathematics' in M. H. Price, ed., *The Development of the Secondary Curriculum* (pub: London, 1986), pp. 103-155; M. H. Price, 'Mathematics in English education 1860-1914: some questions and explanations in curriculum history', *History of Education*, 12(1983), pp. 271-184.

XVI For more details on Dyer, see Olive Checkland, *Britain's Encounter with Meiji Japan, 1868-1912* (Macmillan: Basingstoke, 1989). Graeme Gooday has also stressed the importance of Ayrton's and Perry's Japanese experience for the establishment of practical physics teaching in England. See his 'Teaching telegraphy and electrotechnics in the physics laboratory', *History of Technology*, 13 (1991), pp. 73-114.

INDEX